Aquatic Food Webs

Aquatic Food Webs
An Ecosystem Approach

EDITED BY

Andrea Belgrano

National Center for Genome Resources (NCGR),
Santa Fe, NM, USA

Ursula M. Scharler

University of Maryland Center for Environmental Science,
Chesapeake Biological Laboratory (CBL),
Solomons, MD, USA and Smithsonian Environmental Research Center,
Edgewater, MD, USA

Jennifer Dunne

Pacific Ecoinformatics and Computational Ecology Lab, Berkeley,
CA USA; Santa Fe Institute (SFI) Santa FE, NM, USA; Rocky Mountain
Biological Laboratory, Crested Butte, CO USA

AND

Robert E. Ulanowicz

University of Maryland Center for Environmental Science,
Chesapeake Biological Laboratory (CBL),
Solomons, MD, USA

OXFORD
UNIVERSITY PRESS

OXFORD

UNIVERSITY PRESS

Great Clarendon Street, Oxford OX2 6DP

Oxford University Press is a department of the University of Oxford.
It furthers the University's objective of excellence in research, scholarship,
and education by publishing worldwide in

Oxford New York

Auckland Cape Town Dare es Salaam Hong Kong Karachi
Kuala Lumpur Madrid Melbourne Mexico City Nairobi
New Delhi Shanghai Taipei Toronto

With offices in

Argentina Austria Brazil Chile Czech Republic France Greece
Guatemala Hungary Italy Japan South Korea Poland Portugal
Singapore Switzerland Thailand Turkey Ukraine Vietnam

Oxford is a registered trade mark of Oxford University Press
in the UK and in certain other countries

Published in the United States
by Oxford University Press Inc., New York

British Library Cataloging in Publication Data
(Data available)

Library of Congress Cataloging-in-Publication Data
Aquatic food webs : an ecosystem approach / edited by Andrea Belgrano ... [et al.].
p. cm.
Includes bibliographical references and index.
ISBN 0-19-856482-1 (alk. paper) – ISBN 0-19-856483-X (alk. paper)   1. Aquatic
ecology.   2. Food chains (Ecology)   I. Belgrano, Andrea.
QH541.5.W3A68225 2004
577.6'16–dc22   2004027135
ISBN 0 19 856482 1 (Hbk)   9780198564829
ISBN 0 19 856483 X (Pbk)   9780198564836

10 9 8 7 6 5 4 3 2

Typeset by Newgen Imaging Systems (P) Ltd., Chennai, India
Printed in Great Britain
on acid-free paper by Antony Rowe, Chippenham

CURRENT AND FUTURE PERSPECTIVES ON FOOD WEBS

Michel Loreau

Food webs have been approached from two basic perspectives in ecology. First is the energetic view articulated by Lindeman (1942), and developed by ecosystem ecology during the following decades. In this view, food webs are networks of pathways for the flow of energy in ecosystems, from its capture by autotrophs in the process of photosynthesis to its ultimate dissipation by heterotrophic respiration. I would venture to say that the ecological network analysis advocated by Ulanowicz and colleagues in this book is heir to this tradition. A different approach, rooted in community ecology, was initiated by May (1973) and pursued by Pimm (1982) and others. This approach focuses on the dynamical constraints that arise from species interactions, and emphasises the fact that too much interaction (whether in the form of a larger number of species, a greater connectance among these species, or a higher mean interaction strength) destabilises food webs and ecological systems. The predictions resulting from this theory regarding the diversity and connectance of ecological systems led to a wave of comparative topological studies on the structure of food webs. Thus, the two traditions converge in the search for patterns in food-web structure despite different starting points. This book results from the confluence of these two perspectives, which are discussed in a number of chapters.

Patterns, however, are generally insufficient to infer processes. Thus, the search for explanations of these patterns in terms of processes is still very much alive, and in this search the energetic and dynamical perspectives are not the only possible ones. Biogeochemical cycles provide a functional perspective on food webs that is complementary to the energetic approach (DeAngelis 1992). Material cycles are among the most common of the positive feedback loops discussed by Ulanowicz in his concluding remarks, and may explain key properties of ecosystems (Loreau 1998). The stoichiometry of ecological interactions may further strongly constrain food-web structure (Sterner and Elser 2002; Elser and Hessen's chapter). There has also been considerable interest in the relationship between biodiversity and ecosystem functioning during the last decade (Loreau et al. 2002). Merging the theories that bear upon food webs and the maintenance of species diversity is urgently needed today, and may provide new insights into food-webs structure and ecosystem functioning (Hillebrand and Shurin's chapter).

The structure and functioning of ecological systems is determined not only by local constraints and interactions, but also by larger-scale processes. The importance of regional and historical influences has been increasingly recognised in community ecology (Ricklefs and Schluter 1993). The extent to which they shape food webs, however, has been relatively little explored. The recent development of metacommunity theory (Leibold et al. 2004) provides a framework to start examining spatial constraints on the structure and functioning of local food webs (Melian et al.'s chapter). At even larger time scales, food webs are the result of evolutionary processes which determine their current properties. Complex food webs may readily evolve based on simple ecological interactions (McKane 2004). The evolution of food-web and ecosystem properties is a fascinating topic for future research.

This book provides a good synthesis of recent research into aquatic food webs. I hope this synthesis will stimulate the development of new approaches that link communities and ecosystems.

References

DeAngelis, D. L. 1992. *Dynamics of nutrient cycling and food webs*. Chapman & Hall, London.

Leibold, M. A., M. Holyoak, N. Mouquet, P. Amarasekare, J. M. Chase, M. F. Hoopes, R. D. Holt, J. B. Shurin, R. Law, D. Tilman, M. Loreau, and A. Gonzalez. 2004. The metacommunity concept: a framework for multi-scale community ecology. *Ecology Letters* 7: 601–613.

Lindeman, R. L. 1942. The trophic-dynamic aspect of ecology. *Ecology* **23**: 399–418.

Loreau, M. 1998. Ecosystem development explained by competition within and between material cycles. *Proceedings of the Royal Society of London, Series B* **265**: 33–38.

Loreau, M., S. Naeem, and P. Inchausti. Eds. 2002. *Biodiversity and ecosystem functioning: synthesis and perspectives*. Oxford University Press, Oxford.

May, R. M. 1973. *Stability and complexity in model ecosystems*. Princeton University Press, Princeton.

McKane, A. J. 2004. Evolving complex food webs. *The European Physical Journal B* **38**: 287–295.

Pimm, S. L. 1982. *Food webs*. Chapman & Hall, London.

Ricklefs, R. E., and D. Schluter. Eds 1993. *Species diversity in ecological communities: historical and geographical perspectives*. University of Chicago Press, Chicago.

Sterner, R. W., and J. J. Elser. 2002. *Ecological stoichiometry: the biology of elements from molecules to the biosphere*. Princeton University Press, Princeton.

Contents

Contributors

Daniel Baird, Zoology Department, University of Port Elizabeth, Port Elizabeth, South Africa.

Jordi Bascompte, Integrative Ecology Group, Estación Biológica de Doñana, CSIC, Apdo. 1056, E-41080, Sevilla, Spain. Email: bascompte@ebd.csic.es

Andrea Belgrano, National Center for Genome Resources (NCGR), 2935 Rodeo Park Drive East, Santa Fe, NM 87505, USA. Email: ab@ncgr.org

Ulrich Brose, Technical University of Darmstadt, Department of Biology, Schnittspahnstr. 3, 64287 Darmstadt, Germany.

Robert R. Christian, Biology Department, East Carolina University, Greenville, NC 27858, USA. Email: christianr@mail.ecu.edu

Lorenzo Ciannelli, Centre for Ecological and Evolutionary Synthesis (CEES), Department of Biology University of Oslo, Post Office Box 1066, Blindern, N-0316 Oslo, Norway. Email: lorenzo.ciannelli@bio.uio.no.

Janet T. Duffy-Anderson, Alaska Fisheries Science Center, NOAA, 7600 Sand Point Way NE, 98115 Seattle, WA, USA.

Jennifer A. Dunne, Pacific Ecoinformatics and Computational Ecology Lab, P.O. Box 10106, Berkeley, CA 94709 USA; Santa Fe Institute, 1399 Hyde Park Road, Santa Fe, NM 87501 USA; Rocky Mountain Biological Laboratory, P.O. Box 519, Crested Butte, CO 81224 USA Email: jdunne@santafe.edu.

James J. Elser, School of Life Sciences, Arizona State University, Tempe, AZ 85287, USA. Email: j.elser@asu.edu

Dag O. Hessen, Department of Biology, University of Oslo, P.O. Box 1050, Blindern, N-0316 Oslo, Norway.

Alan G. Hildrew, School of Biological Sciences, Queen Mary, University of London, Mile End Road, London, E1 4NS, UK. Email: A.Hildrew@qmul.ac.uk

Helmut Hillebrand, Institute for Botany, University of Cologne, Gyrhofstrasse 15 D-50931 Köln, Germany. Email: helmut.hillebrand@uni-koeln.de

D.Ø. Hjermann, Centre for Ecological and Evolutionary Synthesis (CEES), Department of Biology University of Oslo, Post Office Box 1066 Blindern, N-0316 Oslo, Norway.

Simon Jennings, Centre for Environment, Fisheries and Aquaculture Science, Lowestoft Laboratory NR33 0HT, UK. Email: S.Jennings@cefas.co.uk

Jeffrey C. Johnson, Institute of Coastal and Marine Resources, East Carolina University, Greenville, NC 27858, USA.

Pedro Jordano, Integrative Ecology Group, Estación Biológica de Doñana, CSIC, Apdo. 1056, E-41080, Sevilla, Spain.

Michio Kondoh, Center for Limnology, Netherlands Institute of Ecology, Rijksstraatweg 6, Nieuwersluis, P.O. Box 1299, 3600 BG Maarssen, The Netherlands. Email: mkondoh@rins.ryukoku.ac.jp

P. Lehodey, Oceanic Fisheries Programme, Secretariat of the Pacific Community, BP D5, 98848 Noumea cedex, New Caledonia.

Mathew Leibold, Section of Integrative Biology, The University of Texas at Austin, 1 University Station, C0930 Austin, TX 78712, USA. Email: mleibold@mail.utexas.edu

Jason S. Link, National Marine Fisheries Service, Northeast Fisheries Science Center, 166 Water St., Woods Hole, MA 02543, USA. Email: jlink@whsunl.wh.whoi.edu

Michel Loreau, Laboratoire d'Ecologie, UMR 7625 Ecole Normale Superieure 46, rue d' Ulm F-75230, Paris Cedex 05, France. Email: loreau@wotan.ens.fr

Joseph Luczkovich, Biology Department, East Carolina University, Greenville, NC 27858, USA.

Neo D. Martinez, Pacific Ecoinformatics and Computational Ecology Lab, P.O. Box 10106, Berkeley, CA 94709 Rocky Mountain Biological Laboratory, P.O. Box 519, Crested Butte, CO 81224 USA.

Carlos J. Melián, Integrative Ecology Group, Estación Biológica de Doñana, CSIC, Apdo. 1056, E-41080, Sevilla, Spain.

Elizabeth T. Methratta, National Marine Fisheries Service, Northeast Fisheries Science Center, 166 Water St., Woods Hole, MA 02543, USA.

James T. Morris, Department of Biological Sciences, University of South Carolina, Columbia, SC 29208, USA. Email: morris@biol.sc.edu

Geir Ottersen, Institute of Marine Research, P.O. Box 1870 Nordnes, 5817 Bergen, NORWAY Current Address: Centre for Ecological and Evolutionary Synthesis, Department of Biology, P.O Box 1050 Blindern, N-0316 Oslo, NORWAY.

Enric Sala, Center for Marine Biodiversity and Conservation, Scripps Institution of Oceanography, La Jolla, CA 92093-0202, USA. Email: esala@ucsd.edu

Ursula M. Scharler, University of Maryland, Center for Environmental Science, Chesapeake Biological Laboratory (CBL), Solomons, MD 20688, USA; Smithsonian Environmental Research Center, Edgewater, MD, USA. Email: scharler@cbl.umces.edu

Jonathan B. Shurin, Department of Zoology, University of British Columbia, 6270 University Blvd. Vancouver, BC V6T 1Z4, Canada.

Nils Chr. Stenseth, Centre for Ecological and Evolutionary Synthesis (CEES), Department of Biology University of Oslo, Post Office Box 1066 Blindern, N-0316 Oslo, Norway.

Dietmar Straile, Dietmar StraileLimnological Institute, University of Konstanz, 78457 Konstanz, Germany. Email: dietmar.straile@uni-konstanz.de

William T. Stockhausen, National Marine Fisheries Service, Northeast Fisheries Science Center, 166 Water St., Woods Hole, MA 02543, USA.

Andrew R. Solow, Woods Hole Oceanographic Institution, Woods Hole, MA 02543, USA. Email: asolow@whoi.edu

Geroge Sugihara, Center for Marine Biodiversity and Conservation, Scripps Institution of Oceanography, La Jolla, CA 92093-0202, USA.

Ross Thompson, Biodiversity Research Centre, University of British Columbia, Vancouver, Canada.

Colin R. Townsend, Department of Zoology, University of Otago, New Zealand.

Robert E. Ulanowicz, University of Maryland, Center for Environmental Science, Chesapeake Biological Laboratory (CBL), Solomons, MD 20688, USA. Email: ulan@cbl.umces.edu

Richard J. Williams, Pacific Ecoinformatics and Computational Ecology Lab, P.O. Box 10106, Berkeley, CA 94709 USA; Rocky Mountain Biological Laboratory, P.O. Box 519, Crested Butte, CO 81224 USA; San Francisco State University, Computer Science Department, 1600 Holloway Avenue, San Francisco, CA 94132 USA.

Guy Woodward, Department of Zoology, Ecology and Plant Science, University College, Cork, Ireland.

Aquatic food-webs' ecology: old and new challenges

Andrea Belgrano

Looking up "aquatic food web" on Google provides a dizzying array of eclectic sites and information (and disinformation!) to choose from. However, even within this morass it is clear that aquatic food-web research has expanded greatly over the last couple of decades, and includes a wide array of studies from both theoretical and empirical perspectives. This book attempts to bring together and synthesize some of the most recent perspectives on aquatic food-web research, with a particular emphasis on integrating that knowledge within an ecosystem framework.

It is interesting to look back at the pioneering work of Sir Alister Hardy in the early 1920s at Lowestoft Fisheries Laboratory. Hardy studied the feeding relationship of the North Sea herring with planktonic assemblages by looking at the species distribution patterns in an attempt to provide better insights for the stock assessment of the North Sea fisheries. If we take a look in his food-web scheme (Figure 1), it is interesting to note that he considered species diversity in both phytoplankton and zooplankton, and also specified body-size data for the different organisms in the food web. Thus, it appears that already almost 100 years ago the concept of constructing and drawing links among diverse species at multiple trophic levels in a network-like fashion was in the mind of many aquatic researchers.

In following decades, researchers began to consider links between food-web complexity and ecological community stability. The classic, and still contentious MacArthur hypothesis that "Stability increases as the number of link increase" (1955) gave rise to studies such as that by Paine (1966) that linked latitudinal gradients in aquatic species diversity, food-web complexity, and community stability.

Following that early MacArthur hypothesis, we find it timely to also ask, *How complex are aquatic food webs?*

The first book on theoretical food-web ecology was written by May (1973), followed by Cohen (1978). Since then, Pimm (1982) and Polis and Winemiller (1996) have revisited some of the ideas proposed by May and Cohen and discussed them in different contexts, and trophic flow models have been proposed and used widely for aquatic and particularly marine ecosystems (e.g. Wulff et al. 1989; Christensen and Pauly 1993). However, recent advances in ecosystem network analysis (e.g. Ulanowicz 1996, 1997; Ulanowicz and Abarca-Arenas 1997) and the network structure of food webs (e.g. Williams and Martinez 2000; Dunne et al. 2002a,b; Williams et al. 2002) in relation to ecosystem dynamics, function, and stability clearly set the path for a new, complementary research agenda in food-web analysis. These and many other studies suggest that a new synthesis of available information is necessary. This new synthesis is giving rise to novel basic research that generalizes across habitats and scales, for example, the discovery of universal scaling relations in food-web structure (Garlaschelli et al. 2003), and is also underpinning new approaches and priorities for whole-ecosystem conservation and management, particularly in marine systems.

Aquatic food-web research is also moving beyond an exclusive focus on taxa from phytoplankton to fish. A new look at the role that marine microbes

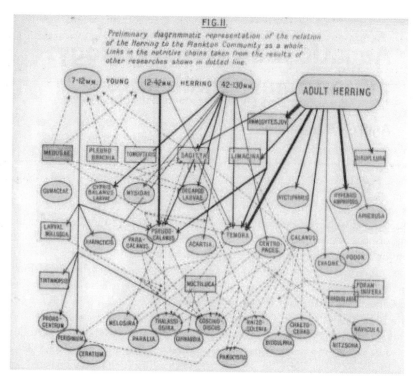

Figure 1 The food web of herring *Clupea harengus* Hardy (1924). From *Parables of Sea & Sky—The life, work and art of Sir Alister Hardy F. R. S.* Courtesy of SAHFOS—The CPR Survey, Plymouth, UK.

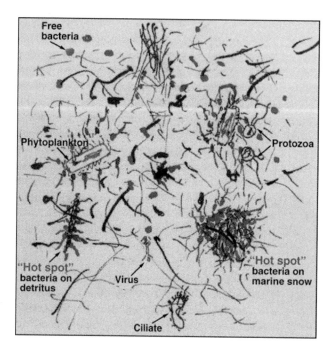

Figure 2 The microbial loop: impressionist version. A bacteria-eye view of the ocean's euphotic layer. Seawater is an organic matter continuum, a gel of tangled polymers with embedded strings, sheets, and bundles of fibrils and particles, including living organisms, as "hotspots." Bacteria (red) acting on marine snow (black) or algae (green) can control sedimentation and primary productivity; diverse microniches (hotspots) can support high bacterial diversity. (Azam, F. 1998. Microbial control of oceanic carbon flux: the plot thickens. *Science* **280**: 694–696.) (See Plate1)

play in the global ocean (Azam and Worden 2004) suggests that oceanic ecosystems can be characterized as a complex dynamic molecular network. The role of microbial food webs (Figure 2—see also, Plate 1—Azam 1998) needs to be considered to understand the nonlinearities underlying the relationship between the pelagic and benthic domains.

Emerging challenges in aquatic food-web research include integrating genomic, biogeochemical, environmental, and economic data in a modeling effort that will elucidate the mechanisms governing the ecosystem dynamics across temporal and spatial scales at different levels of organization and across the whole variety of species diversity, including humans. Aquatic food webs may provide a particularly useful empirical framework for developing and testing an information theory of ecology that will take into account the complex network of interactions among biotic and abiotic components of ecosystems.

Acknowledgments

This work was funded in part or in full by the US Dept of Energy's Genomes to Life program (www.doegenomestolife.org) under the project "Carbon Sequestration in *Synechococcus* sp.: From Molecular Machines to Hierarchical Modeling" (www.genomes-to-life.org).

PART I

Structure and function

Many scientists use food webs to portray ecological communities as complex adaptive systems. However, as with other types of apparently complex systems, underlying mechanisms regulate food-web function and can give rise to observed structure and dynamics. These mechanisms can sometimes be summarized by relatively simple rules that generate the ecosystem properties that we observe.

This section of the book presents and discusses responses of food webs to trophic interactions, transfer efficiency, length of food chains, changes in community composition, the relative importance of grazing versus detrital pathways, climate change, and the effects of natural and anthropogenic disturbances. In addition, research is beginning to incorporate spatial and temporal dimensions of trophic interactions. Along those lines, several of the chapters extend their scope beyond traditional food-web "snapshot" analyses to take into account space and time when assessing changes in food-web structure and species composition.

By comparing food webs from different environments and by encompassing organisms from bacteria to vertebrates, we start to see some common, general constraints that act to shape and change food-web structure and function. These include biological stoichiometry, body-size, and the distribution of interaction strengths. Insights from ecological network analysis also provide new tools for thinking about dynamical and energetic properties of food webs, tools which complement a wide array of more long-standing approaches.

Biosimplicity via stoichiometry: the evolution of food-web structure and processes

James J. Elser and Dag O. Hessen

Introduction

In these days of the vaunted genome and the rest of the proliferating "gnomes" (transcriptome, proteome, metabolome, etc.) and the unveiling of astonishing complex pleiotropy and protein/genome interactions, it may seem headstrong to propose that there is something more complex than the genome currently under study in modern biology. Nevertheless, we propose that the "entangled bank" of food webs, the trophic connections among interacting organisms in ecosystems, is indeed as complex and bewildering as the emerging genome and its products. The complexity increases further when considering the myriad pathways of matter and energy that the species interactions build upon. Consider, for example, a simplified map of central cellular metabolism (Figure 1.1). Here we can see, in basic outline, the key pathways by which energy and key resources are metabolized in maintenance and growth of the organism. Note the complexity of the diagram both in terms of the numbers of nodes and the numbers and types of connections among different components. As shown by the shading, different parts of the overall metabolism can be classified into different functional roles, in this case into 11 categories. Note also that we used the word "simplified." That is, if we were to zoom in on the nucleotide synthesis area of the diagram, more details would emerge, with more nodes (chemical categories) and pathways appearing (you can do this yourself on the Internet at www.genome.ad.jp/kegg/). Yet more magnification, for example,

on purine metabolism would reveal yet more details, finally yielding individual molecules and each individual chemical reaction pathway. The fascinating but intimidating journey just completed should be familiar to food-web ecologists, for whom Figure 1.1(a) (Lavigne 1996) has achieved near-iconic status as a symbol, sure to stimulate uneasy laughter in the audience, of the daunting complexity confronted by food-web ecologists. If we were to follow the metabolic example and zoom in on the northwest Atlantic food web, we would, of course, encounter more and more detail. The node "cod," for example, might resolve itself into larval, juvenile, and mature cod, each connected, by its feeding, in different ways with other parts of the web. Further inspection might then reveal the individual cod themselves, each with a distinct genome and a unique physiological and behavioral repertoire.

How, then, can we deal with this layered complexity in food webs? And how could any connecting thread of simplicity and unifying principles be spotted in this overwhelming complexity? It is our view that, just as the individual molecules in metabolism are the critical level of resolution for the molecular biologist confronting the genome and its products, the level of the individual organism should be of central importance for the food-web ecologist. This is because, just as particular individual molecules (not classes of molecules) are the actual participants in metabolic networks, it is individual organisms (not species, populations, functional groups) that do the actual eating

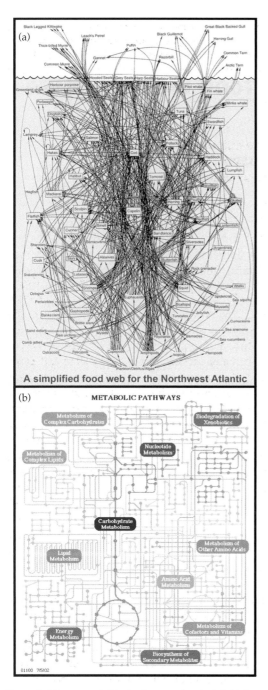

(a)

A simplified food web for the Northwest Atlantic

(b) METABOLIC PATHWAYS

Figure 1.1 Two entangled banks demonstrating the intimidating task that lies before biology. Ultimately, food webs (a) (Lavigne 1996) are the outcome of dynamic interactions among various organisms that acquire resources from the abiotic environment and each other in order to drive their metabolism (b) (www.genome.ad.jp/kegg/) and leave offspring.

(and thus make the trophic connections) in food webs. Thus, organism-focused reasoning based on sound physiological principles is likely to be of great assistance in unraveling food webs. Furthermore, the constituents of food webs are not fixed entities; rather, they are products and agents of continuous evolutionary change. And since evolution operates primarily at the level of individual reproductive success, it seems that evolutionary thinking should play a central role in understanding how and why food webs are shaped the way they are.

Molecular/cell biologists coming to grips with the daunting complexity of the genome and its products (Figure 1.1) have a powerful ally in the fact that each node of a metabolic network is the product (and a reactant) in an enzymatically driven chemical reaction. Thus, all parts of the network must obey strict rules of mass balance and stoichiometric combination in the formation and destruction of the constituent parts. Indeed, these simplifying principles form the basis of various emerging theories through which cell biologists hope to make progress in understanding functional interconnections among genes and gene products in metabolism (e.g. metabolic control theory, Dykhuizen et al. 1987; Wildermuth 2000; stoichiometric network theory, Hobbie et al. 2003; metabolic flux balancing, Varma and Palsson 1994). But why should such powerful tools be left to molecular biologists?

Luckily, food-web ecologists can also take advantage of the considerable traction afforded by the firm laws of chemistry because, just as every node in a biochemical network is a chemical entity, so is every node in a food web. That is, each individual organism forming a connection point in Figure 1.1(b) is an aggregation of biochemicals and chemical elements and is sustained by the net outcome of the coupled biochemical pathways shown in Figure 1.1(a). Thus, the interactions among food-web components in terms of consumption (and the feedbacks imposed by nontrophic relations of excretion and nutrient regeneration) are also constrained by the firm boundaries of mass balance and stoichiometric combination. These principles and their applications are known as "ecological stoichiometry" (Sterner and

Elser 2002), while their more recent extension to the realms of evolutionary biology, behavior, physiology, and cellular/molecular biology are known as "biological stoichiometry" (Elser et al. 2000b). The approach of ecological stoichiometry simplifies the bewildering ecological complexity in Figure 1.1(b) by focusing on key ecological players in food webs and characterizes them in terms of their relative carbon (C), nitrogen (N), and phosphorus (P) demands. In biological stoichiometry, metabolic complexity in Figure 1.1(a) is simplified by focusing on major biochemical pools (e.g. rRNA, total protein, RUBISCO) that determine overall organismal demands for C, N, and P and attempts to connect those biochemical demands to major evolutionary forces operating on each organism's life history or metabolic strategy.

In this chapter we will review some basic principles and highlight some of the most recent findings from the realm of ecological stoichiometry in food webs, to illustrate how a multivariate perspective on energy and chemical elements improves our understanding of trophic relations. More details of these (and other) matters are available in Sterner and Elser (2002); in this chapter we seek to highlight some findings that have emerged since publication of that work. We will then discuss recent movements to integrate stoichiometric study of food webs with the fact that food-web components are evolving entities and that major evolutionary pressures impose functional trade-offs on organisms that may have profound implications for the structure and dynamics of food webs. Our overarching view is that stoichiometric theory can help in integrating food-web ecology and evolution into a more comprehensive framework capable of making a priori predictions about major food-web features from a relatively simple set of fundamental assumptions. In advocating this view we hope to continue to add substance to the vision of food webs offered nearly 100 years ago by Alfred Lotka (1925):

For the drama of life is like a puppet show in which stage, scenery, actors, and all are made of the same stuff. The players indeed 'have their exits and their entrances', but the exit is by way of a translation into the substance of the stage; and each entrance is a transformation scene. So stage and players are bound together in the close partnership of an intimate comedy; and if we would catch the spirit of the piece, our attention must not all be absorbed in the characters alone, but most also be extended to the scene, of which they are born, on which they play their part, and with which, in a little while, they merge again.

Stoichiometric imbalance, "excess" carbon, and the functioning of food webs

Conventional food-web diagrams show binary feeding links among species. Flowchart analyses of food webs go beyond a binary depiction of feeding relations in using dry-weight, energy (Joules), or carbon as a common currency to express the magnitude of particular connections. The advantage of using C-based flow charts is quite obvious since, not only does C account for some 50% of dry mass in most species and taxa, it also allows for inclusion on import and export of inorganic carbon in the same scheme. However, with the realization that the supply and availability of P is a key determinant of the binding, flux, and fate of C in freshwater food webs, in many cases more information may be gained from P-based flow charts. Such a view will also provide a better representation of the recycling of elements, and thus differentiate between "old" and "new" production. In a P-limited system, any extra atom of P will allow for binding of more than 100 atoms of C in autotroph biomass. Due to different elemental ratios in different food-web compartments, conventional flowchart diagrams will normally turn out quite different in terms of C or P (Figure 1.2). Neither is more "true" or correct than the other; instead, each provides complementary information on pools and key processes. Since P is a conservative element that is lost from aquatic systems only by sedimentation or outflow, it will normally be frequently recycled and may thus bind C in stoichiometric proportions a number of times over a season.

In addition to pictures of who is eating whom in food webs, we need to understand the outcome of those feeding interactions for the consumer. "Trophic efficiency" is key aspect of food webs that captures important aspects of this outcome (here we use "efficiency" to refer to the fraction of

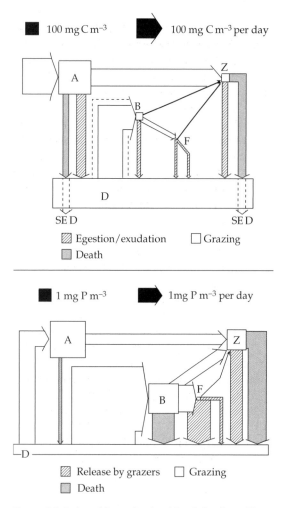

Figure 1.2 Pools and fluxes of carbon (a) and phosphorus (b) in a pelagic food web of a eutrophic lake (data from Vadstein et al. 1989). A: algae, B: bacteria, D: detritus and other kinds of nonliving dissolved and particulate matter, F: heterotrophic flagellates, and Z: metazoan zooplankton. Boxes denotes biomasses, arrows denote fluxes. Note the entirely different size of pools and fluxes for C and P.

energy or C produced by a certain level that is transferred to higher trophic levels). Efficient trophic systems typically have steep slopes from autotrophs at the base of the food web to top predators, and subsequently should also support a higher number of trophic levels than less efficient systems. Typically, planktonic systems are among those with high trophic efficiency, with forests on the extreme low end (Hairston and Hairston 1993; Cebrian and Duarte 1995; Cebrian 1999). Several

explanations may be invoked to explain such patterns but, as argued by Cebrian et al. (1998), surely the high transfer efficiency of C in planktonic systems may be attributed to both high cell quotas of N and P relative to C (high-quality food for reasons given below) coupled with decreased importance of low-quality structural matter like lignins and cellulose that are poorly assimilated. A striking feature of the cross-ecosystem comparison compiled by Cebrian et al. (1998) is the close correlation paths among autotroph turnover rate, specific nutrient (N, P) content, and the trophic efficiency. These associations make perfect sense from a stoichiometric point of view: while consumers are not perfectly homeostatic (cf. DeMott 2003), they have a far closer regulation of elemental ratios in somatic tissue than autotrophs (Andersen and Hessen 1991; Hessen and Lyche 1991; Sterner and Hessen 1994) and their input and output of elements must obey simple mass balance principles. As a general rule, limiting elements are expected to be utilized for growth and transferred in food chains with high efficiency, while non-limiting elements, by definition present in excess, must be disposed of and may be recycled (Hessen 1992; Sterner and Elser 2002). Thus, when feeding on low C : N or low C : P food, a considerable share of N and P may be recycled (Elser and Urabe 1999), while C-use efficiency is high (Sterner and Elser 2002). However, typically autotrophs have higher C : element ratios than consumers (Elser et al. 2000a). Thus, when consumers feed on diets that are high in C : N or C : P, nutrient elements are reclaimed with higher efficiency by the animal (Elser and Urabe 1999), while much of the C is unassimilated and must be egested, excreted, or respired (DeMott et al. 1998; Darchambeau et al. 2003). Since herbivore performance is strongly impaired in these high C : nutrient systems, more C must enter detrital pathways, as is clearly shown by Cebrian's studies.

These factors point to fundamental differences between ecosystems not only with regard to the transfer and sequestration of carbon, but also with regard to community composition and ecosystem function in more general terms. While realizing that pelagic food webs are among the most "efficient" ecosystems in the world, there is certainly a huge scatter in trophic efficiency also among pelagic systems, that is, considerable variation appears when phytoplankton biomass or production is regressed against zooplankton (Hessen et al. 2003). Such scatter may be caused by time-lag effects, external forcing, algal species' composition, and associated biochemistry as well as by top-down effects, but it is also clear that alterations in trophic transfer efficiency due to stoichiometric constraints could be a strong contributor. Thus, understanding food-web dynamics requires understanding the nature and impacts of nutrient limitation of primary production.

Stoichiometry, nutrient limitation, and population dynamics in food webs

Since nutrient limitation of autotroph production only occurs, by definition, when nonnutrient resources such as light are sufficient, a particularly intriguing outcome of stoichiometric analysis in freshwaters, and one that is rather counterintuitive, is that high solar energy inputs in the form of photosynthetically active radiation may reduce secondary (herbivore) production (Urabe and Sterner 1996; Sterner et al. 1997; Hessen et al. 2002; Urabe et al. 2002b). The rationale is as follows: when photosynthetic rates are high due to high light intensity but P availability is low (a common situation in freshwaters), C is accumulated in biomass out of proportion with P. Thus, C:P in the phytoplankton increases, meaning potential reduced C-use efficiency (P-limitation) in P-demanding grazers such as *Daphnia*. The outcomes of such effects have been shown by Urabe et al. (2002b), who applied deep shading that reduced light intensities nearly 10-fold to field enclosures at the Experimental Lakes Area, where seston C:P ratios are generally high (Hassett et al. 1997) and *Daphnia* have been shown to be P-limited (Elser et al. 2001). The outcome was a nearly five-fold increase in zooplankton biomass in unenriched enclosures after the five-week experiment.

However, the negative effects of high algal C:P ratio on zooplankton can be a transient situation and high energy (light) input may eventually sustain a high biomass of slow-growing zooplankton, as demonstrated by long-term chemostat experiments (Faerovig et al. 2002; Urabe et al. 2002a). At a given (low) level of P, high light yields more algal biomass than low light treatments, but with lower food quality (higher C:P). The net outcome will be slow herbivore growth rates at high light, with a higher asymptotic biomass of adults. This is because high growth rate and high reproduction require a diet that balances the grazer's demands in terms of energy, elements, and macromolecules, while a standing stock of (nearly) nonreproducing adults can be sustained on a low-quality diet since their basic metabolic requirements mostly rely on C (energy). This implies a shift from a high to low biomass: production ratio. Eventually, the nutrient constraint in low quality (high autotroph C:P) systems may be overcome by feedbacks from grazers. Such intra- and interspecific facilitation (Sommer 1992; Urabe et al. 2002a) may induce a shift in population dynamics under a scenario of increasing grazing since an increasing amount of P will be available per unit of autotroph biomass due to the combined effect of grazing and recycling (cf. Sterner 1986).

Thus, understanding the biological role of limiting nutrients in both autotrophs and consumers provides a basis for better prediction of how population dynamics of herbivores should respond to changing environmental conditions that alter nutrient supply, light intensity, or other environmental conditions. However, surprisingly little attention in mainstream textbooks on population dynamics has been given to food quality aspects (e.g. Turchin 2003). According to the stoichiometric growth rate hypothesis (described in more detail later) and supported by an increasing body of experimental data (Elser et al. 2003), taxa with high body P-contents commonly have high growth rates and can thus rapidly exploit available resources but are probably especially susceptible for stoichiometric food quality effects. What are the dynamic consequences of this under different conditions of nutrient limitation in the food web? For reasons given above, the a priori assumption would be that predator–prey interactions should be more dynamic when the system sets off with high quality (low C:P) autotroph biomass. Low autotroph C:P will stimulate fast growth of the consumer and relatively high recycling of P for autotroph reuse; the system should therefore rapidly reach an equilibrium where food abundance is limiting to the grazer. On the other hand, a system with high C:P in the autotrophs should have slow grazer response and low recycling of P, yielding sluggish and perhaps erratic dynamics as the system operates under the simultaneous effects of changing food abundance, quality, and nutrient recycling. Indeed, recent models (Andersen 1997; Hessen and Bjerkeng 1997; Loladze et al. 2000; Muller et al. 2001) taking grazer P-limitation and recycling into account clearly demonstrate this kind of dynamic dependency on resource and consumer stoichiometry. As demonstrated by Figure 1.3 (Hessen and Bjerkeng 1997), the amplitudes and periods of autotroph–grazer limit cycles depends both on food quantity and quality. When P:C in the autotroph becomes low, this constrains grazer performance and a high food biomass of low quality may accumulate before the grazer slowly builds up. With assumptions of a more efficient elemental regulation in the autotroph (i.e. lower minimum P:C), limit cycles or amplitudes will be smaller, but the periods will increase.

One intriguing feature of stoichiometric modeling is the potential extinction of the grazer, like a P-demanding *Daphnia*, under a scenario of high food biomass but low food quality (Andersen 1997; Hessen and Bjerkeng 1997). External enrichment of P to the system will also invoke strong shifts in system dynamics due to stoichiometric mechanisms (Andersen 1997; Muller et al. 2001). The relevance for these theoretical exercises for natural systems remains to be tested, however. Clearly the assumption of two compartment dynamics represent an oversimplification, since a considerable share of recycled P and organic C will enter the bacteria or detritus pool, thus dampening the dynamics predicted from the simplified model assumptions.

Thus, one central outcome of stoichiometric theory in consumer–resource systems is deviation from the classical straight Lotka–Volterra isoclines (Andersen 1997; Murdoch et al. 2003). From these analyses, it appears that a combination of inter- and intraspecific facilitation during periodic nutrient element limitation by consumers results in a deviation from straightforward negative density

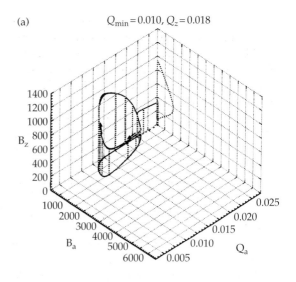

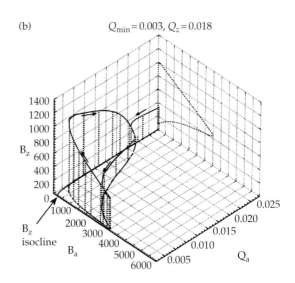

Figure 1.3 Three-dimensional limit cycles for two scenarios with *Daphnia* grazing on algae with different flexibility in their P:C ratio (Q_a). The solid line gives the trajectory, while projections of the three-dimensional trajectory are given on the B_a–B_z plane and the Q_a–B_z plane. B_a: algal biomass (μg C l^{-1}), B_z: grazer (*Daphnia*) biomass (μg C l^{-1}), Q_a: algal P:C (μg P:μg C). In the upper panel, the lower bound of Q_a (Q_{min}) is set to 0.010(a), while Q_{min} in the lower panel is 0.003(b). P:C in the grazer (Q_z) is in both cases fixed at 0.018. By increasing Q_z (higher P:C ratio, lower C:P ratio) slightly in the lower scenario, the grazer will go extinct, and the system will stabilize at a high algal biomass near Q_{min}.

dependence in consumer populations and a much richer array of population dynamics appears. In this way, stoichiometry can provide a logical explanation for Allee effects (positive density-dependence) and hump-shaped curves for density-dependent responses.

Much of the preceding discussion has had herbivores and other primary consumers (e.g. detritivores) in mind. What about the role of stoichiometry higher in food webs? Since metazoans do not vary too much in their biochemical makeup, predators are less likely to face food quality constraints compared with herbivores and especially detritivores (Sterner and Hessen 1994). Fish in general have high P requirements due to investment in bone (Sterner and Elser 2002); this could be seen as another reason, in addition to their large body size, why P-rich *Daphnia* should be preferred prey relative to P-poor copepods. A more important issue is, however, how the predicted dynamics due to stoichiometric mechanisms might be associated with the potential prey susceptibility to predators. A reasonable assumption would be that grazers in low food quality systems would be more at risk for predatory mortality simply because, all else being equal, slow growth would render the population more susceptible to the impacts of any given rate of mortality loss. However, the effects might not quite be so straightforward. For example, fast-growing individuals generally also require high rates of food intake; in turn, more active feeding might increase predation risk (Lima and Dill 1990). Furthermore, there may be some inherent and unappreciated physiological–developmental impacts associated with rapid growth such that overall mortality is elevated in fast-growing individuals, over and above potentially accentuated predation risk (Munch and Conover 2003).

Stoichiometry, omnivory, and the evolution of food-web structure

The fact that different species or taxa have different stoichiometric or dietary requirements has important bearings on the dietary preferences that weave food webs together. Ecologists have commonly generated a coarse classification of species and developmental stages according to their mode of feeding (carnivores, omnivores, herbivores, detritivores, filtrators, raptors, scavengers, etc). We suggest that it might also be useful to adopt a subtler categorization based on dietary, stoichiometric requirements. In fact, in many cases the more specific dietary requirements of a taxon may be the ultimate cause for an organism being a carnivore or a detritivore and, as we discuss below, is probably also an important factor contributing to widespread omnivory among taxa. Hence one could speak about the "stoichiometric niche" of a particular species, in the sense that species (or stages) with high P (or N) requirements would succeed in situations that supply nutrient-rich food compared with species with lower nutrient requirements. For example, for freshwater food webs it has been suggested that when planktonic algae are deprived of P and develop high C:P ratios, P-demanding species like *Daphnia* acutely suffer from "P-starvation" and, probably due to decreased C-use efficiency, become competitively inferior to less P-demanding members of the plankton community like *Bosmina* (DeMott and Gulati 1999; Schulz and Sterner 1999). Thus, the stoichiometric niche space available to *Bosmina* may extend to higher regions of food C:P than in *Daphnia*. But what about the other end of the C:P continuum? Interestingly, in a brand-new stoichiometric wrinkle, recent evidence shows that extremely low C:P may cause decreased growth in *Daphnia* (Plath and Boersma 2001) and the caterpillar *Manduca sexta* (Perkins et al. 2003). While the mechanistic bases of these responses remain obscure, they appear to represent the other side of the stoichiometric niche in the P-dimension.

Another aspect of stoichiometric effects on biodiversity and food-web structure relates to the number of trophic levels and the degree of omnivory. Hastings and Conrad (1979) argued that the evolutionary stable length of food chains would be three, and that the main determinant of the number of trophic levels is the quality of primary production (and therefore, to at least some degree, its C: nutrient ratio) and *not* its quantity, as is often implied in discussions of food-web length. Omnivory may be seen as a compromise between exploiting large quantities of low quality resources at low metabolic cost, or utilizing lower quantities of high-quality food at high metabolic costs. From a stoichiometric point of view, omnivory may be seen as a way of avoiding nutrient deficiency while at the same time having access to a large reservoir of energy. This "best of two worlds" strategy is clearly expressed as life-cycle omnivory, like in crayfish where fast-growing juveniles are carnivorous, while adults chiefly feed on detritus or plants (cf. Hessen and Skurdal 1987).

The fact that organisms can be potentially limited not only by access to energy (carbon) but also by nutrients has obvious implications for coexistence of potential competitors. While this principle has been well explored for autotrophs (e.g. Tilman 1982), the same principle may be invoked for heterotrophs with different requirements for key elements (cf. Loladze et al. 2004). In fact, this will not only hold for interspecific competition, but also for intraspecific competition, since most species undergo ontogenetic shifts in nutrient requirements. Indeed, it now seems that this coexistence principle can be extended to explain the evolution and maintenance of omnivory (Diehl 2003), since utilization of different food resources in species with different nutrient contents promotes and stabilizes feeding diversification.

Biological stoichiometry: the convergence of ecological and evolutionary time

It should be clear from the preceding material that stoichiometric imbalance between food items and consumers has major effects on the dynamics and structure of key points in the food web and especially at the autotroph–herbivore interface. Indeed, the effects of stoichiometric imbalance on herbivores are often extreme and suggest that there should be strong selective pressure to alleviate these impacts. The fact that such impacts nevertheless remain implies that there may be fundamental trade-offs and constraints on evolutionary response connected to organismal stoichiometry. So, why is it that consumer organisms, such as herbivorous zooplankton or insects, maintain body nutrient contents that are so high

that they often cannot even build their bodies from available food? Why do some species seem to be more sensitive to the effects of stoichiometric food quality? In this section we follow the advice of Holt (1995) by describing some recent findings that illuminate some of these evolutionary questions in the hopes that perhaps in the future we will encounter Darwin as well as Lindeman in the reference sections of food-web papers.

Beyond expanding the diet to include more nutrient-rich prey items and thus inducing omnivory as discussed earlier, another obvious evolutionary response to stoichiometrically unbalanced food would be for a consumer to evolve a lower body requirement for an element that is chronically deficient in its diet. Several recent studies emphasizing terrestrial biota have provided evidence for just such a response. Fagan et al. (2002) examined the relative nitrogen content (%N of dry mass, $N:C$ ratio) of folivorous insect species and documented a significant phylogenetic signal in which the recently derived insect group (the "Panorpida," which includes Diptera and Lepidoptera) have significantly lower body N content than the more ancestral groups Coleoptera and Hemiptera which were themselves lower than the still older Lower Neoptera. Their analysis eliminated differences due to body size, gut contents, or feeding mode as possible explanations for the pattern. They noted that the divergence of major insect groups appears to have coincided with major increases in atmospheric CO_2 concentrations (and thus high plant $C:N$ ratio) and hypothesized that clades of insects that emerged during these periods of "nitrogen crisis" in plant biomass were those that had an efficient N economy. Signs of evolutionary response to stoichiometric imbalance in insects is also seen at a finer scale in studies by Jaenike and Markow (2002) and Markow et al. (1999), who examined the body $C:N:P$ stoichiometry of different species of *Drosophila* in relation to the $C:N:P$ stoichiometry of each species' primary host resource. Host foods involved different species of rotting cactus, fruit, mesquite exudates, and mushrooms and presented a considerable range in nutrient content. They showed a significant correlation between host nutrient content and the nutrient content of the associated fly species: taxa specializing on nutrient-rich mushrooms had the highest nutrient contents (in terms of %N and %P) while those specializing on the poorest quality resource (mesquite flux) had the lowest body nutrient contents. Signs of evolutionary response to stoichiometric resource limitation can be found even in the amino acid structure of proteins themselves. Using genomic data for the prokaryote *Escherichia coli* and the eukaryote *Saccharomyces cerevisiae*, Baudoin-Cornu et al. (2001) showed that protein enzymes involved in C, N, or S metabolism were significantly biased in favor of amino acids having low content of C, N, and S, respectively. That is, enzymes involved in uptake and processing of N were disproportionately constituted of amino acids having relatively few N atoms, a response that makes adaptive sense.

These studies suggest that heterotrophic consumers can indeed respond in evolutionary time to reduce the degree of stoichiometric imbalance between their biomass requirements and their often-poor diets. And yet the overall nutrient content (e.g. %N and %P) of herbivorous animals (zooplankton, insects) remains at least 10-times and often 100- to 1,000-times higher than is found in autotroph biomass (Elser et al. 2000*a*). Some fundamental benefit of body nutrient content must exist. The nature of the benefits of higher body nutrient content is perhaps becoming clearer now based on work emerging from tests of the "growth rate hypothesis" (GRH hereafter), which proposes that variation in the C:P and N:P ratios of many organisms is associated with differences in growth rate because rapid growth requires increased allocation to P-rich RNA (Hessen and Lyche 1991; Sterner 1995; Elser et al. 1996; Elser et al. 2000*b*). In this argument, a key life history parameter, growth or development rate, is seen to inherently require increased investment in P-intensive biochemicals, implying a trade-off in that fast-growing organisms are likely to find themselves constrained by an inability to acquire sufficient P from the environment or diet. Work with zooplankton and other organisms such as insects and microbes now makes it clear that the fundamental core of the GRH is correct (Figure 1.4(a)): in

various comparisons involving physiological, ontogenetic, and cross-species comparisons, P-content increases with growth rate (Main et al. 1997; Carrillo et al. 2001; Elser et al. 2003; Makino and Cotner 2003), RNA content also increases with growth rate (Sutcliffe 1970; Vrede et al. 2002; Elser et al. 2003; Makino and Cotner 2003; Makino et al. 2003), and, importantly, increased allocation to RNA quantitatively accounts for the increased P content of rapidly growing organisms (Elser et al. 2003; Figure 1.4(a)). The mechanistic connections among RNA, P, and growth appear to manifest quite directly in how nutrient supply affects herbivore dynamics, as illustrated in a recent study of a plant–herbivore interaction in the Sonoran Desert (Figure 1.4(b)). In this study (Schade et al. 2003), interannual variation in rainfall led to increased soil P supply under wet conditions, which in turn lowered foliar C:P in velvet mesquite trees. Consequently, mesquite-feeding weevils had higher RNA and correspondingly higher P contents, consistent with a P-stimulated increase in growth rate, along with higher population densities on mesquite branches with low foliar C:P. In sum, these findings of tight and ecologically significant associations of growth, RNA, and P contents suggest that major aspects of an organism's ecology and life history have an important stoichiometric component. Given the strong effects of stoichiometric imbalance in trophic relations (Figure 1.3), it seems then that evolutionary adjustments bearing on growth rate will impact on the types of organisms and species that come to dominate key positions in food webs under different conditions.

Thus, better understanding of the genetic basis and evolutionary dynamics of the coupling among growth, RNA, and P may help in understanding how food webs self-organize. Elser et al. (2000*b*) suggested that the genetic basis of variation in growth and RNA (and thus C:P and N:P ratios) lies in the genes encoding for RNA, the rDNA. In particular, they reviewed evidence suggesting that rDNA copy number and length and content of the rDNA intergenic spacer (IGS, where promoter–enhancer sequences are found) were key variables underlying growth rate variation because of their effects on transcriptional capacity and

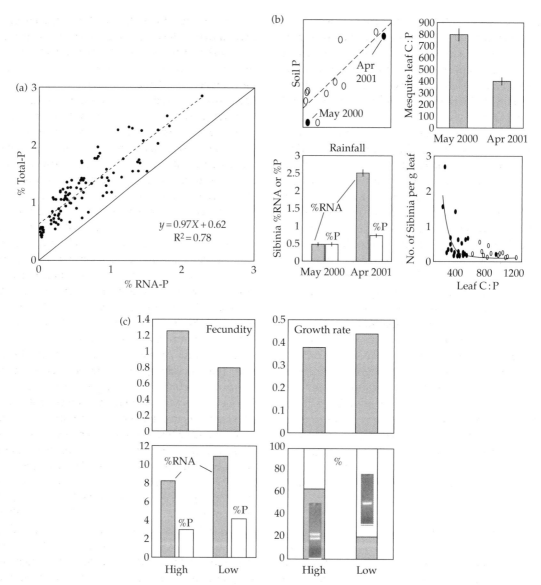

Figure 1.4 From the genome through metabolism to the food web. (a) Variation in allocation to P-rich ribosomal RNA explains variation in the P-content of diverse heterotrophic organisms. The data included in this plot involve data from *E. coli, Drosophila melanogaster,* and various crustacean zooplankton including *Daphnia*. Individual relationships involve interspecific comparisons, changes during ontogenetic development, and changes due to physiological P-limitation. In each case, RNA-rich (and thus P-rich) data were from organisms with relatively fast growth rate. A relationship fit to all of the data combined has a slope of 0.97 and an R^2 of 0.78. Figure from Elser et al. (2003). (b) Transmission of soil P into herbivore ribosomes affects population dynamics. In a study of a desert ecosystem, interannual variation in rainfall led to interannual differences in soil P availability (upper left panel) to velvet mesquite trees (*Prosopis velutina*), leading to a decrease in foliar C:P ratio (upper right). In response, mesquite-feeding weevils (*Sibinia setosa*) had higher RNA and P contents (bottom left) and thus achieved higher population densities (bottom right) when rainfall and soil P supply was higher. Figures from Schade et al. (2003). (c) Artificial selection for five generations on weight-specific fecundity (top left panel) in parthenogenetic *Daphnia pulicaria* produced correlated changes in juvenile growth rate (top right) and in juvenile RNA and P contents (bottom left). Selected lines differed in the relative proportion of individuals carrying both long and short IGS or just the long IGS: high fecundity lines with slow-growing juveniles were dominated by the mixed genotype while low fecundity lines with fast-growing juveniles were dominated by individuals carrying only the long IGS. Figures from Gorokhova et al. (2002).

thus rates of RNA production. Thus, rDNA variation may also be linked to differences in stoichiometric requirements of biota. Emerging data suggest that indeed it is the case that variations in the rDNA can have important stoichiometric ramifications (Gorokhova et al. 2002; Weider et al. 2004).

Of these data, most striking is the study of Gorokhova et al. (2002), who performed an artificial selection experiment on the progeny of a single female of *Daphnia pulex*, selecting on weight-specific fecundity (WSF) of the animals. The treatments responded remarkably quickly to the selection regimen—animals selected for low fecundity were significantly lower than random controls or high fecundity selected animals within three rounds of selection. After five rounds of selection, animals were assayed for juvenile growth rate along with RNA and P contents. The analyses showed that lineages showing low WSF had experienced a correlated inverse response of juvenile growth rate; that is, these females with low WSF produced offsprings that grew faster, offering an opportunity to test the growth rate hypothesis. As predicted by the GRH, these fast-growing juveniles had elevated RNA and P contents compared to the slower-growing counterparts in the control and high-selected lines (Figure 1.4(c)). To determine if these differences had a genetic basis, animals were screened for variation in the IGS of their rDNA. Consistent with the idea above that high RNA (fast growth) phenotypes should be associated with long IGS, a disproportionate number of the (fast-growing, high RNA, high P) animals in the low fecundity treatment carried only long IGS variants while (slow-growing, low RNA, low P) animals in the high WSF and control lines disproportionately carried both long and short IGS variants. Recall that this experiment involved selection on the offspring of a single parthenogenetic female and all offspring throughout were also produced by parthenogenesis. Nevertheless, it appears that functionally significant genetic variation can arise in a few generations even within a "clonal" organism, as is becoming increasingly recognized (Lushai and Loxdale 2002; Loxdale and Lushai 2003; Lushai et al. 2003).

If the results of Gorokhova et al. are confirmed and shown to hold for other biota that comprise food webs, then we would suggest that there is reason to question the traditional distinction between ecological and evolutionary time. That is, ecologists are used to considering species shifts (e.g. during seasonal succession in the plankton) in terms of the sorting out of various ecological transactions such as competition and predation among taxa that have fixed genetic structure on the timescale of the study. The results of Gorokhova et al. suggest that, even with a clonal organism, genetic recombinations with important ecophysiological impacts can arise on time scales that would easily be encompassed, for example, by a single growing season in a lake. This same point of rapid evolution in food webs is also demonstrated by studies of rapid evolution of digestibility-growth trade-offs in rotifer-algae chemostats (Fussmann et al. 2003, Yoshida et al. 2003). Not all evolutionary change requires the slow propagation of small point mutations through the vast protein library that comprises the genome. Indeed it is becoming increasingly obvious that the genome can reorganize quickly via structural mutations that affect regulatory regions and gene copy number and via gene silencing/unsilencing mechanisms such as those resulting from insertion/deletion of transposable elements. It is interesting to note that transposable elements have a significant role in silencing copies of the rDNA in *Drosophila* (Eickbush et al. 1997) and have been identified in the rDNA of various other taxa, including *Daphnia* (Sullender and Crease 2001). To the extent that these reorganizations impact traits that are ecophysiologically relevant (such as those that affect RNA production and thus organismal $C:N:P$ stoichiometry), the shifting genomes of interacting species will need to be incorporated (somehow!) into food-web ecology.

Conclusions

Lotka's ecological play remains to a large degree a mysterious entertainment. There is a long road ahead for food-web ecologists before we can expect to really understand how food webs self-organize and are sustained in their complex interplay with the abiotic environment. So, in some ways this chapter is a plea for a form of a "bio-simplicity" research program in approaching food-web study. That is, Darwin unlocked the key to the origin of the biosphere's bewildering biodiversity with his breathtakingly simple algorithm of variation and selection. Perhaps the complexity we see in food webs is more apparent than real and will also come to be seen to be the product of a simple set of rules ramifying through time and space. Bearing in mind the warning "if your only tool is a hammer then every problem looks like a nail," for the moment we propose that the minimal biosimplicity "rulebox" should include the Darwinian paradigm along with the laws of thermodynamics, mass conservation, and stoichiometric combination. What kind of food-web structures can be built using such tools given the raw materials circulating in the biosphere?

Answering such a question will take some time and a large quantity of cleverness and a certain amount of courage. We can take some solace in the fact that we will be traveling this long road toward understanding of biological complexity with our colleagues who study the genome and its immediate metabolic products. While they might not themselves always realize that high-throughput sequencers and microarray readers do not necessarily result in high throughput of the conceptual insights that might make sense of it all, a time will come (or has it already?) when all biologists, from the geneticists to the ecologists, will lean on each other for insight and necessary data. For guidance at this point we turn to the American poet Gary Snyder whose poem *"For the Children"* begins with what sounds like a familiar problem:

> The rising hills, the slopes,
> of statistics
> lie before us.
> the steep climb
> of everything, going up,
> up, as we all
> go down.

The poem ends with the advice:

> stay together
> learn the flowers
> go light

So, building on this counsel, perhaps we should try to stay together with our colleagues who are confronting the genome, adapting their tools, data, and ideas to better understand our own questions, and perhaps offering some insights of our own. We *do* need to learn the flowers, their names, and the names of those who eat them, and of those who eat those too. But, especially these days when the genomics juggernaut threatens to bury us all in a blizzard of data and detail, perhaps we should try to go light, approach the problems with a simplified but powerful toolbox that includes stoichiometric principles, and see what major parts of the puzzle will yield themselves to our best efforts.

Acknowledgments

The authors are grateful for the support of the Center for Advanced Study of the Norwegian Academy of Letters and Sciences that made this work possible. JJE also acknowledges NSF grant DEB-9977047. We thank M. Boersma and K. Wiltshire for their helpful comments on an early draft.

CHAPTER 2

Spatial structure and dynamics in a marine food web

Carlos J. Melián, Jordi Bascompte, and Pedro Jordano

Introduction

The role of space in population and community dynamics has been recently emphasized (e.g. Hanski and Gilpin 1997; Tilman and Kareiva 1997; Bascompte and Solé 1998). Several models for the coexistence of interacting species in heterogeneous environments have been formulated. These include the energy and material transfer across ecosystem boundaries and its implication for succession and diversity (Margalef 1963; Polis et al. 1997), the geographic mosaic of coevolution (Thompson 1994), the regional coexistence of competitors via a competition–colonization trade-off (Tilman 1994), the random assembly of communities via recruitment limitation (Hubbell 2001), and metacommunities (Wilson 1992). As a general conclusion of these approaches, succession, dispersal, local interactions, and spatial heterogeneity have appeared strongly linked to the persistence of diversity. However, the underlying pattern of ecological interactions in a spatially structured ecosystem and its implications for the persistence of biodiversity remains elusive by the lack of synthetic data (Loreau et al. 2003).

Introducing space and multiple species in a single framework is a complicated task. As Caswell and Cohen (1993) argued, it is difficult to analyze patch-occupancy models with a large number of species because the number of possible patch states increases exponentially with species richness. Therefore, most spatial studies have dealt with a few number of species (Hanski 1983), predator–prey systems (Kareiva 1987), or n-competing species (Caswell and Cohen 1993; Tilman 1994; Mouquet and Loreau 2003). On the other hand, the bulk of studies in food-web structure and dynamics have dealt with either large (but see Hori and Noda 2001) or small (but see Caldarelli et al. 1998) number of species, but make no explicit reference to space (Caswell and Cohen 1993; Holt 1996, 1997). Only a few studies have explored the role of space on a small subset of trophic interacting species (Holt 1997; Melián and Bascompte 2002).

The present study is an attempt to link structure and dynamics in a spatially structured large marine food web. We use data on the diet of 5526 specimens belonging to 208 fish species (Randall 1967) in a Caribbean community in five different habitats (Opitz 1996; Bascompte et al., submitted). First, we analyze structure by addressing how simple trophic modules (i.e. tri-trophic food chains (FCs) and chains with omnivory (OMN) with the same set of species are shared among the five habitats. Second, we extend a previous metacommunity model (Mouquet and Loreau 2002) by incorporating the dynamics of trophic modules in a set of connected communities. Specifically, the following questions are addressed:

1. How are simple trophic modules composed by the same set of species represented among habitats?
2. How does the interplay between dispersal and food-web structure affect species dynamics at both local and regional scales?

Data collection: peculiarities and limitations

The Caribbean fish community here studied covers the geographic area of Puerto Rico–Virgin Islands.

Data were obtained in an area over more than 1000 km^2 covering the US Virgin Islands of St Thomas, St John, and St Croix (200 km^2), the British Virgin Islands (343 km^2), and Puerto Rico (554 km^2). The fish species analyzed and associated data were obtained mainly from the study by Randall (1967), synthesized by Opitz (1996).

Spatially explicit presence/absence community matrices were created by considering the presence of each species in a specific habitat only when that particular species was recorded foraging or breeding in that area (Opitz 1996; Froese and Pauly 2003). Community matrices include both the trophic links and the spatial distribution of 208 fish taxa identified to the species level. Randall's list of shark species was completed by Opitz (1996), which included more sharks with affinities to coral reefs of the Puerto Rico–Virgin islands area, based on accounts in Fischer (1978). Note that our trophic modules are composed only by fishes, and that all fish taxa is identified to the species level, which implies that results presented here are not affected by trophic aggregation.

The final spatially explicit community matrix includes 3,138 interactions, representing five food webs in five habitat types. Specifically, the habitat types here studied are mangrove/estuaries (m hereafter; 40 species and 94 interactions), coral reefs (c hereafter; 170 species and 1,569 interactions), seagrass beds/algal mats (a hereafter; 98 species and 651 interactions), sand (s hereafter; 89 species and 750 interactions), and offshore reefs (r hereafter; 22 species and 74 interactions). To a single habitat 85 species are restricted while 46, 63, 12, and 2 species occupy 2, 3, 4, and 5 habitats, respectively. Global connectivity values (C) within each habitat are similar to previously reported values for food webs (Dunne et al. 2002). Specifically, $C_m = 0.06$, $C_c = 0.054$, $C_a = 0.07$, $C_s = 0.095$, and $C_r = 0.15$.

Food-web structure and null model

We consider tri-trophic FCs (Figure 2.1(a)) and FCs with OMN (Figure 2.1(c)). We count the number and species composition of such trophic modules within the food web at each community. We then make pair-wise comparisons among communities ($n = 10$ pair-wise comparisons) and count the

number of chains (with identical species at all trophic levels) shared by each pair of communities. To assess whether this shared number is higher or lower than expected by chance we develop a null model. This algorithm randomizes the empirical data at each community, yet strictly preserves the ingoing and outgoing links for each species. In this algorithm, a pair of directed links A–B and C–D are randomly selected. They are rewired in such a way that A becomes connected to D, and C to B, provided that none of these links already existed in the network, in which case the rewiring stops, and a new pair of links is selected.

We randomized each food web habitat 200 times. For each pair of habitats we compare each successive pair of replicates and count the shared number of simple tri-trophic FCs and chains with OMN containing exactly the same set of species. Then we estimated the probability that a pair-wise comparison of a random replicate has a shared number of such modules equal or higher than the observed value. Recent algorithm analysis suggest that this null model represents a conservative test for presence–absence matrices (Miklós and Podani 2004).

We calculated the number of tri-trophic FCs, and OMN chains common to all pairs of communities, and compared this number with that predicted by our null model (Figure 2.1(b) and (d)). The coral reef habitat shares with all other habitats a number of FCs and OMN larger than expected by chance ($P < 0.0001$ in all pair-wise comparisons except for the mangrove comparison, where $P < 0.002$ and $P < 0.01$ for FCs and OMN, respectively). Similarly, seagrass beds/algal mats and sand (a/s contrasts) share a significant number of FCs and OMN ($P < 0.0001$). Globally, from the 10 possible intercommunity comparisons, five share a number of modules higher than expected by chance (Figure 2.1(a) and (c) where arrows are thick when the pair-wise comparison is statistically significant, and thin otherwise). This suggests that habitats sharing a significant proportion of trophic modules are mainly composed by a regional pool of individuals.

The average fraction of shared FCs and OMN between habitat pairs is 38%±24.5% and 41%±25%, respectively, which still leaves more than 50% of

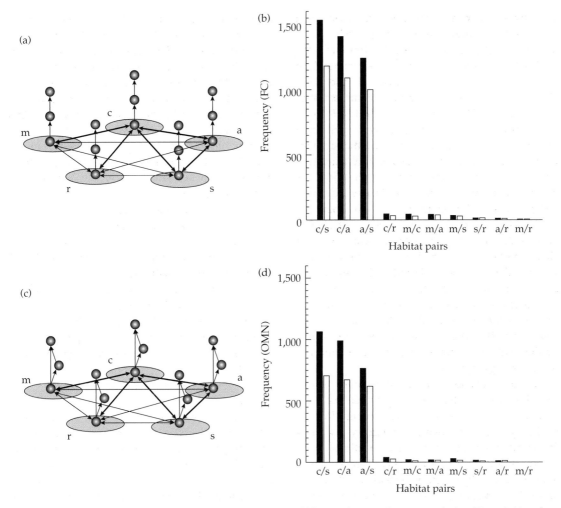

Figure 2.1 The food-web modules studied here are (a) tri-trophic FCs, and (c) OMN chains. Circles represent the five different habitat types. For each habitat pair, the link connecting the two habitats is thick if the number of shared trophic modules is significant, and thin otherwise; (b) and (d) represent the frequency of shared tri-trophic FCs and OMN chains, respectively in all pair-wise community comparisons. Black and white histograms represent the observed and the average expected value, respectively. Habitat types are mangrove/estuaries (m), coral reefs (c), seagrass beds/algal mats (a), sand (s), and offshore reefs (r). As noted, coral reefs (c), share with the rest of the habitats a number of FCs and OMN larger than expected by chance, which suggest a high degree of connectance promoted by dispersal.

different species composition trophic modules between habitats. However, it is interesting to note that 15 species (specifically, herbivorous species from *Blenniidae* and *Scaridae* families, and top species from *Carcharhinidae* and *Sphyrnidae* families) are embedded in more than 75% of trophic modules, which suggests that a small number of species are playing an important role in connecting through dispersal local community dynamics. Note that these highly connected species link

trophic modules across space in larger structures, which suggest a cohesive spatial structure (Melián and Bascompte 2004).

Dynamic metacommunity model

In order to assess the local and regional dynamics of the structure studied, we extend a previous metacommunity model (Mouquet and Loreau 2002, 2003) by incorporating trophic modules (tri-trophic FCs and FCs with OMN) in a set of interacting communities. The model follows the formalism of previous metapopulation models (Levins 1969) applied to the scale of the individual (Hastings 1980; Tilman 1994). At the local scale (within communities), we consider a collection of identical discrete sites given that no site is ever occupied by more than one individual. The regional dynamics is modeled as in mainland–island models with immigration (Gotelli 1991), but with an explicit origin of immigration that is a function of emigration from other communities in the metacommunity (Mouquet and Loreau 2003). Therefore, the model includes three hierarchical levels (individual, community, and metacommunity). The model reads as follows:

$$\frac{dP_{ik}}{dt} = \theta I_{ik} V_k + (1-d)c_{ik}P_{ik}V_k - m_{ik}P_{ik}$$
$$+ R_{ik}P_{ik} - C_{ik}P_{ik}. \qquad (2.1)$$

At the local scale, P_{ik} is the proportion of sites occupied by species i in community k. Each community consists of S species that indirectly compete within each trophic level for a limited proportion of vacant sites, V_k, defined as:

$$V_k = 1 - \sum_{j=1}^{S} P_{jk}, \qquad (2.2)$$

where P_{jk} represents the proportion of sites occupied by species j within the same trophic level in community k. The metacommunity is constituted by N communities. d is the fraction of individuals dispersing to other habitats, and dispersal success, θ, is the probability that a migrant will find a new community, c_{ik} is the local reproductive rate of species i in community k, and m_{ik} is the mortality rate of species i in community k.

For each species in the community, we considered an explicit immigration function I_{ik}. Emigrants were combined in a regional pool of dispersers that was equally redistributed to all other communities, except that no individual returned to the community it came from (Mouquet and Loreau 2003). After immigration, individuals were associated to the parameters corresponding to the community they immigrated to. I_{ik} reads as:

$$I_{ik} = \frac{d}{N-1} \sum_{l \neq k}^{N} c_{il}P_{il}, \qquad (2.3)$$

where the sum stands for all the other communities l. R_{ik} represents the amount of resources available to species i in community k

$$R_{ik} = \sum_{j=1}^{S} a_{ijk}P_{jk}, \qquad (2.4)$$

where a_{ijk} is the predation rate of species i on species j in community k, and the sum is for all prey species. Similarly, C_{ik} represents the amount of consumption exerted on species i by all its predators in community k, and can be written as follows:

$$C_{ik} = \sum_{j=1}^{S} a_{jik}P_{jk}, \qquad (2.5)$$

where a_{jik} is the predation rate of species j on species i in community k, and the sum is for all predator species.

We have numerically simulated a metacommunity consisting of six species in six communities. In each community, either two simple tri-trophic FCs, or two OMN chains are assembled with the six species. The two trophic modules within each community are linked only by indirect competition between species within the same trophic level. We assumed a species was locally extinct when its proportion of occupied sites was lower than 0.01. Mortality rates (m_{ik}) are constant and equal for all species. Dispersal success (θ) was set to 1.

We considered potential reproductive rates to fit the constraint of strict regional similarity, SRS (Mouquet and Loreau 2003). That is, species within each trophic level have the same regional basic reproductive rates, but these change locally among communities. Under SRS, each species within each trophic level is the best competitor in one community. Similarly, we introduce the constraint of strict regional trophic similarity (SRTS). That is, each consumer has the same set of local energy requirements but distributed differently among communities. Additionally, we assumed a direct relationship between the resource's local reproductive rate and the intensity it is predated with (Jennings and Mackinson 2003).

Under the SRS and SRTS scenarios, regional species abundance and intercommunity variance are equal for each of the two species within the same trophic level. Regional abundance in OMN is higher, equal, and lower for top, intermediate, and basal species, respectively. Local abundances differ significantly between the two modules explored. Specifically, when there is no dispersal ($d = 0$) there is local exclusion by the competitively superior species (Mouquet and Loreau 2002). This occurs for the basal and top species in the simple trophic chain. The variance in the abundance of the basal and top species between local communities is thus higher without dispersal for tri-trophic FCs (Figure 2.2(a)).

However, the situation is completely different for OMN. Now, intercommunity variance is very low for both the basal and top species in the absence of dispersal, and dramatically increases with d in the case of the top species. When the communities are extremely interconnected, the top species disappears from the two communities ($P_{ik} < 0.01$), and is extremely abundant in the remaining communities. For intermediate species, increasing dispersal frequency decreases the intercommunity variance, except when d ranges between 0 and 0.1 in FCs (Figures 2.2(a) and (b)).

Finally, we can see in Figure 2.2(b) (as compared with Figure 2.2(a)) that intercommunity variance for high d-values is higher in a metacommunity with OMN. Thus, the interplay between dispersal among spatially structured communities and food-web structure greatly affects local species abundances. The results presented here were obtained with a single set of species parameters. Under the SRS and SRTS scenarios, results are qualitatively robust to deviations from these parameter values.

Summary and discussion

It is well known that local communities can be structured by both local and regional interactions (Ricklefs 1987). However, it still remains unknown what trophic structures are shared by a set of interacting communities and its dynamical implications for the persistence of biodiversity. The present study is an attempt to link local and regional food-web structure and dynamics in a spatially structured marine food web.

Communities in five habitats of the Caribbean have shown significantly similar trophic structures which suggest that these communities are open to

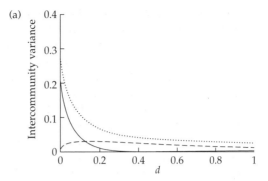

(a)

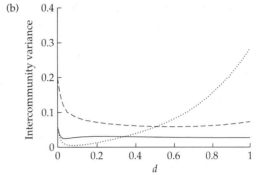

(b)

Figure 2.2 Intercommunity variance in local species abundance for the basal (continuous line), intermediate (broken line), and top (dotted) species as a function of the proportion of dispersal between communities (d). (a) Represents tri-trophic FCs and (b) OMN chains. Parameter values are $m_{ik} = 0.2$, c_{ik} for basal species is 3, 2.8, 2.6, 2.4, 2.2, and 2 from the first to the sixth community, respectively. For intermediate species c_{ik} is 1.5, 1.4, 1.3, 1.2, 1.1, and 1, respectively from the first to the sixth community. Top species reproductive values are 0.8, 0.75, 0.7, 0.65, 0.6, and 0.55, respectively. Predation rates of intermediate and top species j on species i in community k are 0.6, 0.5, 0.4, 0.3, 0.2, and 0.1, respectively. The initial proportion of sites occupied by species i in community k, (P_{ik}) is set to 0.05. As noted, in closed metacommunities, tri-trophic FCs show an extreme variation in local abundances for both the basal and top species ($P_{ik} < 0.01$) in two and three communities, respectively. On the other hand, OMN shows the highest intercommunity variance for high dispersal rates ($d = 1$). The top species becomes unstable, and goes extinct in two local communities ($P_{ik} < 0.01$).

immigration (Karlson and Cornell 2002). It has been recently shown that mangroves in the Caribbean strongly influence the local community structure of fish on neighboring coral reefs (Mumby et al. 2004). Additionally, empirical studies have shown that dispersal among habitats and local species interactions are key factors for metacommunity structure (Shurin 2001; Cottenie et al. 2003; Kneitel and Miller 2003; Cottenie and De Meester 2004), and the persistence of local and regional diversity (Mouquet and Loreau 2003). However, it still remains unclear how the interplay between dispersal and more complex trophic structures affects species persistence in local communities (Carr et al. 2002; Kneitel and Miller 2003).

In the present work, closed communities ($d = 0$) with tri-trophic FCs showed an extreme variation in local abundances for both the basal and top species (Figure 2.2(a)). On the other hand, OMN shows the highest intercommunity variance for high dispersal rates ($d = 1$). The top species becomes unstable, and goes extinct in two local communities (Figure 2.2(b)). Recent empirical studies have shown that increasing dispersal frequency in intermediate species decreases the variance among local communities (Kneitel and Miller 2003), a pattern consistent with theoretical results presented here (see dotted line in Figure 2.2(a) and (b)). Further data synthesis and theoretical work is needed here to integrate the functional links between habitats and the local dynamics of species embedded in food webs.

In summary, the similarity in the trophic modules reported here suggests a strong link among the spatially structured communities. The level of connectivity among these local communities and the type of trophic modules alter local abundance of species and promote local changes in diversity. It still remains unexplored how the results here presented change by the introduction of a larger number of interacting modules in a set of spatially structured communities. Our result predicts a relative stability in the composition of basal species, and a dramatic influence in the abundance of top species depending on the connectivity (i.e. dispersal) among distinct habitats.

Acknowledgments

We thank the editors of this book for inviting us to contribute this chapter. We thank Miguel A. Fortuna and Mayte Valenciano for their useful comments on a previous draft. Funding was provided by the Spanish Ministry of Science and Technology (Grants REN2003-04774 to JB and REN2003-00273 to PJ, and Ph.D. Fellowship FP2000-6137 to CJM).

CHAPTER 3

Role of network analysis in comparative ecosystem ecology of estuaries

Robert R. Christian, Daniel Baird, Joseph Luczkovich, Jeffrey C. Johnson, Ursula M. Scharler, and Robert E. Ulanowicz

Introduction

Assessments of trophic structure through ecological network analysis (ENA) have been done in a wide variety of estuarine and coastal environments. For example, some have used it to compare trophic structures within ecosystems focusing on temporal conditions (Baird and Ulanowicz 1989; Baird et al. 1998) and among ecosystems focusing on spatial conditions (e.g. Baird and Ulanowicz 1993; Christensen 1995). These comparisons have used carbon or energy as the currency with which to trace the interactions of the food webs, although other key elements such as nitrogen and phosphorus have also been used in ENA (Baird et al. 1995; Ulanowicz and Baird 1999; Christian and Thomas 2003). One of the primary features of ENA is that the interactions are weighted. That is, they represent rates of flow of energy or matter and not simply their existence. Other kinds of comparisons have been attempted less frequently. Effects of currency used to track trophic dynamics has received little attention (Christian et al. 1996; Ulanowicz and Baird 1999), and comparisons of ENA with other modeling approaches are quite rare (Kremer 1989; Lin et al. 2001). There is a need to expand the applications of network analysis (NA) to address specific questions in food-web ecology, and to use it more frequently to explain and resolve specific management issues. The NA approach must be combined with other existing methods of identifying ecosystem performance to validate and improve our inferences on trophic structure and dynamics.

Estuaries are excellent ecosystems to test the veracity of the inferences of ENA for three reasons. First, more NAs have been conducted on estuaries than on any other kind of ecosystem. Second, estuarine environments are often stressed by natural and anthropogenic forcing functions. This affords opportunities for evaluating controls on trophic structure. Third, sampling of estuaries has often been extensive, such that reasonable food webs can be constructed under different conditions of stress. Finally, other modeling approaches have been used in numerous estuarine ecosystems. Results of these alternate modeling approaches can be compared to those of ENA to test the coherence of inferences across perspectives of ecosystem structure and function. These conditions set the stage for an evaluation of the status of ENA as a tool for comparative ecosystem ecology.

Comparative ecosystem ecology makes valuable contributions to both basic ecology and its application to environmental management. Given the critical position of estuaries as conduits for materials to the oceans and often as sites of intense human activities in close proximity to important natural resources, ENA has been used frequently for the assessment of the effects of environmental conditions within estuaries related to management. Early in the use of ENA in ecology, Finn and Leschine (1980) examined the link between fertilization of saltmarsh grasses and shellfish production.

Baird and Ulanowicz (1989) expanded the detail accessible in food webs and the consequences of this increased detail in their seminal paper of seasonal changes within the Chesapeake Bay. In 1992, Ulanowicz and Tuttle determined through ENA and field data that the overharvesting of oysters may have had significant effects on a variety of aspects of the food web in Chesapeake Bay. Baird and Heymans (1996) studied the reduction of freshwater inflow into an estuary in South Africa and noted changes in food-web structure and trophic dynamics. More recently, Brando et al. (2004) and Baird et al. (2004) evaluated effects of eutrophication and its symptoms on Orbetello Lagoon, Italy, and Neuse River Estuary, USA, respectively. All of these studies involved comparisons of conditions linked to human impacts.

The first comprehensive review of the methodologies and use of ENA, an associated software NETWRK4, and application in marine ecology was published in 1989 (Wulff et al. 1989). Other approaches to ENA have been developed and applied to food webs. The software programs ECOPATH and ECOSIM have been used throughout the world to address various aspects of aquatic resources management (see www.ecopath.org/for summary of activities; Christensen and Pauly 1993). In parallel with NETWRK4, ECOPATH was developed by Christensen and Pauly (1992, 1995) and Christensen et al. (2000), based on the original work of Polovina (1984). The dynamic simulation module, ECOSIM, was developed to facilitate the simulation of fishing effects on ecosystems (Walters et al. 1997). NETWRK4 and ECOPATH include, to various extents, similar analytical techniques, such as input–output analysis, Lindeman trophic analysis, a biogeochemical cycle analysis, and the calculation of information-theoretical indices to characterize organization and development. However, some analyses are unique to each. There are several differences in the input methodology between the NETWRK4 and ECOPATH software, which lead to differences in their outputs. Heymans and Baird (2000) assessed these differences in a case study of the northern Benguela upwelling system. Environs analysis, developed by Patten and colleagues (reviewed by

Fath and Patten 1999), provides some of the same analyses found in NETWRK4 but includes others based on the theoretical considerations of how systems interact with their environment. Lastly, social NA is beginning to be applied to ecological systems. A software package so used is UCINET (www.analytictech.com/ucinet.htm; Johnson et al. 2001; Borgatti et al. 2002). Although several methods and software packages exist for evaluating weighted food webs, none has been developed and validated to an extent to give a good understanding of the full implications of the variety of results.

We have organized this chapter to address the use of ENA associated with estuarine food webs in the context of comparative ecosystem ecology. Comparisons within and among estuaries are first considered. ENA provides numerous output variables, but we focus largely on five ecosystem-level variables that index ecosystem activity and organization. We address the ability of recognizing ecosystem-level change and patterns of change through the use of these indices. Then we compare several estuarine food webs to budgets of biogeochemical cycling to assess the correspondence of these two facets of ecosystems. Again we use these same indices and relate them to indices from the biogeochemical budgeting approach of the Land–Ocean Interaction in the Coastal Zone (LOICZ) program. How do the two modeling approaches compare in assessing ecosystems? Finally, comparisons of food-web diagrams are problematic if the food webs are at all complex. Recently, visualization tools from biochemistry and social networks have been used to portray food webs. We explore this new approach in the context of intrasystem comparisons.

Estuarine food-web comparisons

We highlight how food webs are perceived to change or remain stable across a variety of conditions. First, we compare systems temporally from intra and interseasonal to longer-term changes. Within a relatively unimpacted ecosystem, food webs may tend to be relatively stable with differences among times related to altered, weather-related metabolism and differential growth, migrations and ontogenetic changes in populations (Baird and Ulanowicz 1989). Human impacts may alter these drivers of change and add new ones. Multiple food-web networks for an ecosystem tend to be constructed under common sets of rules, facilitating temporal comparisons. Then we compare food webs among ecosystems where major differences may exist in the very nature of the food webs. Interpreting such differences is more difficult than intrasystem comparisons and must be viewed with more caution. We have used studies of intersystem comparisons where effort was made by the authors to minimize differences in rulemaking and network structure. Should networks be constructed under different constraints, such as inconsistent rules for aggregation, the interpretation of differences in the NA results is difficult and should be viewed with more caution.

Ecological network analysis provides a myriad of output variables and indices. Each has its own sensitivity to differences in network structure. Generally, indices of population (i.e. at compartment-level) and cycling structure are more sensitive than ecosystem-level indices in terms of responsiveness to flow structure and magnitude of flows (Baird et al. 1998; Christian et al. 2004). Also, because currency and timescale may differ among networks, direct comparisons using different flow currencies are difficult. We focus on five ecosystem-level output variables of ENA, four of which are ratios. These are described in greater detail elsewhere (Kay et al. 1989; Christian and Ulanowicz 2001; Baird et al. 2004). The first adds all flows within a network, total system throughput (TST), and reflects the size, through activity, of the food web. Combinations of flows may be interpreted as occurring in cycles, and the percentage of TST involved in cycling is called the Finn Cycling Index (FCI; Finn 1976). The turnover rate of biomass of the entire ecosystem can be calculated as the sum of compartment production values divided by the sum of biomass (P/B). Networks can be collapsed, mathematically into a food chain, or Lindeman Spine, with the processing of energy or matter by each trophic level identified (Ulanowicz 1995). The trophic efficiency (TE) of each level represents the ingestion of the next level as a percentage of the ingestion of the focal level. The geometric mean of individual level efficiencies is the system's TE. Ulanowicz has characterized the degree of organization and maturity of an ecosystem through a group of information-based indices (Ulanowicz 1986). Ascendency/developmental capacity (A/C) is a ratio of how organized, or mature, systems are, where ecosystems with higher values reflect relatively higher levels of organization. Thus, these five indices can be used to describe both extensive and intensive aspects of food webs. While our focus is on these indices, we incorporate others as appropriate to interpret comparisons.

Temporal comparisons

There are surprisingly few estuarine ecosystems for which food-web networks have been examined during different times. Most networks represent annual mean food webs. We provide a brief review of some for which we have direct experience and can readily assess the focal ecosystem-level indices. These are ecosystems for which NETWRK4 was applied rather than ECOPATH, because of some differences in model construction and analysis (e.g. general use of gross primary production in NETWRK4 and net primary production in ECOPATH). The shortest timescale examined has been for a winter's *Halodule wrightii* ecosystem in Florida, USA (Baird et al. 1998) where two sequential months were sampled and networks analyzed. Seasonal differences between food webs were a central part of the Baird and Ulanowicz (1989) analysis of the food web in Chesapeake Bay. Almunia et al. (1999) analyzed seasonal differences in Maspalomas Lagoon, Gran Canaria, following the cycle of domination by benthic versus pelagic primary producers. Florida Bay, which constitutes the most detailed quantified network to date, has been analyzed for seasonal differences (Ulanowicz et al. 1999). Finally, interdecadal changes, associated with hydrological modifications, were assessed for the Kromme Estuary, South Africa (Baird and Heymans 1996). Table 3.1 shows the five indices for each temporal condition for these ecosystems.

Table 3.1 Temporal changes in ecosystem-level attributes for different estuarine ecosystems

Time period	TST (mg C m^{-2} per day)	FCI (%)	P/B (day^{-1})	TE (%)	A/C (%)
St Marks, intraseasonal					
January 1994	1,900	16	0.037	4.9	36
February 1994	2,300	20	0.041	3.3	32
Neuse, intraseasonal					
Early summer 1997	18,200	14	0.15	5.0	47
Late summer 1997	17,700	16	0.30	4.7	47
Early summer 1998	18,600	16	0.24	3.3	47
Late summer 1998	20,700	16	0.33	4.9	46
Chesapeake, interseasonal					
Spring	1,300,000	24	n.a.	9.6	45
Summer	1,700,000	23	n.a.	8.1	44
Fall	800,000	22	n.a.	10.9	48
Winter	600,000	23	n.a.	8.6	49
Maspalomas Lagoon, interseasonal					
Benthic-producer-dominated system	13,600	18	n.a.	11.4	40
Transitional	12,300	23	n.a.	12.8	38
Pelagic-producer-dominated system	51,500	42	n.a.	8.7	45
Florida Bay, interseasonal					
Wet	3,460	26	n.a.	n.a.	38
Dry	2,330	n.a.	n.a.	n.a.	38
Kromme, interdecadal					
1981–84	42,830	12	0.012	4.5	48
1992–94	45,784	10	0.011	2.8	46

Note: Flow currency of networks is carbon; n.a. means not available.

Ecological network analysis was applied to a winter's *H. wrightii* ecosystem, St Marks National Wildlife Refuge, Florida, USA (Baird et al. 1998; Christian and Luczkovich 1999; Luczkovich et al. 2003). Unlike most applications of ENA, the field sampling design was specific for network construction. From these data and from literature values, the authors constructed and analyzed one of the most complex, highly articulated, time-and site-specific food-web networks to date. Two sequential months within the winter of 1994 were sampled with the temperature increase of 5°C from January to February. Metabolic rates, calculated for the different temperatures and migrations of fish and waterfowl affected numerous attributes of the food webs (Baird et al. 1998). The changes in the focal indices are shown in Table 3.1. Activity estimated by the three indices was higher during the warmer period with >20% more TST, and FCI,

and a 12% increase in P/B. However, organization of the food web (A/C) decreased, and dissipation of energy increased lowering the TE. Although statistical analysis of these changes was not done, it would appear that the indices do reflect perceived effects of increased metabolism.

The food web of the Neuse River Estuary, NC, was assessed during summer conditions over two years (Baird et al. 2004; Christian et al. 2004). The Neuse River Estuary is a highly eutrophic estuary with high primary production and long residence times of water. Temperature was not considered to differ as dramatically from early to late summer, but two major differences distinguished early and late summer food webs. First was the immigration and growth of animals to the estuary during summer, which greatly increased the biomass of several nekton compartments. Second, hypoxia commonly occurs during summer, stressing both nekton and benthos. Hypoxia was more dramatic in 1997 (Baird et al. 2004). Benthic biomass decreased during both summers, but the decrease was far more dramatic during the year of more severe hypoxia. Changes in the ecosystem-level indices were mostly either small or failed to show the same pattern for both years (Table 3.1). A/C changed little over summers or across years. TST, FCI, and TE had different trends from early to late summer for the two years. Only P/B showed relatively large increases from early to late summer. Thus, inferences regarding both activity and organization across the summer are not readily discerned. We have interpreted the results to indicate that the severe hypoxia of 1997 reduced the overall activity (TST) by reducing benthos and their ability to serve as a food resource for nekton. But these ecosystem-level indices do not demonstrate a stress response as effectively as others considered by Baird et al. (2004).

The food web in Chesapeake Bay was analyzed for four seasons (Baird and Ulanowicz 1989). Many of the changes linked to temperature noted for the within-season changes of the food web in St Marks hold here (Table 3.1). TST and P/B are highest in summer and lowest in winter, although the other measure of activity, FCI, does not follow this pattern. However, FCI is a percentage of TST. The actual amount of cycled flow (TST × FCI) does follow the temperature-linked pattern. TE failed to show a pattern of increased dissipation with higher temperatures, although it was lowest during summer. Organization, as indexed by A/C, showed the greatest organization in winter and least in summer. Hence, in both of the aforementioned examples, times of higher temperature and therefore, higher rates of activity and dissipation of energy were linked to transient conditions of decreased organization. These findings are corroborated for spring—summer comparisons of these food webs are discussed later in the chapter.

Maspalomas Lagoon, Gran Canaria, shows, over the year, three successive stages of predominance of primary producers (Almunia et al. 1999). The system moves from a benthic-producer-dominated system via an intermediate stage to a pelagic-producer-dominated system. The analysis of system-level indices revealed that TST and A/C increased during the pelagic phase (Table 3.1). The proportional increase in TST could be interpreted as eutrophication, but the system has no big sources of material input from outside the system. Almunia et al. (1999) explained the increase in A/C as a shift in resources from one subsystem (benthic) to another (pelagic). The FCI was lowest during the benthic-dominated stage and highest during the pelagic-dominated stage, and matter was cycled mainly over short fast loops. The pelagic-dominated stage was interpreted as being in an immature state, but this interpretation is counter to the highest A/C during the pelagic stage. The average TE dropped from the benthic-dominated stage to the pelagic-dominated stage, and the ratio of detritivory to herbivory increased accordingly. Highest values of detritivory coincided with lowest values of TE.

Florida Bay showed remarkably little change in whole-system indices between wet and dry seasons (Ulanowicz et al. 1999; www.cbl.umces.edu/~bonda/FBay701.html). Although system-level indices during the wet season were about 37% greater than the same indices during the dry season, it became apparent that this difference was almost exclusively caused by the change in system activity (measured as TST), which was used to scale the system-level indices to the size of the system. The fractions of A/C and the distribution of the different components of the overhead were almost identical during both seasons. Ulanowicz et al. (1999) concluded that the Florida Bay ecosystem structure is remarkably stable between the two seasons. (FCI was high during the wet season (>26%) but could not be calculated for the dry season since the computer capacity was exceeded by the amount of cycles (>10 billion).)

Lastly, we consider a larger timescale of a decade for the Kromme Estuary, South Africa. Freshwater discharge to this estuary was greatly reduced by 1983 due to water diversion and damming projects, greatly lessening nutrient additions, salinity gradients, and pulsing (i.e. flooding; Baird and Heymans 1996). Can ecosystem-level indices identify resultant changes to the food web? Although there was a slight increase in TST, the trend was for a decrease in all other measures (Table 3.1). However, all of these were decreases of less than 20%, with the exception of TE. This general, albeit slight, decline has been attributed to the stress of the reduced flow regime (Baird and Heymans 1996). The TE decreased during the decade to less than half the original amount. Thus, much less of the primary production was inferred to pass to higher, commercially important, trophic levels. Further, TE of the Kromme under a reduced flow regime was among the smallest for ecosystems reviewed here.

In summary, most temporal comparisons considered were intra-annual, either within or among seasons. Seasons did not have comparable meaning among ecosystems. The Chesapeake Bay networks were based on solar seasons, but Maspalomas Lagoon and Florida Bay networks were not. All indices demonstrated intra-annual change, although the least was associated with A/C. This is to be expected as both A and C are logarithmically based indices. Summer or warmer seasons tend to have higher activity (TST and FCI or FCI × TST), as expected. In some cases this was linked to lower organization, but this was not consistent across systems. We only include one interannual, actually interdecadal, comparison, but differences within a year for several systems were as great as those between decades for the Kromme Estuary. Interannual differences in these indices for other coastal ecosystems have been calculated but with different currencies and software (Brando et al. 2004; others). Elmgren (1989) has successfully used trophic relationships and production estimates to assess how eutrophication of the Baltic Sea over decades of enhanced nutrient loading has modified production at higher trophic levels. Even though the sample size remains small, it appears that intra-annual changes in food-web structure and trophic dynamics can equal or exceed those across years and across different management regimes. Obviously, more examples and more thorough exploration of different indices are needed to establish the sensitivities of ecosystem-level indices to uncertainties in ecosystem condition.

Interecosystem comparisons

Ecological network analysis has been used in intersystem comparisons to investigate the structure and processes among systems of different geographic locations, ranging from studies on estuaries in relatively close proximity (Monaco and Ulanowicz, 1997; Scharler and Baird, in press) to those of estuarine/marine systems spanning three continents (Baird et al. 1991). Perhaps the most extensive comparison has been done by Christensen (1995) on ecosystems using ECOPATH to evaluate indices of maturity. These comparisons are limited, as discussed previously, because of differences in rules for constructing and analyzing networks. We review here some of the estuarine and coastal marine comparisons that have taken into account these issues, beginning with our focal indices.

The geographically close Kromme, Swartkops, and Sundays Estuaries, differ in the amount of freshwater they receive, and consequently in the amount of nutrients and their habitat structure (Scharler and Baird 2003). Input–output analysis highlighted the differences in the dependencies (or extended diets) of exploited fish and invertebrate bait species. Microalgae were found to play an important role in the Sundays Estuary (high freshwater and nutrient input) as a food source to exploited fish and invertebrate bait species, whereas detritus and detritus producers were of comparatively greater importance in the Kromme (low nutrients) and Swartkops (pristine freshwater inflow, high nutrients) Estuaries (Scharler and Baird, in press).

When comparing some indicators of system performance such as TST, FCI, A/C, and TE of the Kromme, Swartkops, and Sundays Estuaries, it revealed an interplay between the various degrees of physical and chemical forcings. The Kromme Estuary is severely freshwater starved and so lacks a frequent renewal of the nutrient pool. Freshets have largely disappeared as a physical disturbance. The Sundays system features increased freshwater input due to an interbasin transfer, and the Swartkops Estuary has a relatively pristine state of the amount of freshwater inflow but some degree of anthropogenic pollution (Scharler and Baird, in press). NA results showed that the Swartkops was more impacted due to a low TST and a high average residence time (ART, as total system biomass divided by total outputs) of material and least efficient to pass on material to higher trophic levels. The Kromme was more self-reliant (higher FCI) than the Sundays (lowest FCI). The Sundays was also the most active featuring a comparatively high TST and low ART. However, a comparatively high ascendency in the Sundays Estuary was not only a result of the high TST, which could have implicated the consequences of eutrophication (Ulanowicz 1995a), but it also featured the highest AMI (the information-based component of ascendency)(Scharler and Baird, in press). The Kromme Estuary had the comparatively lowest A/C, and lowest AMI, and Baird and Heymans (1996) showed that since the severe freshwater inflow restrictions, a decline of the internal organization and maturity was apparent.

Intercomparisons of estuaries and coastal aquatic ecosystems have often focused on other issues in addition to the focal indices of this chapter. One important issue has been the secondary production of ecosystems, which is of special interest in terms of commercially exploited species. As Monaco and Ulanowicz (1997) stated, there can be differences in the efficiency of the transformation of energy or carbon from primary production to the commercial species of interest. By relating the output of planktivorous and carnivorous fish, and that of suspension feeders to primary production, it became apparent that in Narragansett Bay twice as many planktivorous fish and 4.6–7.4 as many carnivorous fish were produced per unit primary production than in Delaware or Chesapeake Bay, respectively. The latter, on the other hand, produced 1.3 and 3.5 as much suspension feeding biomass than Narragansett and Delaware Bay from one unit of phytoplankton production (Monaco and Ulanowicz 1997). This analysis was performed on the diet matrix to quantify a contribution from a compartment (in this case the primary producers) in the network to any other, over all direct and indirect feeding pathways, and is described as part of an input–output analysis in Szyrmer and Ulanowicz (1987).

This approach of tracing the fate of a unit of primary production through the system was also applied by Baird et al. (1991) who calculated the fish yield per unit of primary production in

estuarine and marine upwelling systems. They used a slightly different approach, in that only the residual flow matrices (i.e the straight through flows) were used for this calculation, since the cycled flows were believed to inflate the inputs to the various end compartments. In this study, the most productive systems in terms of producing planktivorous fish from a unit of primary production were the upwelling systems (Benguela and Peruvian) and the Swartkops Estuary, compared to the Baltic, Ems, and Chesapeake (Baird et al. 1991). In terms of carnivorous fish, the Benguela upwelling system was the most efficient, followed by the Peruvian and Baltic (Baird et al. 1991).

Trophic efficiencies have also been used to make assumptions about the productivity of a system. In perhaps the first intersystem comparison using NA, Ulanowicz (1984) considered the efficiencies with which primary production reached the top predators in two marsh gut ecosystems in Crystal River, Florida. Monaco and Ulanowicz (1997) identified that fish and macroinvertebrate catches in the Chesapeake Bay were higher compared to the Narragansett and Delaware Bay, despite its lower system biomass, because the transfer efficiencies between trophic levels were higher. Similarly, transfer efficiencies calculated from material flow networks were used to estimate the primary production required to sustain global fisheries (Pauly and Christensen 1995). Based on a mean energy transfer efficiency between trophic levels of 48 ecosystems of 10%, the primary production required to sustain reported catches and bycatch was adjusted to 8% from a previous estimate of 2.2%.

In the context of the direct and indirect diet of exploited and other species, it can be of interest to investigate the role of benthic and pelagic compartments. The importance of benthic processes in the indirect diet of various age groups of harvestable fish was determined with input–output analysis by Monaco and Ulanowicz (1997). The indirect diet is the quantified total consumption by species j that has passed through species i along its way to j (Kay et al. 1989). They found that benthic processes in the Chesapeake Bay was highly important to particular populations of juvenile and adult piscivores. Indirect material transfer effects revealed that the Chesapeake Bay relied more heavily on its benthic compartments compared to the Narragansett and Delaware Bays and that disturbances to benthic compartments may have a comparatively greater impact on the system (Monaco and Ulanowicz 1997). The pattern changes somewhat with season, as discussed in the section, "visualization of network dynámics" of this chapter.

The shallow Kromme, Swartkops, and Sundays Estuaries were found to rely more on their benthic biota in terms of compartmental throughput and the total contribution coefficients in terms of compartmental input (Scharler and Baird, in press). In terms of carbon requirements, the Kromme and Swartkops Estuaries depended two-third on the benthic components and one-third on the pelagic components, whereas the Sundays Estuary depended to just over half on its benthic components. The Sundays Estuary was always perceived to be "pelagic driven," probably due to the high phyto and zooplankton standing stocks, which are a result of the regular freshwater and nutrient input. By considering not only direct effects, but also all indirect effects between the compartments, the regular freshwater input suppressed somewhat the dependence on benthic compartments, but has not switched the system to a predominantly pelagic dependence (Scharler and Baird, in press).

Indicators of stress, as derived from ENA, have been discussed in several comparative studies. Baird et al. (1991) proposed a distinction between physical stress and chemical stress. The former has in general been influencing ecosystems, such as upwelling systems, for a time long enough so that the systems themselves could evolve under the influence of this type of physical forcing. Freshwater inflow into estuaries similarly determines the frequency of physical disturbance, due to frequent flooding in pristine systems and restrictions thereof in impounded systems. On the other hand, chemical influences are in general more recent through anthropogenic pollution, and the systems are in the process of changing from one response type (unpolluted) to another (polluted) that adjusts to the chemical type of forcing (Baird et al. 1991). With this perspective, Baird et al. (1991)

pointed out that the system P/B ratio is not necessarily a reflection of the maturity of the system, but due to NA results reinterpreted maturity in the context of physical forcing (e.g. the upwelling systems (Peruvian, Benguela) are considered to be mature under their relatively extreme physical forcings, although they have a higher system P/B ratio than the estuarine systems (Chesapeake, Ems, Baltic, Swartkops)).

Comparison of whole-system indices between the Chesapeake and Baltic ecosystems provided managers with a surprise (Wulff and Ulanowicz 1989). The conventional wisdom was that the Baltic, being more oligohaline than the Chesapeake, would be less resilient to stress. The organizational status of the Baltic, as reflected in the relative ascendency (A/C) was greater (55.6%) than that of the Chesapeake (49.5%) by a significant amount. The relative redundancy (R/C) of the Chesapeake (28.1%) was correspondingly greater than that of the Baltic (22.0%), indicating that the Chesapeake might be more stressed than the Baltic. The FCI in the Chesapeake was higher (30%) than in the Baltic (23%). As greater cycling is indicative of more mature ecosystems (Odum 1969), this result seemed at first to be a counterindication that the Chesapeake was more stressed, but Ulanowicz (1984) had earlier remarked that a high FCI could actually be a sign of stress, especially if most of the cycling occurs over short cycles near the base of the trophic ladder. This was also the case in this comparison, as a decomposition of cycled flow according to cycle length revealed that indeed most of the cycling in the Chesapeake occurred over very short cycles (one or two components in length), whereas recycle over loops that were three or four units long was significantly greater in the Baltic. The overall picture indicated that managerial wisdom had been mistaken in this comparison, as the saltier Chesapeake was definitely more disrupted than the Baltic.

Intermodel and technique comparisons

Another modeling protocol was developed under the auspices of the International Geosphere–Biosphere Program (IGBP), an outcome of the 1992

Rio Earth Summit and established in 1993. The aims of the IGBP are "to describe and understand the physical, chemical and biological processes that regulate the earth system, the environment provided for life, the changes occurring in the system, and the influence of human actions." In this context, the Land Ocean Interactions in the Coastal Zone (LOICZ) core project of the IGBP was established. LOICZ focuses specifically on the functioning of coastal zone ecosystems and their role in the fluxes of materials among land, sea, and atmosphere; the capacity of the coastal ecosystems to transform and store particulate and dissolved matter; and the effects of changes in external forcing conditions on the structure and functioning of coastal ecosystems (Holligan and de Boois 1993; Pernetta and Milliman 1995).

The LOICZ biogeochemical budgeting procedure was subsequently developed that essentially consists of three parts: budgets for water and salt movement through coastal systems, calculation of rates of material delivery (or inputs) to and removal from the system, and calculations of rate of change of material mass within the system (particularly C, N, and P). Water and salt are considered to behave conservatively, as opposed to the nonconservative behavior of C, N, and P. Assuming a constant stoichiometric relationship (e.g. the Redfield ratio) among the nonconservative nutrient budgets, deviations of the fluxes from the expected C:N:P composition ratios can thus be assigned to other processes in a quantitative fashion. Using the flux of P (particularly dissolved

inorganic P), one can derive whether (1) an estuary is a sink or a source of C, N, and P, that is $\Delta Y = \text{flux}_{out} - \text{flux}_{in}$, where $Y = C$, N, or P; (2) the system's metabolism is predominantly autotrophic or heterotrophic, that is, $(p - r) = \Delta DIP(C:P)_{part}$, where $(p - r)$ is photosynthesis minus respiration; and (3) nitrogen fixation (nfix) or denitrification (denit) predominates in the system, where $(nfix - denit) = \Delta DIN - \Delta DIP(N:P)_{part}$ (Gordon et al. 1996). A summary of attributes for this modeling approach is shown in Table 3.2.

This section explores the possibility of linkages between the two different methodologies of ENA and LOICZ biogeochemical budgeting protocol. The rationale for this hypothesis is:

The magnitude and frequency of N and P loadings and the transformation of these elements within the system, ultimately affect the system's function. Since we postulate that system function is reflected in network analysis outputs, we infer that there should exist correspondence in the biogeochemical processing, as indexed by the LOICZ approach, and trophic dynamics, as indexed by network analysis outputs.

To do this, we used ENA and LOICZ variables and output results from six estuarine or brackish ecosystems based on input data with a high level of confidence (Table 3.3). We first performed Spearman's and Kendall's correlation analyses between the ENA and LOICZ output results of a number of system indices of the six ecosystems. From the correlation matrices we selected those variables which showed correlation values of 80%

Table 3.2 System properties and variables derived from NA and the LOICZ biogeochemical budgeting protocol used in factor analysis

LOICZ variables	Description of variable/system property
Nutrient loading	*From land to ocean, two macronutrients and their possible origins*
Dissolved inorganic nitrogen (DIN) (mol m^{-2} per year)	Products of landscape biogeochemical reactions
Dissolved inorganic phosphorus (DIP) (mol m^{-2} per year)	Materials responding to human production, that is, domestic (animal, human) and industrial waste, and sewage, fertilizer, atmospheric fallout from vehicular and industrial emissions
ΔDIN (mol m^{-2} per year)	Flux$_{out}$ − Flux$_{in}$
ΔDIP (mol m^{-2} per year)	Flux$_{out}$ − Flux$_{in}$
Net ecosystem metabolism (NEM) (mol m^{-2} per year)	Assumed that the nonconservative flux of DIP is an approximation of net metabolism: $(p - r) = -\Delta DIP(C:P)$
NFIXDNIT (mol m^{-2} per year)	Assumed that the nonconservative flux of DIN approximates N fixation minus denitrification: $(nfix - denit) = \Delta DIN - \Delta DIP(N:P)$

Table 3.3 Ecosystem-level attributes used for comparison of NA results of estuarine food webs with biogeochemical budgeting models

System	Morphology		ENA				
	Volume (m^3)	Area (m^2)	A/C (%)	TST (mgC m^{-2} per day)	FCI (%)	P/B (day^{-1})	TE (%)
Kromme	9.00E+06	3.00E+06	33.7	13,641	26	0.73	6.2
Swartkops	1.20E+07	4.00E+06	28	11,809	44	3.65	4
Sundays	1.40E+07	3.00E+06	43	16,385	20	10.95	2.6
Baltic Sea	1.74E+13	3.70E+11	55.6	2,577	23	29.2	16.2
Cheasapeake	3.63E+10	5.90E+09	49.5	11,224	23	51.1	9
Neuse	1.60E+09	4.60E+08	46.8	11,222	15.4	94.9	4.5

	LOICZ models (mol m^{-2} per year)						
	ΔDIP	ΔDIN	(nfix−denit)	(p−r) NEM	DIN loading	DIP loading	GPP[a]
Kromme	6.80E−03	1.59E−01	4.93E−01	−7.27E−01	3.01E−02	7.30E−04	2.15E+02
Swartkops	−6.25E−02	−1.01E+01	−9.13E+00	6.65E+00	1.40E+00	6.85E−02	1.10E+02
Sundays	4.70E−03	−2.26E−01	−3.02E−01	−5.03E−01	1.43E+00	1.18E−02	8.53E+03
Baltic Sea	−9.41E−03	−1.52E−01	−1.17E−01	5.12E−01	1.33E−01	1.25E−03	2.50E+03
Cheasapeake	−1.42E−02	−5.95E−01	2.78E−01	1.51E+00	4.82E−01	1.00E−02	6.27E+03
Neuse	−1.69E−04	−1.64E−03	4.02E+01	6.55E+00	1.77E−01	1.56E−02	5.04E+03

[a] GPP is in C and was also calculated through stochiometry for N and P. These values are not included here.

and higher, on which we subsequently performed factor analysis. The system properties and the values of the ecosystem properties on which the factor analysis was based are given in Table 3.3.

The output from factor analysis yielded eigen values of six principal components, of which the first four principal components account for 98.4% of the variance between the system properties. The factor loadings for each of the LOICZ and NA variables are given in Table 3.4, and taking +0.7 and −0.7 as the cutoff values, certain variables are correlated with one another and can be interpreted as varying together on these principal factors. The first three principal components explain 87.7% of the variance and none of the factor loadings of the fourth principal component exceeded the cutoff value, so this factor was not considered further. A number of inferences can be made:

1. The first principal component explains 46% of the variance and which includes three LOICZ and three ENA variables. Table 3.4 (under the first principal component) shows that of the LOICZ-derived variables ΔDIN and ΔDIP correlate negatively with DIP loading, which means that the magnitude of DIP loading will somehow affect the flux of DIN and DIP between the estuary and the coastal sea.

Table 3.4 Unrotated factor loadings of the selected system variables listed in Table 3.1

Variables/property	Principal component			
	1	2	3	4
Network variables				
A/C	*0.79*	0.31	0.13	0.51
FCI(%)	*−0.94*	0.04	−0.10	0.19
TST	−0.08	*−0.81*	−0.25	−0.49
P/B(day^{-1})	0.55	−0.02	*0.81*	0.03
Trophic Efficiency (%)	0.23	*0.85*	−0.07	0.46
GPP-C	*0.70*	−0.61	−0.05	0.35
GPP-N	0.69	−0.65	−0.04	0.30
GPP-P	*0.70*	−0.65	−0.03	0.29
LOICZ variables				
DIN loading	−0.47	*−0.76*	−0.20	0.37
DIP loading	*−0.87*	−0.34	0.32	0.17
ΔDIP	*0.89*	0.03	−0.26	−0.37
ΔDIP	*0.94*	0.17	−0.19	−0.20
(nfix−denit)	0.55	−0.10	*0.75*	−0.34
(p−r) NEM	−0.44	−0.19	*0.88*	0.02

Note: Four principal components are extracted (columns 1–4).

The FCI correlates negatively with the A/C and carbon GPP (gross primary productivity) of the ENA-derived properties, and one can thus expect lower FCI values in systems with high A/C.

This inverse relationship has in fact been reported in the literature (cf. Baird et al. 1991; Baird 1998). From the linkage between the ENA and the LOICZ modeling procedure, we can infer from these results that there appears to be a positive correlation among DIN and DIP flux, GPP, and ascendency. Systems acting as nutrient sinks may thus well be positively associated with GPP and ascendency, and such systems are thus more productive (higher GPP) and organized (higher A/C). The data given in Table 3.3 show to some degree that the Baltic Sea, the Chesapeake Bay and the Neuse River Estuary have high A/C associated with their performance as nutrient sinks.

2. Of the variance, 25% is explained by the second principal component, which had high factors scores for one LOICZ-and two ENA-derived variables (Table 3.4). The results would indicate some positive correlation between DIN loading and TST, but both are negatively associated with the TE index (Table 3.3).

3. The third principal component, which accounts for 17% of the variance (Table 3.4) shows positive correlations between two LOICZ variables ((nfix-denit), net ecosystem metabolism $(p - r)$), and one ENA system-level property, the P/B. We can construe from these relationships that the P/B is influenced by the magnitude and nature of one or both of the two LOICZ-derived properties.

The underlying associations are summarized in a scatter plot of the ecosystem positions relative to the first two principal components (Figure 3.1) and a cluster tree (Figure 3.2), which essentially reflects the results from factor analysis presented above. The three systems in the middle of Figure 3.1 occupy a relatively "neutral domain" in the context of their responses to the variability of the NA and LOICZ parameters given on the x- and y-axes, and appears to relate to the analyses of Smith et al. (2003) that a large proportion of the estuaries for which biogeochemical results are available cluster

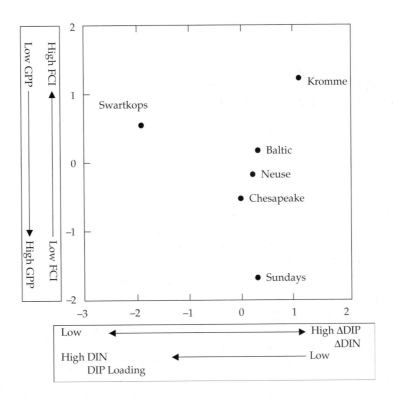

Figure 3.1 System position within the plane of the first two principal components.

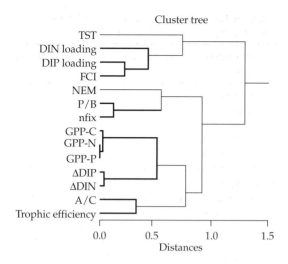

Figure 3.2 Cluster tree of ENA and LOICZ variables.

around neutral values of (p − r)(or NEM) and (nfix − denit). These three systems, namely the Baltic Sea, the Chesapeake Bay, and the Neuse River Estuaries are large in terms of aerial size and volume compared to the three smaller systems (namely the Swartkops, Kromme, and Sundays Estuaries), which are scattered at the extreme ranges of the variables. Table 3.3 shows that the larger systems are bigger in volume and size by 3–5 orders of magnitude, but that the DIN and DIP loadings on a per meter square basis of all six systems fall within in the same range. Other noticeable differences are the shorter residence times of material, the small volume and low rate of fresh water inflows compared with the three bigger systems. Although the scales of the axes in Figure 3.1 are nondimensional, the positions of the various systems reflect the relative order of the four variables plotted on the x- and y-axis, respectively, and corresponds largely with the empirical outputs from ENA and the LOICZ budgeting protocol. Finally, a cluster tree (Figure 3.2), which shows the similarity of the variables using an average clustering of the Pearson correlation r (as a distance measure $= 1 − r$), groups the NA and LOICZ variables in a hierarchical manner. Using a distance of <0.2 as a cutoff, the P/B ratio from NA is closely grouped with the [nfix − denit] of LOICZ, which suggests that overall production and nitrogen balance is linked in these estuaries.

In addition, the FCI from NA and DIP loading from LOICZ vary together as well, which suggests that overall cycling is linked to phosphorous in some way. This result is similar in many ways to the factor analysis above, especially the variables that score highly on principal factors 1 and 3 (Table 3.4).

The fundamental differences between the net flux methodology of LOICZ modeling and the gross flows of material inherent in food-web networks must be kept in mind, but the correlation between the methodologies is encouraging in our search for better understanding of ecosystems function. We should thus emphasize the possible linkages and the complimentary results derived from these methodologies. ENA results have rarely been related to other approaches. Comparisons, such as this, are essential to broaden our understanding of how ecosystems function and are structured in a holistic way.

Visualization of network dynamics

The display of dynamic, complex food webs has been problematic in past, due to the multiple species and linkages that must be rendered. This display limitation has prevented the visualization of changes that occur at the level of the whole food web. Seasonal changes, changes over longer periods of time, impacts due to fishing or hunting, and pollution impacts can all affect food-web structure, but unless this can be quantified and visualized, it is difficult for most to appreciate. Most current approaches involve either simplifying the food web by aggregating species into trophic species and by displaying "wiring diagrams" of the underlying structure. We use network statistical modeling software to analyze the similarities in the food webs and display the results using three-dimensional network modeling and visualization software.

We used the visualization technique described in Johnson et al. (2001, 2003) and Luczkovich et al. (2003) to display a series of food webs of the Chesapeake Bay, originally described by Baird and Ulanowicz (1989). This technique involves arranging the nodes (species or carbon storage compartments) of the food-web network in a three-dimensional space according to their similarity in feeding and predator relationships, as measured by a model called regular equivalence. In the regular equivalence model, two nodes in close position in the three-dimensional graph have linkages to predator and prey nodes that themselves occupy the same trophic role, but not necessarily to the exact same other nodes. Thus, here we visualize the change in trophic role of the compartments in Chesapeake Bay as they change from spring to summer.

In the example we display here, Baird and Ulanowicz (1989) modeled the carbon flow in a 36-compartment food web of the Chesapeake Bay. The model was adjusted seasonally to reflect the measured changes in carbon flow among the compartments. This model was originally constructed using the program NETWRK4. We obtained the input data from the NETWRK4 model from the original study and converted them to text data using a conversion utility from Scientific Committee on Oceanographic Research (SCOR) format (Ulanowicz, personal communication). The carbon flow data in a square matrix for each season was imported into UCINET (Borgatti et al. 2002) to compute the regular equivalence coefficients for each compartment (or node). Due to migrations and seasonal fluctuations in abundance, the model had 33 compartments in spring, 36 in summer, 32 in fall, and 28 in winter. They are listed in Table 3.5 along with their identification codes and seasonal presence and absences. The algorithm for computing regular equivalence (REGE), initially places all nodes into the same class and then iteratively groups those that have similar type of connections to predators and prey. Finally, a coefficient ranging from 0 to 1.00 is assigned to each node, which reflects their similarity in food-web role. These coefficients have been found to have a relationship with trophic level, as well as differentiate the benthos and plankton based food webs (Johnson et al. 2001; Luczkovich et al. 2003). After the REGE coefficients were computed, the matrices for each season were concantentated so that a 144×36 rectangular matrix of the coefficients was created. The combined four-season REGE coefficient matrix was analyzed using a stacked correspondence analysis

Table 3.5 The compartments in the four seasonal models of Chesapeake Bay and their identification numbers

	Compartment name	Spring	Summer	Fall	Winter	Trophic level
1	Phytoplankton	x	x	x	x	1.00
2	Bacteria in suspended POC	x	x	x	x	2.00
3	Bacteria in sediment POC	x	x	x	x	2.00
4	Benthic diatoms	x	x	x	x	1.00
5	Free bacteria	x	x	x	x	2.00
6	Heterotrophic microflagellates	x	x	x	x	3.00
7	Ciliates	x	x	x	x	2.75
8	Zooplankton	x	x	x	x	2.16
9	Ctenophores	x	x	x	x	2.08
10	Sea Nettle		x			3.44
11	Other suspension feeders	x	x	x	x	2.09
12	*Mya arenaria*	x	x	x	x	2.09
13	Oysters	x	x	x	x	2.08
14	Other polychaetes	x	x	x	x	3.00
15	*Nereis sp.*	x	x	x	x	3.00
16	*Macoma spp.*	x	x	x	x	3.00
17	Meiofauna	x	x	x	x	2.67
18	Crustacean deposit feeders	x	x	x	x	3.00
19	Blue crab	x	x	x	x	3.51
20	Fish larvae		x			3.16
21	Alewife and blue herring	x	x	x	x	3.16
22	Bay anchovy	x	x	x	x	2.84
23	Menhaden	x	x	x	x	2.77
24	Shad	x	x			3.16
25	Croaker		x	x	x	4.00
26	Hogchoker	x	x	x	x	3.91
27	Spot	x	x	x		4.00
28	White perch	x	x	x	x	3.98
29	Catfish	x	x	x	x	4.00
30	Bluefish	x	x	x		4.59
31	Weakfish	x	x	x		3.84
32	Summer flounder	x	x	x		3.99
33	Striped bass	x	x	x		3.87
34	Dissolved organic carbon	x	x	x	x	1.00
35	Suspended POC	x	x	x	x	1.00
36	Sediment POC	x	x	x	x	1.00

Source: From Baird and Ulanowicz (1989).

(Johnson et al. 2003), which makes a singular value decomposition of the rows and column data in a multivariate space. We used the row scores (the 36 compartments in each of the 4 seasons) to plot all 144 points in the same multivariate space. The network and correspondence analysis coordinate data were exported from UCINET to a coordinate file so that the food web could be viewed in Pajek (Batagelj and Mrvar 2002). (Note: we have also used real time interactive molecular modeling software Mage for this purpose; see Richardson and Richardson (1992)). Pajek was used to create the printed versions of this visualization.

The three dimensional display of the spring (gray nodes with labels beginning "SP" and end-ing with the node number) and summer (black nodes with labels beginning with "SU" and ending with the node number) food web of the Chesapeake network shows groupings of nodes that have similar predator and prey relationships, so that they form two side groups at the base of the web, and a linear chain of nodes stretching upwards (Figure 3.3). The arrows show the shift in coordinate position from spring to summer (we omit the arrows showing carbon flow here for clarity). The vertical axis in this view (note: normally axis 1 is plotted along the horizontal, but we rotated it here to have high trophic levels at the top) is correspondence analysis axis 1, which is significantly correlated ($r = 0.72$) with the trophic

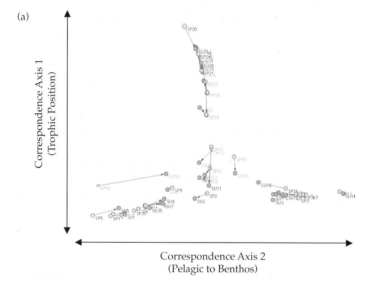

(a)

Correspondence Axis 1
(Trophic Position)

Correspondence Axis 2
(Pelagic to Benthos)

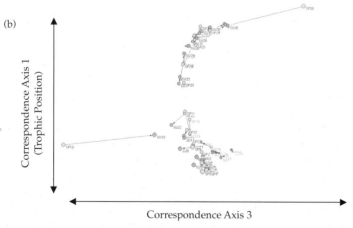

(b)

Correspondence Axis 1
(Trophic Position)

Correspondence Axis 3

Figure 3.3 (a) The food-web network of the Chesapeake Bay in spring (gray "SP" node labels) and summer (black, "SU" node labels), displayed using Pajek. The arrows show the shift in coordinate position from spring to summer. The stacked correspondence analysis row scores were used to plot the positions in three-dimensional space. (b) Another view showing the shift along the first and third axes, which represent trophic position as before and degree of connectedness to the network. The two compartments that were absent from the summer network: sea nettles (10) and other suspension feeders (11) are shown as moving into the center in the summer and becoming connected to the network.

levels (Table 3.5) that were calculated based on annualized carbon flows for each compartment by Baird and Ulanowicz (1989). In Figure 3.3 (a), there is a group at the right side of the base of the web, which is composed of compartments that are associated with detritus or the benthos, including bacteria in the sediment particular organic material (POC; 3), benthic diatoms (4), *Nereis* (15), other polychaetes (14), crustacean deposit feeders (18), and sediment POC (36) (note: the number in the parenthesis is the serial number of each compartment in Table 5). On the left side of the base of this web, there are plankton-associated groups, including phytoplankton (1), free bacteria (5), heterotrophic microflagellates (6), zooplankton (8), and dissolved organic carbon (34), and suspended POC (35). In the center at the base of the web, we find the bacteria in the suspended POC (2), which is midway between the benthic group and the plankton group, due to the fact that these bacteria are important as food of consumers in both groups. Stretching in a near-linear chain above the base are various consumers that are higher trophic levels. Compartments with low effective trophic levels (TL) include oysters (13; TL = 2.08), soft-shelled clams *Mya arenaria* (12; TL = 2.09), bay anchovies (22; TL = 2.77), other suspension feeders (11; TL = 2.09), and menhaden (23; TL = 2.77). Higher on the correspondence analysis vertical axis are compartments catfish (29), white perch (28), spot (27), hogchoker (26), alewife and blueback herrings (21), summer flounder (32), striped bass (33), bluefish (30), shad (24), and larval fish (20).

One way to interpret these visualizations is that those compartments that move the most in the coordinate space show the greatest seasonal change in trophic roles. Species that move downward along axis 1 are consuming more of the primary production or consuming more prey at low trophic levels. This is also shown for the whole system as higher TST and P/B ratios in the summer (Table 3.1), but our visualization shows the contribution of individual compartments to the system-wide changes. Some good examples are other suspension feeders (11), which move downward on the trophic position axis, because they feed more on the phytoplankton

in the summer. This can also be seen in the case of bay anchovy (22), which takes in more zooplankton (8), and spot (27) which increase consumption of "other polychaetes" (14) in the summer (Baird and Ulanowicz 1989). In all of these cases, an increase in consumption of species with lower trophic positions is driving this change in the visualization.

Another interpretation of the coordinate movements is that the species which derive energy from the pelagic zone in the summer are moving toward the center on axis 2. For example, free bacteria (5) and zooplankton (8) move toward the center of the diagram from spring to summer as they increase their consumption of dissolved organic carbon (34) and ciliates (7), respectively, while crustacean deposit feeders (18) move toward the center since they consume less sediment POC (36) in the summer. Thus, the degree to which the whole ecosystem shifts from benthic to pelagic primary production can be easily visualized. This also can be visualized dynamically across multiple seasons. We do not show the other seasons here, but interactively, one can turn off and on a similar display for each season and show that the nodes in the fall and winter move back towards the springtime positions. This can also be done over multiple years, if the data were available, or in varying salinities, temperatures, and under different management schemes.

Conclusions

Estuarine and coastal ecosystems have been locations where numerous studies have incorporated ENA to assess food-web structure and trophic dynamics. ENA also affords a valuable approach to comparative ecosystem ecology. Numerous ecosystem-level indices are calculated and complement indices at lower level of hierarchy. Comparisons of five ecosystem-level indices of food webs over various temporal and spatial scales appeared to correspond with our understanding of levels of development and stress within several estuarine systems. Intra-annual variations in these indices within an ecosystem were equal to or exceeded that for the limited number of cases of interannual comparisons. Interecosystem comparisons are more difficult because of differences in rules for network construction used for different ecosystems, but patterns in calculated indices were consistent with expectations. Finally, two relatively new approaches to understanding estuarine ecosystems, namely models of biogeochemical budgets and visualization tools were compared to the focal indices. The biogeochemical modeling complemented the ecosystem-level network indices, providing an extended assessment of the limited number of ecosystems evaluated. Visualization of food webs is problematic when those food webs are complex. This problem is exacerbated when one wants to compare food-web structures. We demonstrated a relatively new approach to visualizing food webs that enhances one's ability to identify distinctions between multiple conditions. Thus, we evaluated how ENA can be used in comparative ecosystem ecology and offer two new approaches to the discipline.

Acknowledgments

We would like to thank several organizations for the support of this work. We thank the Rivers Foundation and the Biology Department of East Carolina University for support of Dan Baird as Rivers Chair of International Studies during fall 2003. Funding for Christian came from NSF under grants DEB-0080381 and DEB-0309190. Ursula Scharler is funded by the NSF under Project No. DEB 9981328.

CHAPTER 4

Food webs in lakes—seasonal dynamics and the impact of climate variability

Dietmar Straile

Introduction

As a result of increased green-house gases, global surface temperatures increased strongly during the twentieth century and will most likely increase between 1.4–5.8°C within the twenty-first century (Houghton et al. 2001). Early signs of climate-related gradual changes in lake ecosystems have been reported (Schindler et al. 1990; Magnuson et al. 2000; Straile 2002; Livingstone 2003; Straile et al. 2003) and further changes are expected to come. Besides the increase in temperature, the increase in greenhouse gases per se (e.g. Urabe et al. 2003; Beardall and Raven 2004) as well as other associated phenomena, for example, increase in UV-B radiation (e.g. Schindler et al. 1996; Williamson et al. 2002), and their interactions will affect organisms in lake ecosystems (Williamson et al. 2002; Beardall and Raven 2004). This chapter can obviously not cover in detail these issues in global change research; instead the emphasis will be on possible food-web effects of temperature increase and variability. For recent reviews on various effects of climate forcing on lakes see (Carpenter et al. 1992; Schindler 1997; Gerten and Adrian 2002a; Straile et al. 2003).

Studying spatial shifts in species distribution and changes in phenology—although they are important and do undoubtedly occur also in lake food webs (Straile et al. 2003; Briand et al. 2004)—is not sufficient to predict the effects of climate change on ecosystems as they do not consider indirect effects and feedbacks within the food web (Schmitz et al. 2003). Food-web interactions may indeed play a key role for an understanding of the effects of climate change on lakes. Although food-web related impacts of climate change will probably be among the most difficult impacts to predict, they might have also some of the most severe consequences, for example, leading to regime shifts (Sanford 1999; Scheffer et al. 2001a,b).

Studies on lakes provide some of the most complete investigations on the structure, dynamics, and energetics of food webs encompassing organisms from bacteria to vertebrates. The life cycle of organisms in temperate lakes is adapted to a highly seasonal environment. Food-web interactions in these lakes depend on the seasonal overlap of the occurrence of potential prey, competitor, or predator species. This seasonal overlap, that is, the match–mismatch of food-web interactions depends strongly on the seasonal dynamics of the physical environment of lakes such as temperature, light availability, and mixing intensity. Consequently, climate variability influences food-web interactions and hence the structure, dynamics, and energetics of lake food webs. On the other side, a thorough understanding of climate effects on lakes will require information on lake food webs and—as will be shown below—on the seasonal dynamics of lake food webs. Hence, this chapter is separated into three sections: The section titled "The trophic structure of aquatic food webs" examines the trophic structure of lake food webs, especially of lake food webs as observed during the summer season. The section "Seasonal sucession" analyses the seasonal development of lake food webs, and hence provides the more dynamical viewpoint necessary for examining the effects of climate on the food webs of lakes (section "Climate variability and plankton food webs"). I will illustrate some aspects of lake food-web structure with examples from Lake Constance, which food web has been rather well studied during recent years (see Box 4.1).

The trophic structure of aquatic food webs

The most successful models on food-web regulation in pelagic lake food webs are based on the assumption that food webs can be aggregated into trophic chains, that is, that trophic levels act dynamically as populations. These assumption has its origin in the famous "green world" hypothesis of Hairston, Smith, and Slobodkin (1960) (hereafter referred to as HSS) although aquatic systems were not discussed by HSS. According to HSS, the relative importance of competition and predation alternates between trophic levels of food webs such that carnivores compete, herbivores are controlled by predation, which frees producers from predation and results into competition at the producer trophic level. Similar ideas developed more or less simultaneously in freshwater ecology. Already in 1958, Hrbáček (1958) provided seminal observations on trophic level dynamics in lakes: he noted the relationship between fish predation, the reduction of large-bodied herbivores, and the resulting decrease in transparency and increase in phytoplankton density in lakes. An increasing number of subsequent studies used the aggregation of food webs into trophic levels as a tool to analyze food-web interactions especially in plankton food webs. The theory of trophic level dynamics was further developed into the concepts of the trophic cascade (Carpenter et al. 1985) and biomanipulation (Benndorf et al. 1984; Shapiro and Wright 1984). Food chain theory (Oksanen et al. 1981) formally extended HSS "green world hypothesis" to more than three trophic levels based on the proposed relationship between primary productivity and food-chain length. The trophic cascade has received strong support in both mesocosm experiments (Brett and Goldman 1996) as well as in whole lake manipulative

experiments (Carpenter and Kitchell 1993). Biomass relationships in a suite of 11 Swedish lakes (Persson et al. 1992) also largely corrobated the predictions of HSS and the food-chain theory in systems with three trophic levels (algae, zooplankton, planktivorous fish) and with four trophic levels (algae, zooplankton, planktivorous fish, piscivorous fish).

The success of HSS in aquatic systems is evidence that the complexity of aquatic food webs can indeed be aggregated into trophic levels (Hairston and Hairston 1993). Nevertheless, this assumption has hardly been tested explicitly with data from food webs. Recently, Williams and Martinez (2004) used highly resolved carbon flow models to provide a first test: based on mass-balanced carbon flow models, they estimated the trophic position of web components by computing their food-chain length and their relative energetic nutrition trough chains of different length (Levine 1980). They suggested that most species can be assigned to trophic levels and that the degree of omnivory appears to be limited. To provide another test of this idea, seasonally resolved carbon flow models, which were established for the pelagic food web of Lake Constance (Box 4.1) can be used. To calculate trophic positions in the Lake Constance food-web basal trophic positions of phytoplankton and detritus/DOC were set to one. The latter implies that bacteria, which derive their nutrition from the detritus/DOC pool, were assigned to a trophic position of two. As a consequence, two different flow chains can be recognized: a grazing chain starting from phytoplankton and a detritus chain starting from detritus, respectively bacteria (Figure 4.2(a)). However, as (a) in Lake Constance, bacterial production is considerably lower than primary production, and (b) one group, that is, heterotrophic nanoflagellates (HNF) are the major consumers of bacterial production, a strong impact of the detritus chain on the trophic position of consumers is only evident for HNF. All other groups rely energetically directly (ciliates, rotifers, herbivorous crustaceans) or indirectly (carnivorous crustaceans, fish) on the grazing chain (but see Gaedke et al. (2002) for the dependence of consumers on phosphorus). Trophic positions in the Lake Constance carbon flow models were not distributed homogenously

Box 4.1 The pelagic food web of Upper Lake Constance

Lake Constance is a large (500 km²) and deep ($z_{max} = 254$ m) perialpine lake in central Europe, which has been intensively studied throughout the twentieth century. The lake consists of the more shallow Lower Lake Constance, and the deep Upper Lake Constance (Figure 4.1(a)). Due to its deep slope the latter has a truly pelagic zone, which seems to be energetically independent from littoral subsidies. Like many other temperate lakes, Lake Constance went through a period of severe eutrophication starting in the 1930s and culminating in the 1960s/1970s (Bäuerle and Gaedke 1998 and references therein). Beginning with the 1980s total phosphorus concentrations declined again. However, the response of the plankton community to oligotrophication was delayed. The pelagic food web of Upper Lake Constance during the oligotrophication period has been analyzed within several years of intensive sampling (Bäuerle and Gaedke 1998). Different food-web approaches, that is, body-mass size distributions (Gaedke 1992, 1993; Gaedke and Straile 1994b), binary food webs (Gaedke 1995), and mass-balanced flow networks (Gaedke and Straile 1994a,b; Straile 1995; Gaedke et al. 1996; Straile 1998; Gaedke et al. 2002) were applied to a dataset consisting of five, respectively eight (Gaedke et al. 2002) years of almost

weekly sampling. A special strength of this dataset is that it encompasses both, the classical food chain as well as the microbial food web. For carbon flow models the pelagic food web was aggregated into eight different compartments (Figure 4.1(b)), of which five can be assigned to the "classical food chain," that is, phytoplankton, rotifers, herbivorous crustaceans, carnivorous crustaceans, and fish, and three to the microbial loop, that is, bacteria, heterotrophic nanoflagellates, and ciliates (Figure 4.1(b)). In addition flows between these eight compartments and the detritus/DOC (dissolved organic carbon) pool were considered (exudation of phytoplankton, egestion and excretion of consumers, DOC uptake by bacteria). To analyze seasonal changes in carbon flows data were subdivided into up to 10 seasonal time intervals per year lasting between 14 and 102 days. For all seasonal time intervals mass-balanced carbon and phosphorous flows were established and further processed with the techniques of network analyses (Ulanowicz 1986). The results shown here are based on 44 different mass-balanced food-web diagrams for different seasonal time intervals from the study years 1987 to 1991 (Straile 1995, 1998).

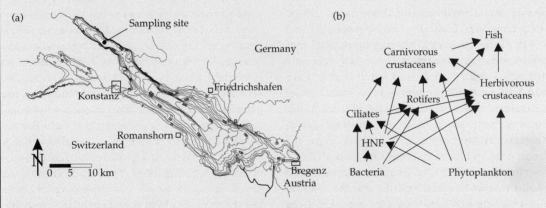

Figure 4.1 (a) Map of Lake Constance, and (b) aggregation of the Lake Constance pelagic food web into eight trophic guilds. Cannibalistic food-web interactions were considered for ciliates, rotifers, and carnivorous crustaceans, but are not shown here. Also not shown are the flows between these compartments and the detritus/DOC pool.

(Figure 4.2(b)). Rather, the distribution seems to be three-modal with peaks around trophic levels of two, three, and below four. This is especially remarkable as the inclusion of the microbial food-web with bacteria considered as second

trophic level will cause an overestimation of trophic level omnivory, for example, herbivores do consume bacterivorous HNF (trophic level two to three, Figure 4.2(a)). Interestingly the distribution of trophic positions is not symmetrical around the

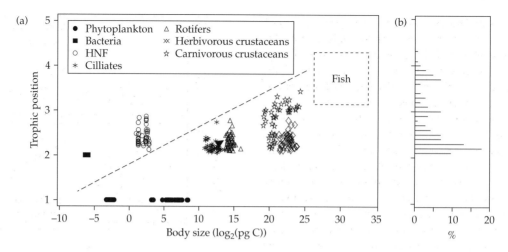

Figure 4.2 (a) Relationship between trophic positions in the carbon flow models and mean body size of trophic guilds; (b) percentage distribution of trophic positions (trophic positions of phytoplankton, bacteria, and HNFs not included).

different trophic levels, but rather skewed to the right, which might reflect the tendency of omnivorous consumers to gain most of their energy on the lowest trophic level on which they feed (Hairston and Hairston 1993).

Regarding the relationship of trophic position with body size, within both, detritus chain, and the grazing chain trophic position increases with body size. However, predominantly herbivorous groups do cover a large body size range between 2^{10} and 2^{24} pg C (Figure 4.2(a)), that is, more than four orders of magnitude. This data support a recently assembled binary food web for Tuesday Lake which also provides information on body size of species and in which a strong overall relationship between body size and trophic position (prey-averaged trophic position calculated from the binary food web) was observed (Cohen et al. 2003).

In many cases lake food webs hence provide text book examples for HSS and the trophic cascade. However, most of these examples are taken from the food webs of lakes during the summer situation. As climate variability will affect lakes throughout the season, it is necessary to consider seasonal variability of lake food webs.

Seasonal succession

The small size of planktonic organisms is the cause why it is necessary to consider seasonal succession

in a discussion about climate effects on plankton food webs. Due to the small size and consequently high intrinsic growth rates of the food-web components, successional processes taking centuries in terrestrial systems will take place every year anew in the plankton. Environmental variations experienced by the plankton community in a lake during one year are considered to be scale analogous of gross climate change since the last glacial maximum (Reynolds 1997). Different successional stages are usually not considered together when examining terrestrial food webs, rather than food webs of for example, grasslands, shrubs, and forests are studied separately. This might suggests that there is in the pelagic zone of a specific lake not one food web, but rather a succession of different food webs. In fact, the classical paper on plankton succession in lakes, the PEG (Plankton Ecology Group) model (Sommer et al. 1986) distinguishes between 24 successional stages, in which the relative importance of abiotic forcing through physical and chemical constraints and of biotic interactions, that is, competition and predation, differ (Sommer 1989). Consequently, seasonal variability in driving forces will result into seasonal differences in for example, species composition, diversity, interactions strength between food-web components, and finally food-web configurations.

Food-web structure including the number of functionally relevant trophic levels changes during the season. During winter, primary productivity of algae in many lakes will be strongly limited by the availability of light, for example, due to deep mixing and/or the formation of ice cover, both preventing algal blooms. During this time period, growth of algae is constrained by abiotic conditions and not by the activity of herbivores. Reynolds (1997) considers this successional state as an aquatic analog to bare land. Only with ice thawing and/or the onset of stratification in deep lakes algal blooms can develop with subsequent growth of herbivores. Within a rather short time of several weeks different herbivore population increase in size and finally often control algae. Within this so-called clear-water phase (Lampert 1978) algal concentration is strongly reduced and water transparencies rise again to values typical for winter. During this successional phase the food web is functionally a two-trophic level system. However, the "green world" returns due to—at least partially—increased predation pressure on herbivores due to fish, whose larvae surpass gape limitation and/or due to invertebrate predators (Sommer 1989). In addition to predatory control of herbivores, the world in summer is also prickly and tastes bad (Murdoch 1966) as for example, large and spiny algal species develop which are difficult to ingest for herbivores. Hence the system has developed from a one level system in early spring where herbivores are not abundant enough to control phytoplankton development into a complex three–four level system in summer, which collapses again toward winter.

This development of trophic structure is also evident for the trophic positions of the different compartments of the Lake Constance flow model (Figure 4.3). The trophic position of herbivorous crustaceans remains rather constant throughout the year, at approximately two (Figure 4.3(a)). However, the trophic positions of carnivorous crustaceans and of fishes do increase seasonally (Figure 4.3(b) and (c)). This increase of both groups is due to the increase in trophic position of carnivorous crustaceans. During winter and early spring, carnivorous crustaceans consist of cyclopoid copepods. Although considered as carnivorous at least as more ontogenetically more advanced stages, there is simply not enough

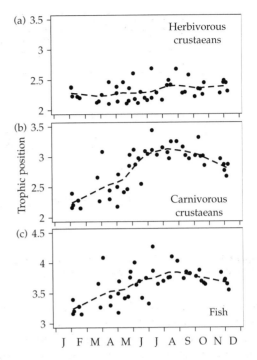

Figure 4.3 Seasonal changes in trophic position of herbivorous zooplankton, carnivorous zooplankton, and fish. Trophic positions are shown on the time axis at the midpoints of the respective time intervals. The dotted line represents a loess fit.

herbivorous production to sustain a carnivorous feeding mode for a major invertebrate group (Straile 1995). As a consequence, "carnivorous crustaceans" do rely largely on phytoplankton production, that is, they are in fact herbivorous during winter and early spring. Only from late spring/early summer onwards, herbivore production is large enough to sustain energetically populations of carnivorous crustaceans. This is also the time when more specialized carnivorous crustaceans, the cladocerans *Bythotrephes longimanus* and *Leptodora kindtii* develop from resting eggs and build up important populations in the planktonic community. Hence, the food web develops energetically from a three-trophic level system in winter/early spring to a four-trophic level system in summer. This is not due to the addition of a new top predator, as fish are present during the whole year, but due to the combined effects of diet switches of taxa, that is, cyclopoid copepods, and the addition of new intermediate predators, that is, *Bythotrephes* and *Leptodora*. This does not imply that the pelagic food web during summer can be considered dynamically as a simple four-trophic level system as among other things spatial (Stich and Lampert 1981) and size refuges (Straile and Hälbich 2000) of prey species at different trophic levels seem to blur cascading trophic interactions in Lake Constance.

However, despite these seasonal differences in food-web composition, structure, and dynamics, the season-specific pelagic food webs are clearly interconnected temporally. This is either due to the species which manage to persist during all food-web configurations within the year in the open water, or due to the existence of diapause and or dormancy stages. The latter allow species—given sufficiently correct cues for the start and termination of diapause—to exploit a specific season, respectively food-web configuration to ensure long-term persistence in a specific lake. These time-travelers (Hairston 1998) allow for the existence of a specific summer food web without the need for yearly new colonization events from other lakes. In addition, the latter is not really a possibility, as food webs in nearby lakes will be at a similar successional stage. This is also an interesting contrast to terrestrial system where a landscape mosaic of different successional stages offers the possibility of a new colonization. Recent research has shown that resting stages are important components of the life cycle of nearly all plankton taxa, for example, for phytoplankton (Hansson 1996), ciliates (Müller and Wünsch 1999), rotifers (Hairston et al. 2000), and crustaceans (Hairston and Cáceres 1996; Cáceres 1998; Jankowski and Straile 2004). Furthermore, bet hedging strategies even allow for multiyear dynamics and the persistence of species even when reproduction fails within specific years (Cáceres 1998).

On the other hand, lake food webs in the various successional stages or even within several successional cycles might be connected through the presence of long-living organisms, which are for example, top predators. This can have important dynamical consequences when for example, long-lived piscivores suppress planktivorous fish over several years (Post et al. 1997; Sanderson et al. 1999; Persson et al. 2003). Persson et al. (2003)

provide evidence for shifts in trophic cascades caused by intrinsically driven population dynamics of top predators including size-specific cannibalism. This is an important case study showing that cannibalism and life-cycle omnivory can have important food-web consequences culminating in a trophic cascade, that is, less zooplankton and more phytoplankton biomass during high summer when large-sized "gigantic cannibals" produced high numbers of young-of-the-year fish (Persson et al. 2003). However, even for long-lived animals there is a need to adjust their life cycle to the seasonal environment, that is, to seasonal changes in the food-web structure. Any analysis of climatic forcing of pelagic food-web hence will need to consider fast changes in food-web structure occurring during seasonal succession and the adaptation of species to cope with this variability.

Climate variability and plankton food webs

Changes in climate may result into changes in the timing and duration of successional stages (e.g. Straile 2002) as well as in modifications of food-web structure within distinct successional stages (e.g. Weyhenmeyer et al. 1999). Climate warming has been shown to increase the stratification period in lakes (Livingstone 2003) which may in turn intensify nutrient limitation and phytoplankton competition. Species need to adapt their life cycles, for example, the timing of reproduction of long-lived species, the timing of critical life history shifts, for example, the timing of metamorphosis of copepod species, and the timing of diapause initiation and termination to temporal shifts in food-web structure, respectively food-web shifts. Match or mismatch of specific life-history events with the successional state of the food web may crucially influence population dynamics of species. This has been suggested for, for example, the interaction of fish larvae and their zooplankton prey (Cushing 1990; Platt et al. 2003) and the hatching of over-wintering eggs (Chen and Folt 1996). The latter authors suggest that fall warming may result into the maladaptive hatching of overwintering eggs of the copepod *Epischura lacustris* possibly resulting into the loss of the species from the lake.

As a consequence of the strong seasonal dynamics within lake food webs, climate effects will be highly season-specific. Interestingly, and in contrast to many food-web studies, which consider the summer situation, there is a wealth of studies reporting ecological climatic effects of climate forcing in lakes during winter and early spring periods. Although climate forcing is also important in summer (see below), this might point to a special importance of winter meteorological forcing for temperate lake food webs. It is hence important to ask why this is the case: as often, there is not a single explanation, but meteorological, physical, and biological reasons contribute to this observation: (1) there are strong changes and there is a high interannual variability in winter meteorological forcing, (2) there is a high susceptibility of lake physics to meteorological forcing in the winter half year, (3) there is a high susceptibility of species physiology and life history and hence of successional dynamics to changes in lake physics during the winter half year.

1. Strong changes and high variability in winter meteorological forcing—the largest warming on our planet during the twentieth century has occurred (1) during winter and (2) over Northern Hemisphere land masses. Likewise, large-scale climatic oscillations such as the North Atlantic Oscillation (NAO) and the El Nino Southern Oscillation (ENSO) do have their strongest teleconnections to northern hemispheric meteorology during winter (Hurrell 1995; Rodionov and Assel 2003). In contrast, summer warming was not as pronounced during the twentieth century, and climate forcing during summer seems to be more controlled by local and regional factors, but less under the control of large-scale climate oscillations.
2. Lake physics will be especially sensitive to climate variability during the winter half year. During this time period important mixing events do occur and lakes may or may not be covered by ice. Physical conditions such as the presence/absence and intensity of mixing and the presence/absence and duration of ice cover may be highly sensitive—depending on lake morphology, latitude, and altitude—to changes in winter meteorology. For example, ice cover duration in European and North

American lakes is associated with the NAO, respectively ENSO (Anderson et al. 1996; Livingstone 2000; Straile et al. 2003). Winter severity influences the mixing intensity in deep lakes, thereby influencing nutrient distributions (Goldman et al. 1989; Nicholls 1998; Straile et al. 2003) and oxygen concentrations in the hypolimnion (Livingstone 1997; Straile et al. 2003). In contrast, stratification is usually strong during summer and interannual meteorological variability might result into more or less steep temperature gradients but unlikely in mixing (but see George and Harris (1985) for more wind-exposed lakes). However, warmer summer temperatures may reduce mixing events in shallow polymictic lakes.
3. Species physiology and life history and hence plankton succession is highly sensitive to physical factors during the winter half year (Sommer et al. 1986). Ice cover duration (Adrian et al. 1999; Weyhenmeyer et al. 1999), mixing intensity, and timing (Gaedke et al. 1998) will have a strong impact on phytoplankton bloom formation. Temperature effects on growth rates of zooplankton and fish seem to be especially important during spring. During the phytoplankton spring bloom zooplankton growth is unlikely to be limited by food concentration. Consequently, temperature limits zooplankton growth rates (Gerten and Adrian 2000; Straile 2000; Straile and Adrian 2000). Similar arguments have been put forward regarding growth of whitefish larvae (Eckmann et al. 1988). As zooplankton is abundant in late spring—given that there is not a mismatch situation—fish growth depends on water temperatures. However, higher winter temperatures may also have negative impacts on consumers due to enhanced metabolic requirements at low food abundance. This has been suggested to have caused a decline of *Daphnia* abundance in high NAO winters in Esthwaite Water (George and Hewitt 1999). Additionally, competition with the calanoid copepod *Eudiaptomus,* which has lower food requirements and increased in abundance during winter in high NAO years might have contributed to the decline of *Daphnia* in Esthwaite Water.

Another important reason for the importance of winter is that winter represents a population bottleneck for many species. For example, in temperate lakes, the abundance of plankton is strongly reduced during winter, and populations rely on resting stages to overwinter (see above). Fitness benefits might be associated with the ability to overwinter in the plankton as opposed to produce resting stages. For example, the dominance of the cyanobacteria *Planktothrix* in Lake Zurich in recent years has been attributed to a series of mild winters resulting in less deep water mixis (Anneville et al. 2004).

Also winter is often associated with reduced consumption and starvation of juvenile fishes influencing fish life history (Conover 1992) and population dynamics (Post and Evans 1989). Furthermore severe winters associated with long-lasting ice cover and oxygen deficiency can cause winterkill of fish species with consequences for fish community composition (Tonn and Paszkowski 1986) and lower trophic levels (Mittelbach et al. 1995).

However, also summer food webs are affected by climate variability. This is due either to summer meteorological forcing or indirect food-web mediated effects of winter/early spring meteorological forcing (Straile et al. 2003). For example, temperature effects on the growth rate of the cladoceran genus *Daphnia* will change the timing of phytoplankton suppression, that is, the timing of the clear-water phase in early summer (Straile 2000). Higher temperatures associated with a positive phase of the NAO resulted into higher *Daphnia* growth rates and earlier phytoplankton depression in Lake Constance, in addition the duration of phytoplankton suppression increased (Straile 2000). Similar shifts in the timing of the clear-water phase were observed in a number of European lakes (Straile 2000, 2002; Anneville et al. 2002; Straile et al. 2003). Overall warming trends since the 1970s resulted into an advancement of the clear-water phase of approximately two weeks in central European lakes (Straile 2002). This phenological trend of a predator–prey interaction is of similar magnitude as phenological shifts in, for example, plant, insect, and bird populations (Walther et al. 2002; Straile 2004). In shallow lakes, the spring clear-water phase is of crucial importance for the growth and summer dominance of macrophytes (Scheffer et al. 1993). Hence, higher

growth rates of daphnids during spring and an earlier onset of the clear-water phase has been suggested to promote a regime shift in shallow lakes from a turbid phytoplankton dominated to a clear macrophyte-dominated state. This might be an example on how even small temporal changes in the food-web interaction between phytoplankton and daphnids may result into large-scale ecosystem changes in lakes. However, the database used by Scheffer et al. (2001*b*) is confounded by management effects and cannot be used as a strong support of their hypothesis (van Donk et al. 2003).

Differences in winter meteorological forcing can also change the nutrient availability for phytoplankton due to changes in deepwater mixing or runoff with subsequent consequences for annual primary productivity. For example, primary production in Lake Tahoe and Castle Lake was related to winter mixing depth and precipitation, respectively (Goldman et al. 1989). Likewise, decreased summer transparency and increased epilimnetic pH and O_2 concentrations indicated that the phytoplankton summer bloom was more intense after mild winters in Plußsee (Güss et al. 2000). Food-web mediated effects of climate variability in summer will also occur when fish numbers are affected by winter severity due to starvation or oxygen deficiency under ice (see above).

Summer meteorological forcing can also have significant effects on lake food webs. For example, in Lake Windermere, interannual differences in June stratification linked to the position of the Gulf stream (George and Taylor 1995) are related to the growth of edible algae and finally to *Daphnia* biomass (George and Harris 1985). Also, copepod species have been shown to increase in abundance as a result of warm summers possibly due to development of an additional generation (Gerten and Adrian 2002*b*).

Effects of climate variability will also change the relative importance of species within successional stages. For example, phytoplankton species differ in respect to sedimentation and/or light limitation. Climate-related changes in stratification hence will favour different species. For example, in Müggelsee (Adrian et al. 1999) and Lake Erken (Weyhenmeyer et al. 1999), mild winters with less ice cover in response to high NAO years favored diatoms species over other phytoplankton. Diatoms need turbulent conditions to prevent them from sinking out of the euphotic zone, a condition not met under ice. In contrast, the dinoflagellate *Peridinium* can already build up a large population size under clear ice (Weyhenmeyer et al. 1999). Likewise, in Rostherne Mere, the intensity of summer mixing is crucial for which phytoplankton species will dominate: extremely stable stratification leading to the dominance of *Scenedesmus*, with increasing mixing favoring *Ceratium/Microcystis* and finally *Oscillatoria* dominance (Reynolds and Bellinger 1992). The establishment of a more stable stratification is of special concern since stratification may often favor buoyant species including toxic cyanobacteria over sedimentating species (Visser et al. 1996).

Also zooplankton growth rates will show different responses to interannual temperature variability. Based on a literature survey of laboratory data, Gillooly (2000) suggested that larger zooplankton species should respond more strongly to increasing water temperatures than small ones. According to his survey, difference in generation time between the large cladoceran *Eurycercus lamellatus* and the rotifer *Notholca caudata* is reduced from 88 days at 5°C to 15 days at 20°C suggesting that large zooplankton should be correspondingly favored with warming. A hypothesis that remains to be tested with field data (Straile et al., in prep.). Also the life history of zooplankton species is of importance. Population growth rate of parthenogenetically reproducing rotifers and cladocerans exceeds those of copepods which reproduce sexually and do have a long and complex ontogeny. As a consequence, daphnids should be able to benefit more strongly from spring warming than copepods. This can be demonstrated by examining interannual variability of daphnid and copepod biomass in relation to interannual variability to vernal warming. Both taxa are able to build up higher biomasses in years with early vernal warming and consequently high May water temperatures in Lake Constance (Figure 4.4). However, the effect is much more pronounced for daphnid biomass, which increases several orders of magnitude. Interestingly, an increase in spring temperatures seems to have a similar effect on the

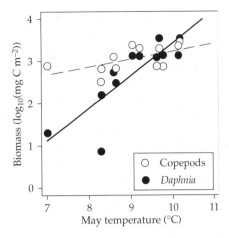

Figure 4.4 Response of *Daphnia* and copepod biomass in May to interannual differences in vernal warming as expressed as average May water temperatures (0–20 m water depth). Both taxa show a significant relationship with temperature ($n = 12$, $r = 0.84$, $p < 0.001$ for daphnids, and $n = 12$, $r = 0.57$, $p < 0.06$ for copepods).

relative abundances of daphnids and copepods in spring as an increase in food availability due to eutrophication (Straile and Geller 1998). Eutrophication as well as temperature increase seems to favor taxa with high intrinsic growth and developmental rates.

To summarize, due to the fast changes in food-web structure in pelagic systems, and the importance of food-web interactions, it is important to adopt a food-web approach when studying climate effects on lake ecosystems. Plankton systems seem to be especially suitable to study indirect food-web mediated effects due to the small size of the organisms concerned, their fast growth rates, and the relative promptness in which indirect effects will occur. On the other hand, food-web variability due to meteorological forcing, that is, so-called "natural experiments," can also be considered as an important tool to analyze food-web regulation. The highly seasonal signal of climate effects on lake food webs suggest that food-web studies in lakes need to consider seasonal variability. In addition, more attention should be paid to the winter period. Many studies in the past seem to have neglected the winter due to the admittedly overall lower biological activity during winter. Instead, the winter period may be a critical one for lake functioning, and for climate effects on lake functioning as it seems that the winter/early spring situation might set the stage for much what will come during the vegetation period. When adopting a seasonal approach, more emphasis should also be paid to time and season as resources. Resources, which will undoubtedly change with climate change and to which organisms have to fit their life cycle.

Acknowledgments

Data acquisition was mostly performed within the Special Collaborative Program (SFB) 248 "Cycling of Matter in Lake Constance," supported by Deutsche Forschungs-gemeinschaft. I thank all people contributing to the establishment of the food-web dataset of Lake Constance. The manuscript benefited from comments by K.-O. Rothhaupt.

CHAPTER 5

Pattern and process in food webs: evidence from running waters

Guy Woodward, Ross Thompson, Colin R. Townsend, and Alan G. Hildrew

Introduction: connectance webs from streams

Early depictions of food webs were usually simple and consisted of few taxa and a limited number of links. These "connectance" webs seemed to support models suggesting that the dynamical stability of food webs was inversely related to their complexity, and thus the prediction that simple webs should predominate in nature (May 1972, 1973; Cohen 1978). Pioneering descriptions of food webs from running waters proved no exception (Figure 5.1(a); see Hildrew 1992), though more latterly streams have become particularly prominent in the new generation of high quality, detailed food-webs that challenge the received wisdom that complexity is the exception rather than the rule (e.g. Tavares-Cromar and Williams 1996; Benke and Wallace 1997; Townsend et al. 1998; Williams and Martinez 2000; Schmid-Araya et al. 2002a,b).

Many of the coarse groupings (e.g. "algae" or "meiofauna") used previously to lump supposedly similar "trophic species" are now separated into distinct taxa in these more detailed webs. Such research shows that a large number of samples are required to characterize complex food-webs, not least because many predators have empty guts and species rank-abundance curves follow log-normal or geometric distributions (Tokeshi 1999; Woodward and Hildrew 2001). Yield-effort curves for the detection of species and links suggest that small sample sizes are often insufficient to capture the true complexity of real food webs (e.g. Thompson and Townsend 1999). Most importantly, greater sampling effort and the improved resolution of small-bodied organisms has revealed recurrent and complex patterns in stream food-webs including the ubiquity of omnivory, feeding loops and cannibalism, high linkage density, a decline in connectance with increased species richness, and long food chains (Schmid-Araya et al. 2002a,b; Woodward and Hildrew 2002a).

The number of species and links described in the food web of Broadstone Stream (southern England), currently the most detailed of any lotic system, increased by almost an order of magnitude from its earliest incarnation, as resolution was improved among smaller species. The initial description (Figure 5.1(b); Hildrew, Townsend, and Hasham 1985) contained 24 species and 109 links, and described the trophic links of the numerous but small-bodied predatory chironomids for the first time in any food web. Ten years later, predatory links to the microcrustacea were resolved (Lancaster and Robertson 1995), producing a web of intermediate completeness (Woodward and Hildrew 2001; Figure 5.1(c)). This was followed by the inclusion of the soft-bodied meiofauna and algae (Figure 5.1(d); Schmid-Araya et al. 2002a), and there are now 131 species and over 841 links described (Woodward, et al. in press). Although Broadstone Stream seems like a complex system, its acidity actually renders it relatively depauperate (e.g. fish and specialist grazers are absent, and there are few algae), and other streams are certainly much more complex, potentially yielding further surprises. For instance, a single small, sandy stream in central Germany, the

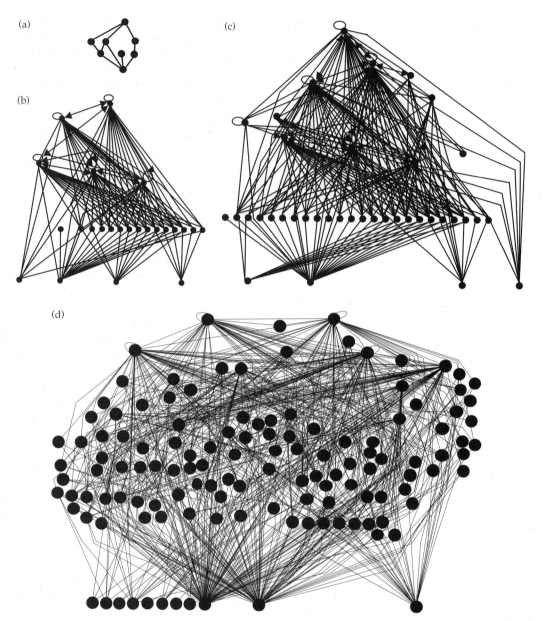

Figure 5.1 Connectance food webs from the early and more recent stream literature: (a) Early stream food-web (redrawn from Cohen 1978) (b) Initial connectance web from Broadstone Stream (after Hildrew et al. 1985) (c) Intermediate resolution web from Broadstone Stream (after Woodward and Hildrew 2001) (d) Highly resolved Broadstone Stream food-web (after Schmid-Araya et al. 2002*a*).

Breitenbach, has over 1,000 species of invertebrates (Allan 1995). Nonetheless, the sample size of stream food-webs where individual feeding links have been resolved, while still very small, is increasing (e.g. Tavares-Cromar and Williams 1996; Townsend et al. 1998; Schmid-Araya et al. 2002*b*; Thompson and Townsend 2003), and encouraging progress has been made over the last 20 years in revealing the true complexity of trophic networks in streams.

Small-bodied species are not the only ones prone to underestimation; many top predators, such as semi-aquatic birds, are comparatively rare, mobile, and often ignored by stream ecologists (Steinmetz et al. 2003). In addition, links between the main energy sources in stream food-webs, algae and detritus of terrestrial or aquatic origin plus associated heterotrophic microbes, still remain largely unresolved (Hildrew 1992). Probably the best-characterized links within the microbe-detritus "sub-web" are those between leaf-litter and hyphomycete fungi (e.g. Gessner and Chauvet 1994). Unraveling the individual feeding links between undoubtedly heterogeneous basal food resources and their nonpredatory consumers remains a major challenge for stream ecologists.

Food web patterns related to body size

Stream food-webs often contain reciprocal predation links, in accord with the "niche" food-web models of Warren (1996) and Williams and Martinez (2000) but not the earlier "random" or "cascade" models (e.g. Cohen et al. 1993a). These may involve body-size effects, including the seasonal ontogenetic reversals that occur when generations overlap, so that large individuals of "small" species are able to eat small individuals of "large" species (Woodward and Hildrew 2002a). Moreover, both the trophic status and niche width of secondary consumers can be determined mainly by body size rather than taxonomy, with similar-sized predators occupying almost the same position within the food web, regardless of their species identity (Woodward and Hildrew 2002a).

Other powerful influences of body size on food web structure and dynamics are the partitioning of food resources among predators, via indirect horizontal (i.e. potentially competitive) links (Cohen et al. 1993a; Woodward and Hildrew 2002a) and the expansion of prey size range with predator size (Allan 1982; Woodward and Hildrew 2002a). The latter occurs when small prey remain vulnerable to growing predators that are increasingly able to take larger prey. This "nested hierarchy" of feeding links confers an obvious size-related structure to food webs (Cohen et al. 1993a). A pattern is produced, known as "upper triangularity", such

that if a web is arranged into a matrix of consumers and resources ranked in order of increasing body-size, feeding links are overrepresented, relative to a random distribution, in the upper part of the triangle above the leading diagonal (Cohen et al. 1993a). The situation might be reversed in host–parasite webs, where consumers are smaller than their resources, but little is known about the importance of parasites in stream food webs.

We might expect that the size difference between predators and prey ultimately becomes so great that the feeding links are broken; thus, piscivorous fish are unlikely to feed directly on rotifers, which therefore occupy a lower-size refugium (the "size-disparity hypothesis" (Hildrew 1992; Schmid-Araya et al. 2002a)). This suggests that stream food webs might be partially compartmentalized, such that, although upper triangularity could exist within "subwebs" (perhaps delimited by portions of the community size-spectrum, for example, meiofauna, macrofauna, megafauna), these subwebs might have relatively few connections between them. Size disparity between the top and bottom of food webs might thus also account for the decline in web connectance with increasing species richness (this last partly attributable to the inclusion of the small-bodied meiofauna) apparent in a sample of stream food webs (Schmid-Araya et al. 2002b), though verifying such patterns is fraught with methodological difficulty.

Cohen et al. (2003) have demonstrated recently the potential importance of body size as a structuring force in aquatic food-webs by examining relationships within a three-way data matrix: they found strong trivariate correlations between the presence/absence of links, mean body-size, and the mean abundance of each species. Small species were abundant, but low in the food web, with large species being rarer and higher in the food web and also possessing a greater number of links. Although these patterns have yet to be examined explicitly in a range of systems, they are true of the Broadstone Stream food web (Woodward et al. in press). In addition, many of the component univariate or bivariate patterns they report have already been described in stream communities, such as inverse relationships between body-size and abundance (Schmid et al. 2000) and upper triangularity

(Williams and Martinez 2000; Schmid-Araya et al. 2002a). Cohen et al.'s (2003) findings may therefore apply to many food webs in running waters. We return to body size effects in later sections.

From pattern to process

Community and ecosystem approaches to quantifying stream food webs

The analysis of connectance webs, which represent equally species and the links between them, focuses on patterns in structure (e.g. links per species; predator–prey ratios) rather than processes (e.g. energy flux, interaction strengths). Trivial interactions are inevitably overemphasized and *vice versa*

(Benke and Wallace 1997). While web structure and the strength of links clearly interact, and a knowledge of both is necessary to understand function, there is now a shift in emphasis to the construction of quantitative, and thus hopefully more realistic, food webs (e.g. Hall et al. 2000; Benke et al. 2001; Woodward et al. in press).

Food-web ecologists working on streams have made recent progress by using semi-quantitative measures of "interaction strength" or by detailed analysis of subsets of whole community webs (e.g. Benke and Wallace 1997; Woodward and Hildrew 2002a; Figure 5.2). However, the sample of such webs is very small and interaction strength has been expressed in a variety of ways, hindering comparisons (indeed, Berlow et al. 2003 list no less

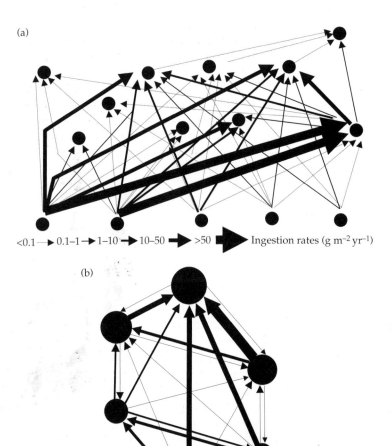

(a)

$<0.1 \rightarrow 0.1–1 \rightarrow 1–10 \rightarrow 10–50 \rightarrow >50$ Ingestion rates (g m^{-2} yr^{-1})

(b)

Figure 5.2 Approaches to quantifying food webs: (a) the flux of organic matter and the trophic basis of production among consumers in a subset of a stream food-web (redrawn after Benke and Wallace 1997), (b) body size, trophic status, and feeding loops within the predator subweb of Broadstone Stream, with links expressed as per capita consumption of "prey" by "predators" (% of each "prey" population consumed per "predator" individual per 24 h). The area of the circles is proportional to $\log_{10}$ mean body-mass (redrawn after Woodward and Hildrew 2002a).

than 11 separable uses of the term in the ecological literature). Attempts to quantify webs have focused either on population/community dynamics (e.g. Power 1990; Wootton et al. 1996) or on the flux of energy or matter (the "ecosystem approach" e.g. Benke and Wallace 1997), with virtually none combining the two (but see Hall et al. 2000). The former approach can itself be divided into questions about overall community dynamics (for instance, how the distribution of interaction strength may affect community stability), and those seeking predictions about the dynamics of constituent species populations (Berlow et al. 2003). The ecosystem approach has often been used to view food webs from a mass–balance perspective in which, for example, the patterning and strength of flow pathways have been analyzed to assess indirect effects, trophic and cycling properties, and the organizational status of the system (e.g. Christian et al. Chapter 3, this volume, review several such studies from marine systems).

Interaction strengths defined or measured in the different ways are not necessarily related, so comparisons between quantified webs constructed using the different approaches should be made with caution. For instance, while a single prey species might account for much of a predator's production, the population size of the prey may itself be unaffected. Conversely, a predator may consume a large proportion of the production of one prey species, while that prey contributes little to the overall production of the predator (a situation that can arise where predators are subsidized by alternative food sources). This creates the potential for indirect effects, such as apparent competition, whereby prey compete for "enemy-free space" rather than for more conventional resources (Holt 1977). The potential for diffuse, apparent competition can be gauged from prey-overlap graphs. Those from Broadstone Stream (Woodward and Hildrew 2001) show that virtually every prey species shares at least two predator species with virtually every other prey species, with many taxa being preyed on by all of the dominant predators (Figure 5.3). The strength of these indirect prey–prey interactions is extremely difficult to measure but, if they prove generally prevalent, they could be a major driver of community dynamics. The blurring of the traditional boundaries between competition and predation that occurs in real food webs (e.g. intraguild predation and cannibalism), allows for a variety of complex interactions to arise,

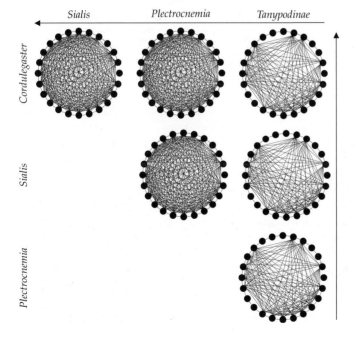

Figure 5.3 Prey overlap graphs derived from the Broadstone Stream food-web. Links between prey join species that share pairs of the four predator taxa in the matrix. The arrows point in the direction of increasing predator body-mass. There is high overlap, especially when comparing pairs of predators of similar size (e.g. *Sialis* and *Cordulegaster*), suggesting a high degree of trophic redundancy in the food web and the potential for diffuse apparent competition among prey (constructed from the food-web data matrix in Woodward and Hildrew 2001).

including apparent competition, apparent mutualism, and even potentially neutral effects of predators on their prey, and a given predator can have mixed trophic impacts on the different members of a web (Ulanowicz and Puccia 1990; Woodward and Hildrew 2002c).

The population/community approach: the complexity-stability debate revisited

Recent theoretical advances have allowed the development of more sophisticated mathematical models that predict better the *dynamics* of food webs (Polis 1998). In the case of the population/community approach, there is increasing support for the argument that the complexity (i.e. many species and/or links) commonly observed in nature can increase the stability of a food web (de Ruiter et al. 1995; McCann et al. 1998; McCann 2000; Neutel et al. 2002). McCann et al. (1998) challenged conventional wisdom by demonstrating mathematically that complexity could enhance stability—*if* most links were weak. Intriguingly, some of the earliest models also demonstrated this possibility (see May 1972) but this was largely ignored in favor of models that assumed strong interactions and showed an inverse relationship between complexity and stability.

Most dynamic food-web models constructed prior to those of McCann et al. (1998) were based on the assumptions that interactions between species were strong (and often uniformally distributed) and that feeding relationships were linear (e.g. Type I rather than sigmoid Type III functional responses) (Polis 1998). An increasing body of evidence from real systems suggests that most links in food webs are "weak" while fewer are "strong" (McCann 2000), and this seems true regardless of what measure of interaction strength is used (Berlow et al. 2003). In this context, experiments in streams have often demonstrated that, although predators can have strong impacts on a few prey species, they have much weaker, or negligible, direct effects on many others (e.g. Woodward and Hildrew 2002a).

The dominance of terrestrial detritus at the base of many stream food-webs means that a large number of primary consumer–resource interactions will be donor-controlled. Supporting this, several studies have reported the absence of trophic cascades in detritus-based stream food-webs, where omnivorous predators have had negative effects on two adjacent trophic levels (e.g. Hildrew 1992; Usio 2000; Rosemund et al. 2001; Woodward and Hildrew 2002a,b). This differs markedly from the situation in algal-based systems, in which cascades have been recorded, where consumers typically exert strong top-down control on their resources (e.g. Power 1990; Townsend 2003), which may also be responding simultaneously to the bottom-up effects of nutrients or light (Hillebrand 2002). The presence of a large number of weak or donor-controlled interactions within a food web, such as those in detritus-based streams, is thought to increase its stability (McCann 2000), and may serve to buffer any potentially strong top-down effects of pairwise interactions between predators and primary consumers (Schmid-Araya et al. 2002a). The idea that a diffuse network of pathways within a diverse community can dampen strong, pairwise interactions, thereby increasing dynamic stability, is of course not new, having been proposed by Charles Elton almost 80 years ago (Elton 1927) and reiterated by MacArthur (1955).

These suggestions are supported by the results of a large-scale experiment in which leaf-litter inputs to a detritus-based stream were manipulated experimentally: the strength of links (in terms of per unit biomass consumption of prey by predators) increased when leaf-litter was excluded, suggesting that a large reservoir of detritus may indeed weaken top-down effects and increase stability (Hall et al. 2001). Subsidies from external energy sources are in general common in stream food-webs, such as terrestrial invertebrates that fall into the stream (e.g. Nakano et al. 1999), CPOM and FPOM derived from terrestrial leaf-litter (e.g. Usio 2000; Usio and Townsend 2001; Woodward and Hildrew 2002a) and marine-derived biomass from carcasses of anadromous salmon that die after spawning (e.g. Wipfli and Caouette 1998).

Refugia may also serve to weaken predator–prey interactions and thus stabilize food webs (Closs 1996), and the physical complexity of streams may explain why trophic cascades appear to be rarer than in simpler habitats, such as the pelagic zone of lakes, where physical refugia are scarce and strong cascading interactions can arise. Stabilizing, density-dependent Type III functional responses can arise where prey populations are constrained by the availability of physical refugia, such as interstitial spaces between particles in the stream bed. At low prey population density, the relative availability of refugia is high and predators may be unable to drive prey to extinction. As prey density increases, however, *per capita* vulnerability increases. There is now plenty of evidence from streams of density-dependent mortality and the stabilizing effect of refugia (e.g. Hildrew and Townsend 1977; Townsend 1989; Hildrew et al. 2004).

Measuring interaction strength and food-web dynamics in practice

The term "interaction strength" is used loosely in ecology (Berlow et al. 2003). However, by expressing carbon (or matter) flux in terms of g C (or dry mass) m^{-2} per year and *per capita* consumption of individuals m^{-2} per year, common baselines can be obtained for characterizing links, while also approximating to the units used in most models (Cohen et al. 1993*b*). Detailed gut contents analysis (GCA) can be used to assess both *per capita* consumption rates (e.g. Hildrew and Townsend 1982; Speirs et al. 2000) and energy fluxes (e.g. Benke and Wallace 1997; Hall et al. 2000). Nevertheless, large numbers of individuals need to be sampled because GCA provides only a snapshot of a predator's diet, which can vary in time, while most guts are empty or contain few prey items, some of which are difficult to quantify (e.g. filamentous algae, soft-bodied prey, biofilm).

Stable Isotope Analysis (SIA) uses the signature of stable isotopes of carbon and nitrogen to reveal, respectively, the overall basal resources of the web and the trophic position (height) of a species within the web. This technique thus has the major advantages that it is sensitive to what is assimilated by consumers, rather than what is simply ingested, and integrates feeding over a longer period. However, SIA mixing models cannot be used to construct the complex, multispecies food-webs that can be described using GCA, partly because the calculations soon become algebraically intractable. Consequently, the main strength of SIA is that it provides a complementary, rather than alternative, approach to GCA, which is most commonly used to quantify ingestion (e.g. Hall et al. 2001).

Despite the lack of methodological standardization among studies, a similarly skewed distribution of "link strength" occurs in a wide range of terrestrial and aquatic food webs, irrespective of which measures are used (e.g. de Ruiter et al. 1995; Benke and Wallace 1997; Wootton 1997; Emmerson and Raffaelli 2004; Woodward et al. in press). To quantify the impact of a predator on a prey species (i.e. the interaction coefficient in a Lotka–Volterra model, a_{ij}) and vice versa (i.e. a_{ji}), we need to know how much of a predator's production is accounted for by that prey species, and what proportion of the prey's production is ingested by the predator. Then a_{ij} can be estimated by dividing ingestion of a prey species by the mean annual abundance of a predator or expressed as per unit biomass effects. Both halves of the community (Jacobian) matrix can therefore be calculated from secondary production and ingestion data, as long as information is available on the assimilation and gross production efficiencies of the consumers. Measuring secondary production and ingestion rates is laborious, as it requires extensive population surveying and exhaustive GCA. However, it does provide an approximate measure of "interaction strength," *in lieu* of data obtained via the even more logistically challenging, and indeed next to impossible, route of attempting to characterize intergenerational population dynamics (as used in Lotka–Volterra type models) following pairwise manipulations of the entire multispecies assemblage of a real food web.

Stream ecologists are presently making concerted efforts to unite the currently disparate ecosystem and community approaches to food webs by combining detailed GCA with SIA. A promising avenue for further progress is provided by the relatively new field of ecological stoichiometry (Chapter 5, this Volume). Essentially, because any given species must keep elemental ratios (e.g. $C:N:P$) in its body tissues within certain narrow limits, changes in the ratios among food resources will determine which consumers will be most successful and this, in turn, will determine species composition and population dynamics and interactions in the web.

Stoichiometric techniques have been used most extensively in the pelagic zone of lakes, where results suggest that an imbalance in $C:N:P$ ratio can have profound consequences for plankton population dynamics, even to the extent of creating alternative equilibria, whereby systems are dominated either by algae or by rooted macrophytes (Elser and Urabe 1999). To date, very few authors have employed stoichiometric approaches to the study of $C:N:P$ ratios in lotic systems, although many studies have examined either $N:P$ or $C:N$ ratios. For instance, the rate of N turnover in the tissues of primary consumers increases as the $C:N$ ratio declines, suggesting that the latter may act as a surrogate for biological activity with regard to N flux in streams (Dodds et al. 2000). A recent study demonstrated how global climate change could alter energy flux in detrital food chains in streams: raised CO_2 reduced the quality of the basal resources by increasing the $C:N$ ratio which, in turn, suppressed microbial activity and the rate of decomposition (Tuchman et al. 2003).

Overall, because terrestrial leaf-litter is a poor quality food resource, being high in C and poor in P and N, there is considerable scope for nutrient recycling in streams to be strongly influenced by consumer-driven dynamics (Vanni 2002; Woodward and Hildrew 2002c). Large-scale switches, along the river corridor, from a resource base dominated by terrestrial detritus (very C enriched) to one dominated by algae/aquatic macrophytes should, in theory, have strong effects upon nutrient uptake and storage by the consumers within the web. A change in $C:N:P$ ratio might have dramatic impacts on stream food-webs as productivity increases (e.g. as P increases during eutrophication). Indeed, Cross et al. (2003) recently demonstrated marked differences in the $C:N:P$ ratio of CPOM, driven by variations in microbial conditioning, across streams that spanned a gradient of nutrient enrichment. Intriguingly, they also found P enrichment in the body-tissues of consumers, suggesting a degree of deviation from the homeostasis usually assumed in stoichiometric models (*sensu* Sterner 1995).

Spatial and temporal variation in food web pattern and process

Variability in space

Riverine food webs vary in both their structural and dynamic properties over the three spatial dimensions and with time (Woodward and Hildrew 2002c), and much of this variation is patchy and scale-dependent (Townsend 1989). Marked shifts can occur in both web topology and the strength of feeding interactions, even at the microhabitat scale (Lancaster 1998; Stone and Wallace 1996).

The ecosystem size hypothesis (Cohen and Newman 1988) states that as the size of a habitat increases so too does species richness, habitat availability, and habitat heterogeneity; these are all factors that may contribute to reduced omnivory and greater dietary specialization, with consequent increases in food-chain lengths (Post et al. 2000; Woodward and Hildrew 2002c). This hypothesis has rarely been tested, but a study of 18 food webs, collected using standardized methods, found a weak positive relationship between ecosystem size (estimated as wetted area of a 30-m reach) and mean food-chain length (Thompson and Townsend, in press).

The question of spatial variation is also relevant at still larger scales. The River Continuum Concept (RCC), developed for North American river systems with headwaters flowing through deciduous forest, postulated variations along a river in habitat conditions, resource quality and quantity, and in ecological patterns (Vannote et al. 1980). Any longitudinal changes in the energy base and its associated fauna should have important consequences for food-web structure. Despite distinct downstream variation in the nature of the basal resources (changing from coarse to fine particulate matter), Rosi-Marshall and Wallace (2002) reported remarkable consistency in overall food-web structure, though not in energy flow along specific links in their study stream.

An important, though largely unexplored, aspect of spatial scale in stream food webs relates to the enormous range in body size of the constituent organisms and the consequent differences in the spatial "grain" with which they perceive their environment (Stead et al. 2003). Food webs attempt to integrate information on groups ranging from bacteria (for whom meaningful biological scales may be measured in millimeters) to fish, mammals, and birds (which may operate over scales of meters to many kilometers). It has been suggested that food webs should be constrained by the home range of the top predator in the system (Cousins 1996). For many streams this would require inclusion of information appropriate to the large scales over which predatory fish forage. The dispersal of adult insects along streams and between catchments is also likely to be important in determining local food web structure in streams (Woodward and Hildrew 2002c).

In general, the definition of appropriate boundaries is a major challenge for food-web ecologists. In streams, the limit is usually drawn at the aquatic–terrestrial interface, but in reality this "ecotone" is leaky to materials and energy, and the stream food-web is firmly embedded within that of the neighboring landscape (Ward et al. 1998). While lateral links with the adjacent terrestrial system have long been recognized as important in studies of stream food-webs (e.g. Hynes 1975; Junk et al. 1989), attempts to quantify the energetic importance of terrestrial subsidies to streams are more recent. Allen's classic (Allen 1951) study of a brown trout stream identified "Allen's paradox"— the observation that in-stream invertebrate prey production was apparently insufficient to provide for the observed trout production. This paradox was resolved in part by realization of the importance of terrestrial invertebrates to trout diet (Huryn 1996) and stream food-webs in general (e.g. Nakano et al. 1999). The use of stable isotopes (e.g. Rounick and Hicks 1985) has also proved useful in characterizing the relative importance of biomass generated in the catchment and in the stream.

Several large-scale manipulative experiments have attempted to elucidate the role of terrestrial inputs by excluding allochthonous material. Exclusion of leaf-litter from a North Carolina stream resulted in the loss of a trophic level from the stream food-web (Hall et al. 2000), while similar experiments excluding terrestrial invertebrates from Japanese streams evoked a trophic cascade, with an increase in algal biomass when fish switched from terrestrial prey to stream invertebrates (Nakano et al. 1999). Comparative studies of streams receiving a variety of types of litter, as compared with those receiving a single type, also suggest that the heterogeneity of terrestrial subsidies may be an important factor in determining food-web characteristics (Thompson and Townsend 2003).

There are also food-web linkages to the landscape in the "reverse" direction. For instance, aquatic insects accounted for around 60% of the carbon assimilated by riparian spiders close to two New Zealand streams (Collier et al. 2002) and aquatic prey supports populations of many riparian birds (e.g. Ormerod and Tyler 1991). In the Pacific northwest of North America nitrogen derived from carcasses of migrating salmon is assimilated by riparian plants, incorporated into terrestrial litter, and ultimately contributes to stream productivity (Milner et al. 2000).

Expansion of streams and rivers across the aquatic/terrestrial ecotone during high flow events is also known to be important in determining in-stream food webs. The lateral "flood pulse" is proposed to be a major force controlling productivity and biota in floodplains (Junk et al. 1989). Young and Huryn (1997) showed that dissolved organic carbon varies along the length of New Zealand's Taieri River, with the highest concentrations where the river flows through floodplain reaches. A similar role for floodplains in maintaining and subsidizing river food-webs has been suggested for the tropics by Winemiller (1990, 1996). The consequences for food-web structure of such energy subsidies from the land have been little explored and warrant future attention. As a note of caution, however, Bunn et al. (2003) point to accumulating evidence from stable isotopes that terrestrial carbon provides little energy to aquatic food webs in arid-zone flood plain rivers, despite the sometimes enormous inputs. Rather, animals on Australian floodplain pools were labeled mainly with C derived from aquatic microalgae, despite the latter's apparent paucity.

Most studies of stream food-webs have been carried out over relatively limited spatial scales, and usually in an arbitrarily determined reach, while experimental studies have commonly resorted to the use of microcosms, mesocosms, or in-stream channels. While such studies may measure adequately the changes and processes that operate on small scales, they cannot incorporate properly species that operate on large spatial scales (particularly fish). Experiments at a large scale, such as the whole catchment manipulations of litter input at Coweeta, USA (Hall et al. 2000) and in Japan (Nakano et al. 1999), are particularly notable exceptions.

Variability in time

Food webs have been constructed over a wide range of different timeframes: ranging from data gathered on a single day to collated webs that represent decades of cumulative research effort (Hall and Raffaelli 1993). Time is critical because temporal variation in community structure and functioning is inevitable on one or more timescales, particularly in dynamic systems such as streams. Overall, underlying temporal variation in stream food-webs means that summary food webs gathered over one or many years are likely to misrepresent true (instantaneous) food-web structure. This is because species may be grouped together that never actually occur at the same time, thus producing anomalously low values of connectance in the summary web (Thompson and Townsend 1999).

A number of studies of detritus-based streams have shown considerable seasonal variability in community structure (Winemiller 1990; Closs and Lake 1994; Tavares-Cromar and Williams 1996), due in part to seasonal patterns in the inputs of terrestrial leaf-litter. Thus, autumn pulses of leaf-litter may be important in determining the synchrony of life-cycles evident in many Northern Hemisphere streams, which in turn influences food-web structure. However, seasonal variation in stream food-webs is also evident in temperate regions with relatively muted cycles. For instance, Thompson and Townsend (1999) reported seasonal food-web patterns in New Zealand streams that had no autumnal pulse of leaf-litter, modest seasonality in rainfall and a fauna of invertebrates with predominantly asynchronous life-cycles.

Seasonality in the life history of invertebrates can affect food-web structure via ontogenetic shifts. In Broadstone Stream, pulses of invertebrate recruitment in autumn and summer resulted in marked changes in size spectra through time (Woodward and Hildrew 2002a). For predators, which tend to be relatively long lived, this entailed changes in diet as a series of prey moved through a temporal "window" of vulnerability, in which they were detectable as prey but had not yet either achieved an upper size refugium or had emerged as adults. Woodward and Hildrew (2002a) showed that the diet of predators changed with ontogeny, with that of their prey, and with variations in the abundance of prey. In Broadstone this produces a much simpler web structure in winter than summer, and a much more complex summary web than is ever realized at one time (Schmid-Araya et al. 2002a). In contrast, the asynchrony of invertebrate life histories in New Zealand streams means that a common core of invertebrates is represented through the year across a range of size classes (Thompson and Townsend 1999).

Stream systems are characterized by a relatively high degree of physical disturbance that can vary seasonally. Highly variable and/or unpredictable discharge in small streams acts as a disturbance of variable frequency and intensity (Poff and Allan 1995) and can modify population, community, and ecosystem processes (Townsend 1989; Death and Winterbourn 1995). The pattern of bed-disturbance can also dictate species composition and richness, and the distribution of species traits (Death and Winterbourn 1995; Townsend et al. 1997). Townsend et al. (1997) found that insect species that are small-bodied, mobile as adults, relatively streamlined, and generalist in habitat preference were more common in highly disturbed streams. Those characters were grouped into those that conferred resistance (the ability to survive spates) and those that conferred resilience (the ability to recover quickly after spates).

Although few studies have investigated the influence of disturbance on food-web structure, Townsend et al. (1998) found that, in 10 streams that differed in bed disturbance regime, the number of dietary links per species was negatively related to intensity of disturbance. Jenkins et al. (1992)

hypothesized that disturbance could also affect food-chain length, because species higher in the food chain are rarer and are more likely to be lost by chance from the system during a disturbance event, but stream studies have not provided unequivocal data in support of this (Townsend et al. 1998).

The persistence of food-web structure has been relatively poorly studied on timescales greater than a single year (e.g. Winemiller 1990; Closs and Lake 1994; Tavares-Cromar and Williams 1996 and above). Bradley and Ormerod (2001) followed invertebrate communities in eight Welsh streams over 14 years. They found that the persistence of communities was related to the occurrence of climatic cycles, with mild, wet winters destabilizing communities. A long-term study (>20 years) showed that Broad-stone Stream (Woodward et al. 2002) has maintained a broadly consistent food-web structure over very long periods, overlain with a gradual shift in response to a rise in pH since the 1970s.

Anthropogenic effects on stream food-webs

The pervasive effects of human activities on running waters, from local to global scales, are not usually analyzed in terms of stream food-webs per se (see Malmquist et al. (in press) for a recent view of the prospects for the running water environment in general). Here we address three specific issues in relation to stream food-webs: catchment land-use change, species introductions-invasions, and larger-scale environmental impacts such as air-borne pollutants and climate change.

Land-use change

Land-use effects are among the most profound influences on stream food-webs at both local and landscape spatial scales (Woodward and Hildrew 2002c). Deforestation for agricultural development, and afforestation with exotic, plantation conifers, can alter geomorphology (e.g. Leeks 1992), water chemistry (e.g. Hildrew and Ormerod 1995) and the disturbance regime (e.g. Fahey and Jackson 1997) over large scales. Conversion of forest to agriculture reduces shading of the channel, increases the supply of nutrients, and enhances algal productivity.

Plantation forests often shade the channel profoundly, reducing algal productivity, but may increase the supply of organic matter.

Lindeman (1942) predicted more trophic levels (longer food chains) in productive systems, and data from New Zealand grassland and forested streams (Townsend et al. 1998; Thompson and Townsend, unpublished) showed a weak, positive relationship between productivity and food-chain length (as well as the total number of trophic links and links per species). Grassland food-webs ("high productivity settings") had more algal species, displayed greater internal connectance and had longer food chains, while forest food-webs ("low productivity settings") tended to be based on detritus, with short food chains and low prey: predator ratios (Figure 5.4), patterns that have also been described in detritus-based food-webs from Canada (Tavares-Cromar and Williams 1996), Australia (Closs and Lake 1994), and the United States (Thompson and Townsend 2003). Thompson and Townsend (2003) also compared streams in exotic pine forests in southern New Zealand with those at two locations in eastern North America and found that food-web structure was essentially identical, providing support for the occurrence of ecological equivalence with taxonomically different species performing the same functional roles.

Invasion of stream food-webs

The deliberate translocation of fish for sport fisheries provides many examples of invasion (e.g. Townsend 1996). Others have occurred through accidental introductions (e.g. Schreiber et al. 2003), or as a result of a relaxation of physicochemical constraints (e.g. Woodward and Hildrew 2001). Not all invaders persist (e.g. Woodward and Hildrew 2001), but those that do may occupy an unoccupied trophic niche but not exert strong effects on prey (e.g. Wikramanayake and Moyle 1989), and thus have weak effects on food-web structure or processes. Others occupy a similar niche to an existing species and replace it (e.g. Dick et al. 1993), but may feed at a different rate and thus alter ecosystem processes (e.g. Power 1990; McIntosh and Townsend 1995). In extreme cases,

(a)

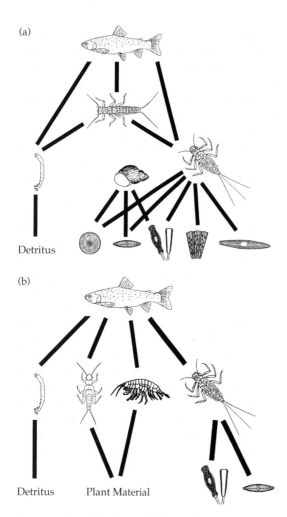

Detritus

(b)

Detritus   Plant Material

Figure 5.4 Generalized food-web diagrams from (a) grassland and (b) forested streams showing differences in food-web architecture (Thompson and Townsend unpublished).

an invader may drive vulnerable competitors or prey to extinction or have widespread consequences that resonate through the food web.

Despite attempts to model the effects of invasions on food webs (e.g. Mithen and Lawton 1986), empirical studies are rare. The impact of brown trout (*Salmo trutta*) has been investigated by comparing New Zealand streams with and without this invader. The equivalent native species were galaxiid fishes, which are lost through predation and competition from trout (Townsend 2003). The presence of trout elicits changes in prey behavior,

and algal consumption is reduced as grazing invertebrates become more nocturnal (McIntosh and Townsend 1995).

Trout do not appear to influence the structural properties of the food web (i.e. as assessed by the population/community approach), because in general they simply replace galaxiids (Townsend et al. 1998), but in terms of ecosystem processes they cause major changes in the allocation of biomass and the flux of materials. Galaxiids consume a moderate amount of invertebrate production, leaving sufficient grazing invertebrates to exert top-down pressure on algal biomass. After invasion, however, a trophic cascade is induced, whereby trout consume a larger fraction of invertebrate production, thus reducing grazing pressure on algae, and resulting in an accumulation of algal biomass (McIntosh and Townsend 1995) and higher primary productivity (Huryn 1998).

In Broadstone Stream, the population of the dragonfly *Cordulegaster boltonii*, previously very rare, irrupted probably as a result of reduced acidity (Woodward and Hildrew 2001), and food-web patterns were recorded before, during and after this invasion. The larva is large, voracious and, by virtue of mouthpart morphology, able to feed on a wide variety of prey items. The invasion by *C. boltonii* increased mean food-chain lengths, web complexity and omnivory but top-down control of prey abundance was observed for only a few, particularly vulnerable, prey species. No species has been lost, probably reflecting the importance of abundant refugia in the heterogeneous streambed, though a competing top predator might be deleted in the longer term (Woodward and Hildrew 2001). It is of interest that both *S. trutta* and *C. boltonii* are large, polyphagous top predators that are able to achieve "size refugia" from predation by all except larger conspecifics (Huryn 1996; Woodward and Hildrew 2002*a*), characteristics that have been described for certain other successful stream invaders (Power 1990; Charlebois and Lamberti 1996).

Air-borne pollutants and climate change

In general, the study of the effects of pollutants and toxins on aquatic communities has been limited largely to ascertaining lethal doses for inverte-brates and fish, although a limited number of stream studies have reported reduced species diversity (e.g. Winner et al. 1980) and more general effects on ecosystem functioning (e.g. Molander et al. 1990). Food-web studies are an appropriate unit of study for improving understanding of the wider consequences of pollutants in the environ-ment (see Clements and Newman (2002) for a wider discussion of community-level ecotoxicology). The best known in this context are the air-borne acidifying pollutants that affect large parts of the globe.

Acidification of streams affects the biota and influences the concentration and toxicity of aluminum and other metals (Hildrew and Ormerod 1995; Friberg et al. 1998). Acid rain reduces stream pH in areas with a susceptible geology, and conifer plantations can further exacerbate the effects of acidification (Hildrew and Ormerod 1995). Many fish species are intolerant of low pH and their absence from acid streams allows large, generalist predatory invertebrates to proliferate (e.g. Hildrew et al. 1984; Hildrew 1992), resulting in greater connectance and a predominance of diffuse links, which may enhance stability and reduce the likelihood of trophic cascades (Schmid-Araya et al. 2002a,b).

Acidification has sometimes been associated with a reduction in the productivity of algae (Ledger and Hildrew 2000a) and bacteria (Edling and Tranvik 1996), with consequences for nutrient recycling (Mulholland et al. 1987). Sutcliffe and Hildrew (1989) describe an apparent switch from an algivore-dominated to a detritivore-dominated community associated with low pH, a pattern confirmed by Lancaster et al.'s (1996) analysis of acid-stream food webs. Specialist invertebrate grazers, such as mayflies and snails, are usually missing from acid streams, which are generally less productive than circumneutral systems. However, Ledger and Hildrew (2000a,b; 2001) showed that the suite of acid-tolerant "detritivor-ous" species that remain grow well on algae and are effective grazers, indicating that ecosystem processes are maintained despite a loss of biodi-versity, and that trophic links in stream food-webs can be somewhat flexible.

Global climate change potentially has an even more pervasive and larger scale effect on food-web structure and function. Mean river temperature in Europe probably increased by approximately 1°C during the twentieth Century (Webb 1996) and similar trends are predicted elsewhere (Matthews and Zimmerman 1990; Hogg and Williams 1996; Keleher and Rahel 1996). The effects of the poten-tial consequent increase in productivity and nutrient flux in streams are unknown. In addition, global warming may drive species with narrow thermal tolerances locally extinct (Matthews and Zimmerman 1990). For example, salmonid habitat may become limited to higher altitudes (Keleher and Rahel 1996), with an overall reduction in the area available to fish (Mohseni et al. 2003). Higher temperature can also cause reduced invertebrate density, earlier onset of maturity, and altered sex ratios (Hogg and Williams 1996). Higher atmo-spheric CO_2 can reduce flux rates in detrital food chains via an increase in the C:N ratio (Tuchman et al. 2003). Global changes in the El Nino Southern Oscillation (ENSO) and the North Atlantic Oscil-lation (NAO), and more generally in patterns of rainfall and drought, also have the potential to alter profoundly food-web structure and functioning (Puckridge et al. 2000; Bradley and Ormerod 2001). For example, Closs and Lake (1994) showed increases in food web size, prey:predator ratio, and food-chain length as a stream food web reas-sembled after a drying event.

Biodiversity and ecosystem function in stream food-webs

Recognition of the implicit links between food-web structure and ecosystem processes has under-pinned recent attempts to position food-web eco-logy within the context of the debate about biodiversity and ecosystem function (Paine 2002; Petchey et al. 2004), with significance for both ecological theory and applied ecology (i.e. dealing with the consequences of biodiversity loss). Many classic studies on biodiversity-ecosystem function have focused on horizontal (i.e. competitive) interactions among terrestrial primary producers (e.g. Lehman and Tilman 2000). More recently, however, aquatic ecologists have focused attention on vertical (i.e. predatory) as well as horizontal interactions (e.g. Downing and Leibold 2002; Paine 2002; Petchey et al. 2004).

In temperate forested streams, detrital food chains predominate and decomposition processes are the means not only by which energy and organic matter are incorporated into stream food-webs but are also pathways for the uptake of toxins that can impair ecosystem function (Bird and Schwartz 1996). Consequently, leaf-litter breakdown has been the primary focus of such research in streams (although predation and filtra-tion rates have also been examined more recently; see Jonsson and Malmqvist 2003). Most studies have measured how decomposition is affected by manipulation of either species richness of detriti-vores (macrofaunal shredder) or of microbial decomposers. Shredders are often the dominant processors, accounting for up to 75% of the mass loss of coarse particles (Hieber and Gessner 2002), although microbial breakdown seems to become increasingly important with decreasing latitude (Irons et al. 1994). Many stream studies have shown positive effects of invertebrate species richness on decomposition (e.g. Jonsson et al. 2001; Jonsson et al. 2002), others have reported no relationship or found that species identity, rather than richness per se, was the main driver (Jonsson and Malmqvist 2003). This idiosyncratic pattern of results resembles that found in marine ecosystems (Emmerson et al. 2001).

A number of factors will determine the nature of the relationship between biodiversity and decom-position. First, there appears to be a high degree of "trophic redundancy" within stream food-webs, as suggested by the prevalence of generalist, as opposed to specialist, feeding modes (Mihuc 1997; Woodward and Hildrew 2002a), resulting in com-plex, reticulate food webs with a high proportion of taxa with similar consumers and resources (Benke and Wallace 1997; Williams and Martinez 2000; Woodward and Hildrew 2001). Even in quantified webs, intraguild variation in diets may be slight, particularly among similar-sized con-sumers (Woodward and Hildrew 2002a). Trophic equivalence among species might explain why curves plotting the increase in a specified eco-system process often begin to saturate at quite low species richness (e.g. from three species onwards) in stream systems.

Because the extent of dietary overlap tends to increase with resource availability (Nakano et al. 1999; Schmid and Schmid-Araya 2000; Woodward and Hildrew 2002a), the degree of redundancy in stream food-webs, and hence the dependence of ecosystem processes on biodiversity, is likely to vary both spatially and temporally. Redundancy in the food web may be further enhanced by "trophic plasticity" (one source of dietary generalism), whereby individual species alter their feeding behavior in response to resource availability, so that they are able to exploit alternative food sources when available. The example of grazing being taken over by detritivores in species-poor acidic streams may be a case in point (see above: Ledger and Hildrew 2000a,b; 2001). Further, a 7-year litter exclusion experiment revealed that shredders and microbes responded by shifting their feeding from leaves to wood when leaf litter was excluded (Eggert and Wallace 2003).

Trophic plasticity is also evident across "guilds" among the primary consumers: for instance, fine particulate and dissolved organic matter bound up in epilithic biofilms can provide a significant energy source to taxa that are traditionally viewed as algal grazers, as revealed by the uptake of iso-topically labeled biofilms in cave streams by snails (Simon et al. 2003). Moreover, extensive sharing of resources can also occur across trophic levels in streams: many predators are extremely omnivor-ous and can consume significant quantities of algae, macrophytes, and detritus (Usio 2000; Schmid-Araya et al. 2002a; Woodward and Hildrew 2002a). A high level of redundancy in stream food-webs accords with the "insurance hypothesis" of ecosystem function (e.g. Walker 1992; Naeem 1998), in that trophically equivalent taxa may be lost without any obvious effects on process rates, at least in the short term, because the remaining taxa are able to compensate for changes in community composition.

It is worth noting that all studies of biodiversity and ecosystem function in streams to date have involved manipulation of a single trophic level (usually the primary consumers, rather than the basal resources). However, ecosystem processes, such as leaf breakdown, can be influenced by both top-down and bottom-up processes, which can act simultaneously (e.g. Rosemond et al. 2001). It is dangerous to generalize when higher trophic levels have been ignored because there are many examples of indirect, predator-mediated impacts on links between the primary consumers and the basal resources in streams (e.g. Power 1990; Flecker and Townsend 1994; Huryn 1998; Nakano et al. 1999). Members of the primary consumer guild in streams vary in their vulnerability to predation (e.g. Woodward and Hildrew 2002b,d) and the identity of species included in experiments may skew results, even to the extent of changing the sign of the relationship between biodiversity and ecosystem function (Petchey et al. 2004). Petchey et al. (2004) modeled the consequences for ecosystem function of species loss in multitrophic systems and found that function was affected not only by the number of species that were lost, but also by the trophic status of the deleted species and whether omnivory was present in the web.

Another potentially fundamental flaw inherent in many biodiversity-ecosystem function experiments is the equal weighting of the species represented (either as numbers or biomass), even though in real communities geometric or log-normal species rank-abundance curves are the rule rather than the exception (Tokeshi 1999). The lack of realism associated with the equal weighting of species in biodiversity-ecosystem function experiments may be analogous to the equal ranking and inadequate sampling of species and links in early connectance food webs, where unrealistic patterns were modeled repeatedly despite the inevitably spurious results (e.g. the supposed rarity of omnivory) (Hall and Raffaelli 1993; Polis 1998). A recent study (Dangles and Malmqvist 2004) has highlighted the pitfalls of this over-simplistic approach, and revealed the importance of incorporating more realistic levels of dominance in biodiversity experiments.

Finally, the random assortment of species, typical of biodiversity-ecosystem function experiments, is unrealistic because community assembly is often nonrandom, as demonstrated by the predictability of community (i.e. taxonomic) composition in streams in response to environmental gradients (e.g. Hämäläinen and Huttunen 1996: Wright et al. 2000). In a revealing experiment on the crucial role of species identity, a "fixed" sequence in the loss of species of shredders associated with two perturbations, acidification and eutrophication, resulted in a different response in the rate of leaf breakdown, as compared to the effects of "random" losses (Jonsson et al. 2002). Indeed, species losses from systems are not random but are determined by differential vulnerability to environmental factors (e.g. pollutants, disturbance) or trophic effects (e.g. invasion/expansion of a predator) and life history characteristics (slow growth rate, low fecundity). Studies exploring the consequences of the predicted loss of species, in the order of their known vulnerability to particular changes, are the most likely to offer useful insight into the way that real systems function and can be managed.

Species that are particularly prone to extinction because of traits such as body size also may contribute disproportionately to stream food-web structure and dynamics, with attendant influences on ecosystem function (Petchey et al. 2004). Fish can have particularly strong effects in stream food-webs, and may induce trophic cascades that alter both decomposition and primary production (e.g. Power 1990; Flecker 1996). However, fish are not the only important "keystone" taxa; the larger members of each guild in a food web are frequently the main drivers of species interactions and ecosystem processes in streams. For instance, leaf breakdown is often dominated by the larger macroinvertebrate shredders such as crayfish and amphipods (Usio and Townsend 2001; Dangles et al. 2004). Similarly, strong top-down effects on primary production are often associated with large grazers, such as snails. Secondary production has also been regarded as driven mainly by larger individuals, particularly within taxa but also between taxa (e.g. hydropsychid caddis among filter-feeders (Benke and Wallace 1997); stoneflies among invertebrate predators (Benke et al. 2001). Large invertebrate predators, such as dragonflies (Woodward and Hildrew 2002b) and omnivorous crayfish (Usio 2000), can also have relatively strong impacts on prey populations, whereas the often far more abundant small species, such as tanypod midge larvae, generally appear to have much weaker or negligible effects (Woodward et al. in press).

Conclusion

Food webs are complex, and stream food-webs are no exception with even relatively "simple" depauperate systems, such as acid streams, containing many hundreds of feeding links. However, most of these links may be energetically and dynamically trivial: increasingly, theoretical advances and improved empirical data suggest that most links are weak, and that this may stabilize even very complex webs (McCann 2000; Neutel et al. 2002). The ubiquity of generalist feeding in streams creates high levels of redundancy within webs, whereby many species are functionally similar, and may also give rise to ecological equivalence among webs, whereby different assemblages perform the same functional role (e.g. at low pH snail and mayfly grazers are replaced by stoneflies, which can act as facultative grazers and maintain herbivory). Although some individual species can have very powerful effects on community structure and ecosystem processes (e.g. trophic cascades induced by fish can have profound impacts on primary production), it seems that many are relatively weak interactors which, if removed from a system, have little discernible effect because their role is assumed by another, trophically similar, species. The few biodiversity-ecosystem function experiments that have been carried out to date with stream assemblages have shown that, for a given process rate, saturation often occurs rapidly as species richness increases, lending support to the "insurance hypothesis" of ecosystem function, which predicts that high species redundancy may serve to buffer the effects of disturbance. However, as progressively more species are lost this buffering capacity will eventually be impaired, and large changes in ecosystem function can occur when certain species (e.g. top predators; keystone species) are lost. Arguably the most pressing challenge at present is to be able to predict the consequences of species loss from lotic systems: to do this we will need to unite the community and ecosystem approaches to food-web ecology, to identify species traits associated with vulnerability to extinction, and then to use this information to perform multitrophic experiments across a range of scales, in order to develop more realistic, predictive mathematical models.

PART II

Examining food-web theories

Ideally, food-web theory results from the rigorous and creative interplay between qualitative or quantitative models, including appropriate null or random models, and observed data collected from field and lab, along gradients or from experiments. The following chapters touch on some important aspects of food-web theory, including the statistical analysis of community food webs for testing for patterns of regularities, such as constant connectance; testing parametric models as the cascade model; and ecological network analysis and the patterns of information theory based metrics associated with network size and function.

In aquatic systems, the interplay between models and data is particularly crucial given the tension between the preservation and exploitation of many aquatic taxa. Theory can play an important role in conservation, and food-web theory in particular is helping to underpin new whole-ecosystem approaches to marine resource management. For example, analysis of abundance–body mass relationships in marine food webs is providing new ways to predict species abundance, food-web structure, and trophic transfer efficiency in the presence and absence of fisheries exploitation, as well as characterizing changes in those properties related to environmental variability.

CHAPTER 6

Some random thoughts on the statistical analysis of food-web data

Andrew R. Solow

Introduction

For the purposes of this chapter, the term food web is taken to mean the binary predator–prey links among the species (or other elements) in a biological community. The description and analysis of food-web data has a long history, dating back in the United States at least to the landmark paper of Forbes (1887). One line of food-web research has been the search for general patterns or regularities in the structure of observed food webs and the construction of parsimonious models that reproduce these regularities. It is not clear that treating food-web structure *in vacuo* is either efficient or sensible (Winemiller and Polis 1996). If structural regularities are present in food webs, then presumably they are present for a reason. The most obvious possibility is that certain structural features have beneficial implications for the dynamics of the community. The connection between food-web structure and community dynamics has been pursued for decades by Robert May (1973), Stuart Pimm (1982), and many others. This suggests that it may be profitable to change the question from "What regularities are present in observed food webs?" to "What food-web structures have beneficial implications for community dynamics and are these structures present in observed webs?" Be that as it may, this chapter will focus exclusively on the analysis of food-web data.

To search for general patterns and to construct models, it is necessary to have a reasonably large collection of observed food webs. A significant contribution in this area was the compilation by Joel Cohen and Frédéric Briand of a freely disseminated database called ECOWeB containing well over 100 published food webs. On the basis of these observed food webs, Cohen and his co-workers identified a number of regularities and went on to propose a simple one-parameter model of an *S*-species food web. This and related work is discussed in the collected articles by Cohen et al. (1990). The model proposed by Cohen and Newman (1985) is called the cascade model and is discussed in further detail below.

Neo Martinez and others questioned the degree of resolution of the food webs in the ECOWeB collection (Martinez 1992, 1993). Upon analyzing a smaller number of more highly resolved food webs, Martinez found that the regularities identified in the ECOWeB data did not survive improved resolution and proposed an alternative model—the constant connectance model. The enterprise did not end there. New and improved food-web data have been published on a regular basis (e.g. Link 2002; Schmid-Araya et al. 2002), new regularities have been discovered (e.g. Solow and Beet 1998; Camacho et al. 2002), and new models have been proposed (e.g. Williams and Martinez 2000).

The purpose of this chapter is to discuss some statistical issues that arise in the analysis of food-web data. There have been many informal statistical analyses of food-web data. The relatively small existing body of *formal* statistical work in this area has focused on assessing the goodness of fit of the cascade model (Solow 1996; Neubert et al. 2000) or testing proposed regularities (e.g. Murtaugh 1994;

Murtaugh and Kollath 1997; Murtaugh and Derryberry 1998). This chapter focuses on statistical issues in modeling food-web data. The current practice in this area is, first, to identify regularities in the data and, second, to show via simulation that a proposed model either does or does not exhibit the same behavior. Although it can be useful, from a statistical perspective, this approach is ad hoc. Not only are its statistical properties unknown, but regularities do not (usually) constitute a complete food-web model in the sense described below. The main goal here is to discuss a more formal approach to modeling.

Likelihood methods for food-web models

Consider a community consisting of S species s_1, $s_2, \ldots, s_S$. The term species is used loosely here. It is common in food-web analysis to refer to trophic species. A trophic species is defined as a set of one or more species with the same predators and prey. The food web for this community can be represented by an S-by-S binary predation or adjacency matrix $a = [a_{jk}]$ where:

$$a_{jk} = 1, \text{ if } s_k \text{ preys on } s_j$$
$$= 0, \text{ otherwise.} \qquad (6.1)$$

This predation matrix is viewed as a realization of a matrix random variable A. A random variable is completely characterized by its distribution. Thus, a food-web model is complete if it completely specifies the distribution of A. As noted, unless a set of descriptive features can be shown to completely characterize the distribution of A, it does not constitute a complete description. Let f_θ be the probability mass function of A under a particular food-web model, where θ is the unknown (possibly vector-valued) parameter of the model. Unlike a collection of regularities, $f_\theta(a)$ is a complete food-web model. Also, let $a_1, a_2, \ldots, a_n$ be the observed predation matrices of a random sample of n food webs of sizes, $S_1, S_2, \ldots, S_n$, respectively.

Statistical inference about the parameter θ can be based on the likelihood function:

$$L(\theta) = \prod_{i=1}^{n} f_\theta(a_i). \qquad (6.2)$$

The likelihood and statistical inference based upon it are discussed in the monograph by Azzalini (1996). What follows is a very brief review of the basic theory. The maximum likelihood estimate $\hat{\theta}$ of θ is found by maximizing $L(\theta)$ (or its logarithm) over θ. The likelihood can also serve as a basis for testing the null hypothesis H_0: $\theta = \theta_0$ that the parameter θ is equal to a specified value θ_0 against the general alternative hypothesis H_1: $\theta \neq \theta_0$. Specifically, the likelihood ratio test statistic is given by:

$$\Lambda = 2(\log L(\hat{\theta}) - \log L(\theta_0)). \qquad (6.3)$$

Provided n is large enough (and other regularity conditions hold), under, H_0, Λ has an approximate χ^2 distribution with degrees of freedom given by the dimension $\dim(\theta)$ of θ, so that H_0 can be rejected at significance level α if the observed value of Λ exceeds the upper α-quantile of this distribution. A $1 - \alpha$ confidence region for θ is given by the set of values of θ_0 for which H_0: $\theta = \theta_0$ cannot be rejected at significance level α. Finally, to test the overall goodness of a fitted model (an exercise inferior to testing against an alternative model), if the fitted model is correct, then $2 \log L(\hat{\theta})$ has an approximate χ^2 squared distribution with $\dim(\theta)$ degrees of freedom.

Some examples

As an illustration, consider the problem of fitting the cascade model of Cohen and Newman (1985). Under the cascade model, there is an unknown ranking $r = (r(s_1)r(s_2) \cdots r(s_s))$ of the S species in a food web and the elements of A are independent with:

$$\text{prob}(A_{jk} = 1) = 0, \quad \text{if } r(s_j) \geq r(s_k)$$
$$= \gamma/s, \quad \text{otherwise,} \qquad (6.4)$$

for an unknown constant γ with $0 \leq \gamma \leq S$. That is, under the cascade model, species can prey only on other species of lower rank and the probability that they do so is inversely proportional to the size of the food web. A consequence is that the expected number of links increases linearly with web size S, and not with S^2 as would be expected if predation probability is independent of S. Note that, if the species are relabeled so that, $r(s_j) = j$, then the matrix A will be upper triangular. If no such relabeling is possible, then the food web contains a loop, which is forbidden under the cascade model. In terms of the previous notation, for a single food web, under the cascade model, the parameter θ includes the ranking r and the constant γ. The likelihood is identically 0 if $a_{r(s_j)\ r(sk)} = 1$ for any pair of species s_j, s_k with $r(s_j) \geq r(s_k)$. Any ranking with this feature is inadmissible. Let y be the sum of the elements of a. For any admissible ranking, the likelihood is given by the binomial probability:

$$f_\theta(a) = \left(\binom{\binom{S}{2}}{y} \right) (\gamma/S)^y (1 - \gamma/S)^{\binom{S}{2} - y}. \qquad (6.5)$$

Importantly, this likelihood is the same for all admissible rankings. The overall likelihood for a collection of n independent webs—each with at least one admissible ranking—is just the product over n terms each with the binomial form (6.5). However, if even a single food web in this collection has no admissible ranking, then the overall likelihood is identically 0.

Solow (1996) used likelihood methods to fit and test the cascade model to 110 ECOWeB food webs. First, the cascade model was fit by maximum likelihood. The estimate of γ is 4.11, which is close to the value of 4 proposed by Cohen. However, the goodness of the fitted cascade model is easily rejected. One possible explanation is that the systematic part of the model—namely, a predation probability *exactly* inversely proportional to S—is incorrect. To explore this possibility, Solow (1996) fit a more general model under which predation probability is inversely proportional to S^β. Cohen (1990) referred to this as the superlinear-link-scaling model. The maximum-likelihood estimate of the parameter β is 0.87, which is close to the value of 1 under the cascade model. However, the goodness of this fitted model is also easily rejected. An alternative explanation of the lack of fit of the cascade model is that the predation probability for food webs of size S is not fixed, but varies from food web to food web, resulting in extra-binomial variation in the number of links. To explore this possibility, Solow (1996) assumed that the predation probability $P(S)$ for a food web of

size S is a random variable with mean $p(S) = \gamma/S$ and variance:

$$\text{Var } P(S) = \delta \, p(S)(1 - p(S)). \qquad (6.6)$$

Under this model, which is due to Williams (1982), the expected number of links is the same as under the cascade model, but the variance is inflated by the factor $1 + (S - 1)\delta$. Using this model, Solow (1996) showed that the smallest value of δ that explains the extra-binomial dispersion in the number of links is around 0.023. To put this result into context, under this model, the greatest variability in predation probability occurs when $S = 8$, in which case the expected predation probability is around 0.5 with standard deviation around 0.08. Murtaugh and Kollath (1997) also found overdispersion in food-web data.

Like the cascade model, almost all food-web models include a ranking of species. However, the cascade model is special in the sense that the likelihood is the same for all admissible rankings. In statistical terminology, under the cascade model, for any admissible ranking, the sufficient statistic y is ancillary for the ranking. As a result, in fitting the cascade model, it is not necessary to estimate the true ranking of the species. This congenial property is not shared by even the simplest alternatives to the cascade model. For example, Neubert et al. (2000) tested the cascade model against several alternative models in which predation probability varied with the rank of the predator, the rank of the prey, or both. In this application, rank was determined by body mass. Consider the first possibility, which Neubert et al. (2000) referred to as predator dominance—see, also, Cohen (1990). If the rank of each species is known, then the problem amounts to testing whether the proportion of potential prey that is actually consumed is constant across predators. This can be done through a standard χ^2 test for homogeneity of probabilities in a 2-by-S contingency table. Briefly, the two rows in the table represent "eaten" and "not eaten," the S columns represent the species according to rank, and the entries in the table represent the number of potential prey that are and are not eaten by each species (so, for example, the sum of the entries in column j must be $j - 1$). Of course, to construct this table, it is necessary to know the rank of each species. Neubert et al. (2000) used the species size as a proxy for rank, which is certainly a reasonable approach (Cohen et al. 1993). However, the problem is more difficult if the species ranks are not known, for in that case, the number of potential prey that is not eaten by each species is unknown.

Under the assumption that species can only prey on species of lower rank, the observed food web a determines a partial ordering of the species. For example, suppose that, in a four-species food web, s_1 is observed to prey on s_2 and s_3 is observed to prey on s_4. If these are the only observed species, then there are six possible complete rankings: in brief, but obvious notation, $(1>2>3>4)$, $(1>3>2>4)$, $(3>1>2>4)$, $(3>1>4>2)$, $(3>4>1>2)$, and $(1>3>4>2)$. These are called the linear extensions or topological sortings of the partial ordering $(1>2, 3>4)$. An algorithm for generating the linear extensions of a partial ordering is given in Varol and Rotem (1981). This algorithm is relatively simple, but not computationally efficient. A more efficient, but less simple, algorithm is given in Pruesse and Ruskey (1994).

To give an idea of the statistical complications arising from an unknown ranking, return to the test for predator dominance of Neubert et al. (2000). The following approach—which is not directly based on the likelihood—could be used when the species ranking is not known. First, identify the linear extension that gives the largest value χ^2_{max} of the χ^2 statistic in the corresponding 2-by-S contingency table. This is the ranking that most favors predator dominance. The value χ^2_{max}

will serve as the test statistic. Second, randomize the location of the 1s in the upper triangular part of the observed predation matrix. This produces an equally likely realization of the cascade model conditional on the number of links. Third, as before, find the maximum value of the χ^2 statistic over the linear extensions of the partial order imposed by the (randomized) feeding pattern. Fourth, repeat the randomization procedure a large number of times and estimate the significance level (or p-value) by the proportion of randomized food webs with values of χ^2_{max} that exceed the observed value.

Conclusion

To an outsider, the recent history of food-web research—at least, the kind of food-web research discussed in this chapter—seems absolutely remarkable. The availability of the ECOWeB food webs led to a burst of activity involving the identification of structural regularities and the development of models. However, virtually none of this work—some of which is quite elegant—survived the realization that the ECOWeB webs were gross caricatures of nature. Ecologists are now left with a relatively small number of what appear—but are not guaranteed—to be more realistic food webs. It seems that some sorting out is now in order to help ensure that future research in this area lies along the most profitable lines. An admittedly small part of this sorting out ought to focus on proper ways to approach statistical inference in food-web models.

Analysis of size and complexity of randomly constructed food webs by information theoretic metrics

James T. Morris, Robert R. Christian, and Robert E. Ulanowicz

Introduction

Understanding the interrelationships between properties of ecosystems has been a central theme of ecology for decades (Odum 1969; May 1973). Food webs represent an important feature of ecosystems, and the linkages of food-web properties have been addressed specifically (Pimm 1982; Paine 1988). Food webs consist of species networks, and the effect on ecosystem processes of the imbedded species diversity has become a major focal point for research (Tilman 1997; McGrady-Steed et al. 2001; Chapin et al. 2000). In recent years the properties of food webs have been placed into a network perspective, but a major limitation to addressing these properties is simplification associated with data availability and software limitations (Paine 1988; Cohen et al. 1993). One major simplification involves aggregation of species into guilds or trophic species to reduce the number of taxa to be considered (Ulanowicz 1986a). Sizes of published food-web networks are less than 200 taxa (Martinez 1991) when binary feeding relationships (i.e. diet item or not) are considered, and no more than 125 taxa (Ulanowicz et al. 1998) when feeding relationships are quantified (i.e. as fraction of total diet). We have constructed hypothetical food webs with weighted feeding relationships that overcome the limitation of aggregation. Food webs ranged in size up to 2,426 taxa.

Ulanowicz has developed several metrics of system growth and development based on information theory (Ulanowicz 1986, 1997). They are predicated on the importance of autocatalysis (i.e. positive feedback loops) during growth and development. Autocatalysis has both extensive and intensive properties. The extensive metric (characteristic of size) is the total amount of flow within the system, that is total system throughput (TST). When the intensive metric (characteristic of complexity) of flow diversity is scaled by TST, the result is called the system's developmental capacity. Flow patterns become increasingly constrained as a system develops and matures, a characteristic represented by average mutual information (AMI). When AMI has been scaled by TST, the ensuing quantity is called the system ascendency. These metrics, and related ones, provide ways to gauge and characterize systems and changes within systems. For example, eutrophication appears to increase system ascendency via an increase in TST that more than compensates for a drop in AMI (Ulanowicz 1986b). This quantitative definition distinguishes simple enrichment (which does not induce a drop in AMI) from eutrophication (which does).

The objectives of this study were twofold. One objective was to assess the effects of different assumptions used to construct the hypothetical webs. A second objective was to evaluate how recent theoretical metrics of food-web organization and complexity (Ulanowicz 1986a, 1997) respond to two key variables that vary with ecosystem development, namely web size and function. Size was examined in terms of the number of taxa and amount of processing (i.e. TST). Function was evaluated on the basis of the number and distribution of connections among taxa. We evaluated the interrelationships among size, connectivity, and network metrics of both hypothetical and empirical webs.

Methods

Food-web construction

We generated a large diversity of food webs for analysis by creating thousands of randomly constructed networks varying in size (number of taxa) and distribution of connections. Random, donor-controlled food webs were generated by populating a transfer matrix (A) with random coefficients that were drawn from realistic probability sets, randomly partitioning exogenous inputs (f) among primary producers (gross production), and solving for the biomass vector (x) for the steady-state condition:

$$dx/dt = 0 = f + Ax, \tag{7.1}$$

$$x = -A^{-1}f, \tag{7.2}$$

where f represents a vector of inputs (gross primary production, autochthonous inputs). Inputs to all structured webs (Figure 7.1) were divided into gross primary production (GPP) and

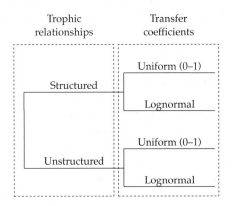

Figure 7.1 Terminology and classification of hypothetical food web construction. Realistic trophic structure was imposed on structured webs (Figure 7.2), while the taxa in unstructured webs were randomly connected without regard to trophic identity. Structured webs had no constraints on the relative numbers of taxa within each trophic level. Transfer coefficients were drawn either from lognormal or uniform distributions.

allochthonous import of organic matter. GPP was set at $1,000 \, \text{kcal} \, \text{m}^{-2}$ per year. Allochthonous import was $100 \, \text{kcal} \, \text{m}^{-2}$ per year. Total input to unstructured webs was fixed at $1,100 \, \text{kcal} \, \text{m}^{-2}$ per year. These standard conditions facilitated comparison of networks.

The food webs were donor-controlled in the sense that the flow from one taxon to the next was proportional to the biomass of the donor taxon. The model structure was purely descriptive and not meant to have any predictive capability of temporal dynamics. Rather, every feasible food web generated is a description of possible energy flows within a hypothetical network constructed from realistic principles. We posit that real food webs lie within the state space of these randomly generated food webs, provided that the sample size of hypothetical webs is sufficiently large.

Solving for the steady state (equation 7.2) from a randomly generated transfer matrix can result in negative biomass numbers for any number of taxa. Such webs were identified as being nonfeasible and were discarded. Hypothetical food webs were generated until a sufficiently large population of feasible (positive biomass for all taxa) webs were obtained to make a convincing analysis of their properties and trends. Four kinds of networks are considered here (Figure 7.1) and the rules are described below.

We constructed networks using different sets of rules concerning the nature of connections. Every structured food web consisted of "taxa" that were each defined as belonging to a group of primary producers, primary, secondary, and tertiary consumers, or detritivores, plus an organic matter or detritus compartment (Figure 7.2; plate 2). For each web that was generated, the number of taxa within each group was determined randomly. By this method we generated feasible webs that consisted of as few as 7 or as many as about 2,200 taxa.

For each structured web we generated three types of connections among food-web members (Figure 7.2). These included mandatory flows from all taxa to organic matter or detritus, mandatory flows from lower to higher trophic levels, and nonobligatory flows among taxa, such as a flow from a secondary consumer to a primary

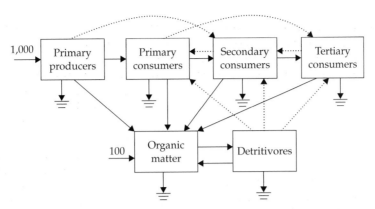

Figure 7.2 This figure shows the basic architecture of structured food-webs constructed with realistic trophic relationships. The energy input to every randomly constructed food web was standardized at $1,000\,kcal\,m^{-2}\,yr^{-1}$ of GPP and $100\,kcal\,m^{-2}\,yr^{-1}$ of exogenous organic matter input. Solid arrows (——→) denote mandatory flows of energy and their direction. For example, every primary consumer in the web is made to consume at least one primary producer. Dotted arrows (·····▶) denote flows that are possible, but not mandatory. The rules for making these connections are discussed in the text.

consumer. Consumer taxa belonging to the same trophic group were not connected. With flows of the second and third kind, a random number generator was used to determine the presence of a flow between two specific nodes. Mandatory flows from lower to higher trophic levels defined the trophic positions of taxa. For mandatory flows from lower to higher trophic levels, such as flow from primary producers to herbivores, the probability of flow from a lower to a higher-level taxon was assumed to be inversely proportional to the number of taxa within the lower trophic level. For example, with four primary producers in the food web, there was a 25% probability that any one of them flowed to any one of the herbivores. Unit random numbers were drawn for each pair of donor and recipient taxa to make the connections, subject to the requirement that every higher-level taxon had to feed on at least one lower level taxon. Connections of the third kind, or non-obligatory flows as from a tertiary to a secondary consumer, were made using a unit random number drawn for each possible pair of taxa after first drawing a unit random number that set the overall connectivity of the web. For each one of these possible connections, a connection or flow was made if the value of a unit random number was less than the overall connectivity. This procedure generated food webs that ranged from sparsely to densely connected.

The structure we have described above is general, and definitions of trophic components and flows are flexible. For example, loss from the organic matter compartment can be defined as either a respiratory flow and/or an export from the system depending on how the microbial community is defined. If detritivores include microbes, then export from organic matter is best defined as a loss from the system rather than a respiratory loss. At the level of the whole web, the entirety of living components can be thought of as including microbes that are parasitic or saprophytic on multicellular organisms from neighboring trophic levels.

A second type of web architecture, termed unstructured (Figure 7.1), was purely random in construction. There were no rules of trophic structure of any kind to govern the presence or absence of flows, except that the distribution of GPP was restricted to a group of taxa representing between 5% and 75% of total taxa in the web as determined by random number. Every taxon, including "primary producers," had an equal probability of feeding on any other taxon as determined by the overall connectivity of the web, which was determined by a uniformly distributed random variable ranging from 0.2 to 0.7 In addition, these unstructured networks had no detritus component to accumulate and degrade the products of mortality. These networks were simply random distributions of respirations and of exogenous inputs (gross production) and feeding among nodes.

Respiration rates

Weight-specific respiration rate coefficients were randomly chosen from uniform distributions spanning ranges representing realistic rates taken from the literature. The respiration of primary producers consisted of two parts: dark respiration and photorespiration. Photorespiration was accounted for by subtracting a fraction equivalent to 25–50% of the rate of GPP. The rate of photorespiration was randomly chosen for each primary producer. A weight-specific rate of dark respiration ranging from 40% to 70% of plant weight per year was also randomly chosen for each primary producer (Lambers 1985; Landsberg 1986). Heterotrophs (consumers and detritivores) had specific respiration rates ranging from 19% to 3,400% per year of their biomass, which is a range that spans a wide spectrum of body size (Gordon et al. 1972). For each heterotoph, a unit random number was drawn to assign a respiration rate within the appropriate range. Based on numerous sources (Morris and Lajtha 1986; Schomberg and Steiner 1997; Asaeda and Nam 2002; Van Santvoort et al. 2002; Qualls and Richardson 2003; Salovius and Bonsdorff 2004), detritus was given an annual decay rate ranging from 6.6% to 164%. In the case of the unstructured networks, the specific respiration rate assigned to each taxon was like that of the heterotrophs.

Transfer coefficients

Transfer coefficients representing energy flow among taxa were either drawn from a population of unit random numbers (0–1) or from a lognormal with $\mu = -2.3$ and $\delta = 2$, where the mean of the distribution is given by $\exp(\mu + \delta^2/2)$. The lognormal distribution allows for coefficients greater than one as would be the case for biomass turnover rates greater than once annually. The two alternative methods of generating transfer coefficients were applied to the structured webs as well as to the unstructured webs.

Network analysis

For every feasible food web generated, we computed total system biomass, the number taxa within each trophic group, the net production, and total respiration within each trophic group, the number of connections, total system throughput, and various metrics derived from information theory (Ulanowicz 1986, 1997). These included average mutual information or AMI, flow diversity (SI), total system throughput (TST), full development capacity ($Cd = TST \times SI$), and ascendancy ($TST \times AMI$).

Average mutual information is inversely proportional to the degree of randomness in a system and is a measure of the predictability of flow. Adopting the convention that flows can be represented by a matrix T_{ij} (row-column order) in which flow moves from a species in column j to a species on row i, then AMI (bits) was computed as:

$$AMI = \sum_i \sum_j p(T_{ij}) \log_2[\{p(T_{ij})/p(T_j)\}/p(T_i)],$$

$$(7.3)$$

where T_{ij} is the flow from species j to species i, $p(T_{ij})$ is the joint probability given by:

$$p(T_{ij}) = T_{ij}/TST, \qquad (7.4)$$

$$p(T_i) = \sum_j p(T_{ij}), \text{and } p(T_j) = \sum_i p(T_{ij}),$$

and where TST is total system throughput:

$$TST = \sum_i \sum_j |T_{ij}| \qquad (7.5)$$

The Shannon flow diversity (SI) is based on the individual joint probabilities of flows from each species j to each species i:

$$SI = \sum_i \sum_j [-p(T_{ij}) \log_2(p(T_{ij}))]. \qquad (7.6)$$

Empirical food webs

The hypothetical data were compared against data consisting of biomass and flow distributions of 31 empirically derived food webs from various ecosystems. The empirical webs were analyzed using the same suite of metrics as described above. However, the biomass and flows of the empirical food webs were first scaled so as to transform the total GPP of the empirical webs to a constant 1,000 units across all webs in order that they would be consistent with the GPP of hypothetical webs and to remove GPP as a variable. The procedure was to scale the input vector f and absolute flows by the quotient $1000/\Sigma f$.

Results

Food-web dimensions

Food webs that have been described empirically are necessarily aggregated to various degrees and typically overlook connections or flows that are minor. The average number of taxa in the set of empirical food webs was 39 ± 36 (± 1 SD). The largest empirical food web contained 125 taxa; the smallest contained 4. Empirical food webs averaged 350 ± 563 (± 1 SD) total connections and 5 ± 5 connections per taxon (Table 7.1). The minimum and maximum numbers of connections per web were 4 and 1,969. Minimum and maximum connections per taxon were 1 and 15.7.

The accounting of flows or connections among the taxa that comprise empirical food webs is likely to miss rare items in an organism's diet, and we arbitrarily defined rarity as any flow that constitutes 5% or less of an organism's diet. Thus, major flows were greater than 5% of an organism's diet. Using this definition of a major flow, the theoretical maximum number of major flows per taxon approaches 20, provided the flows into a taxon are equal in magnitude. The theoretical minimum number of major flows is zero as, for example, when there are greater than 20 uniform flows. Among empirical food webs the number of major flows per taxon averaged 2.3 ± 1, and ranged from 1 to 4.5.

Hypothetical webs had a mean size of 827 ± 444 taxa per web, with a maximum size of 2,426 and a minimum of 7. The probability of generating a feasible (biomass of all taxa must be positive), structured web declined precipitously at sizes exceeding 500 taxa (Figure 7.3) and was independent

of the means of generating transfer coefficients (uniform or lognormal). The probability distribution was described by a sigmoid: $Y = 1/\{1 + \exp(.02X–11)\}$, where Y is the probability and X is the number of taxa. For webs with 500 taxa the probability of successfully generating a feasible random web was 0.5. For webs with 1,000 taxa the probability declined to 0.0001.

For each web generated, we calculated the actual number of connections as well as the maximum possible number of connections, which depends on size as well as the specific distribution of taxa among trophic groups (Figure 7.4). Among structured webs, the total number of connections per web averaged $280,763 \pm 324,086$, with a maximum of 1,885,606 and a minimum of 23. The number of connections per taxon averaged 263 ± 220, with a maximum of 1050 and a minimum of 2. The set of hypothetical webs varied greatly in the total number of connections and connections per taxon, irrespective of size (number of taxa) (Figure 7.4). Both sparsely and densely connected webs were generated, even among the largest sizes. The average number of major (flows greater than 5% of the total input flows) flows per taxon in the set of hypothetical webs was similar to the empirical webs and averaged 2.1 ± 1.2 per taxon. This may be an artifact of a process that produces a distribution like that of the broken stick model (MacArthur 1957), but the distribution of flow stems ultimately from the transfer coefficients in a donor-controlled fashion, and their distribution does not resemble a broken stick distribution (Figure 7.5). On the other hand, GPP imposes a limit on flow in a manner similar to the resource space that MacArthur (1957) argued should set a limit on the distribution of species. Ulanowicz (2002) used an information- theoretic homolog of the May–Wigner stability criterion to posit a maximal connection per taxon of about 3.01 (which accords with these results).

The values of individual fractional flows into each taxon as a percentage of the total flow into each taxon were computed for each food web generated, excluding gross photosynthesis, by type of web construction. For each food web generated, a frequency distribution of these fractional flows was computed, and then the global average

Table 7.1 Summary statistics from empirical ($n = 31$) and hypothetical food-webs. Statistics were computed from 800 samples each of the lognormal and uniform variety of unstructured webs and 641 and 1006 samples of structured lognormal and uniform webs, respectively

Variable	Web type	Mean	Standard deviation	Minimum	Maximum
Number of connections	Empirical	5.1	4.8	1.0	15.8
per taxon	Unstructured lognormal	349.7	242.6	1.3	1025.3
	Unstructured uniform	358.8	247.2	2.0	1049.7
	Structured lognormal	179.7	156.5	3.2	849.0
	Structured uniform	172.6	138.6	4.2	829.6
Number of major connections	Empirical	2.3	1.0	1.0	4.5
per taxon	Unstructured lognormal	3.1	0.8	1.5	4.9
	Unstructured uniform	0.8	0.9	0.0	5.1
	Structured lognormal	2.9	0.5	1.0	4.0
	Structured uniform	1.7	0.8	0.5	5.4
TST (kcal m^{-2} yr^{-1})	Empirical	3,412	662	2,175	4,684
	Unstructured lognormal	18,120	11,311	2,209	50,934
	Unstructured uniform	14,051	7,992	2,299	36,381
	Structured lognormal	4,117	801	2,382	11,250
	Structured uniform	4,052	637	2,912	9,133
AMI	Empirical	1.6	0.3	1.0	2.0
	Unstructured lognormal	3.3	0.6	1.0	4.5
	Unstructured uniform	1.6	0.4	0.9	2.5
	Structured lognormal	2.7	0.4	1.2	4.0
	Structured uniform	2.0	0.3	1.2	2.9
Flow diversity (bits)	Empirical	4.0	1.0	1.8	5.7
	Unstructured lognormal	13.4	3.2	1.3	17.5
	Unstructured uniform	15.0	3.7	2.0	19.8
	Structured lognormal	9.2	1.7	3.2	14.7
	Structured uniform	10.5	1.7	4.2	16.5
Full capacity	Empirical	13,905	5,001	3,935	25,104
	Unstructured lognormal	273,323	205,619	2,892	889,544
	Unstructured uniform	236,335	167,057	4,824	720,361
	Structured lognormal	39,146	15,008	7,736	165,117
	Structured uniform	43,379	13,731	13,059	143,630
Ascendency as a fraction	Empirical	0.41	0.07	0.31	0.64
of full capacity	Unstructured lognormal	0.26	0.04	0.19	0.76
	Unstructured uniform	0.12	0.05	0.05	0.52
	Structured lognormal	0.30	0.03	0.23	0.38
	Structured uniform	0.20	0.04	0.11	0.29

frequency distribution of fractional flows was computed by the type of web (Figure 7.5). Regardless of the type of web construction, the overwhelming majority of flows were less than 5% of the total flow into each taxon. Random webs without realistic structures had the highest percentage of these small flows; 98% ±6 (±1 SD) of flows were less than 5% of the total. Among webs with structure, 95% ±9 of flows were of a magnitude less than 5% of the total.

The distributions of transfer coefficients were also computed and averaged over all webs within each class of web construction to determine if they differed from the distributions expected on the basis of the random number generator. If the distributions differed from the expected, then the web structures must be selecting the distribution. Webs constructed of transfer coefficients drawn from the uniform, unit distribution had frequencies exactly as expected (Figure 7.5). Webs constructed of transfer coefficients drawn from lognormal distributions also did not differ from the expected. Note that transfer coefficient values greater than 1 were grouped into one category for analysis, which accounts for the upturn in frequency on the tail of the distribution. Among webs constructed of lognormally distributed transfer coefficients, 12.5% ±0.3 of coefficient values were greater than one.

Total system throughput

Total system throughput, which is simply the sum of all flows among the taxa, including respirations and exogenous flows (i.e. GPP, losses and allochthonous inputs), was bounded within the structured webs. Among the unstructured webs, TST increased with web size without limit, at least within the range of the food-web sizes generated (Figure 7.6; Plate 3). This increase in TST, seemingly without limit and in defiance of the laws of physics, was a consequence of the high degree of flow reciprocity that occurs among unstructured webs constructed without realistic trophic relationships. TST increased by approximately 15 and 21 kcal m^{-2} per year per taxon in unstructured uniform and unstructured lognormal webs, respectively. TST was limited among structured webs, as finite inputs dictate a finite TST (Ulanowicz 1997), at least when food webs have realistic structures and thermodynamic constraints. TST among structured webs was bounded by upper and lower asymptotes. These asymptotes were sensitive to GPP (not shown), which in these examples was always set to 1,000 kcal m^{-2} per year. Mean TST among structured webs was about 4,000 kcal m^{-2} per year (Table 7.1) or four times GPP. That this mean is very near to four times GPP tells us that the mean effective number of trophic levels (= TST/GPP) is about 4, which is consistent with real food-webs. Of course, the structured webs were designed with 4 trophic levels (Figure 7.2), but it does

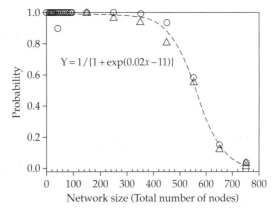

$$Y = 1/\{1 + \exp(0.02x - 11)\}$$

Figure 7.3 Probability of generating a feasible, random network with realistic structure using transfer coefficients drawn from either uniform (△) or lognormal (○) distributions.

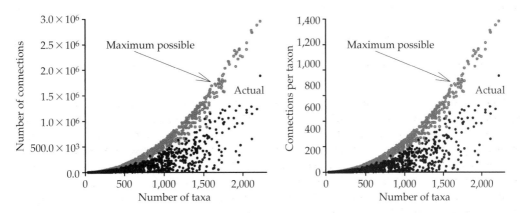

Figure 7.4 Density of connections of structured, hypothetical food webs as a function of the number of taxa. Shown are the maximum possible number of connections (○), determined by the greatest number of feasible connections for a particular distribution of taxa, and the corresponding actual number of connections (●) for approximately 6,800 networks (see also Plate 2).

support the choice of coefficients used to construct the hypothetical webs. The mean TST (normalized to GPP = 1,000 kcal m^{-2} per year) of the empirical webs was $3,400 \pm 662$ kcal m^{-2} per year (Table 7.1).

The pattern of TST when plotted against the average number of major (>5% flow) connections per web was quite different between the two types of unstructured webs and between the unstructured and structured webs (Figure 7.6). Among unstructured normal webs, TST was greatest at low connectivity and declined toward a lower asymptote as connectivity increased. In contrast, the TST of unstructured, lognormal webs was bifurcated (Figure 7.6). For example, for webs with

2 connections per taxon, TST was either greater than 40,000 or less than 10,000 kcal m^{-2} per year, while at the highest connectivity, TST focused on a single cluster less than 10,000 kcal m^{-2} per year. This suggests a strong and peculiar control by the structure of the web over its function. TST of structured and empirical webs was independent of the connectivity.

Flow diversity

Flow diversity (SI) among empirical and hypothetical webs increased with size (taxa) (Figure 7.7; Plate 4). Unstructured, uniform webs had the highest SI, averaging 15 ± 3.7 bits, empirical webs

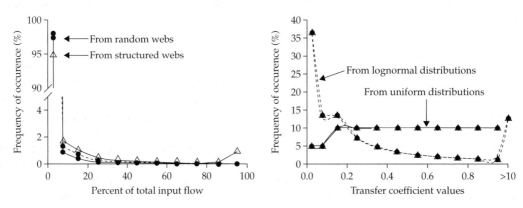

Figure 7.5 Average frequency distribution of flow strengths into taxa expressed as fractions of the total flow into each taxon (left). Average frequency distribution of transfer coefficients (right). Note that the transfer coefficient is equivalent to the fraction of the standing stock of a donor taxon that flows into a recipient taxon in a unit of time in steady state.

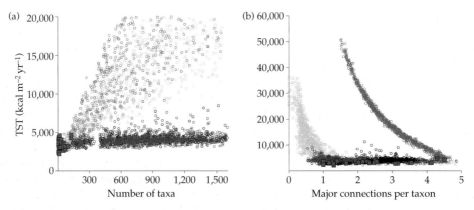

Figure 7.6 TST of different types of hypothetical food-webs as a function of (a) the number of taxa and (b) of the average number of major connections (flows >5% of inputs) per taxon. Hypothetical webs were either unstructured in design with uniformly ○ or lognormally ○ distributed transfer coefficients, or structured in design with uniformly ● or lognormally ○ distributed transfer coefficients. Empirical webs are denoted by ■ (see also Plate 3).

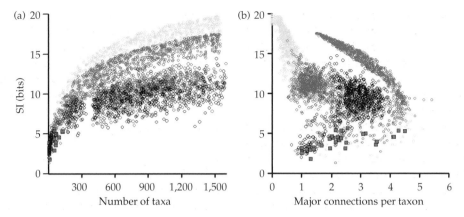

Figure 7.7 Flow diversity (SI) of different types of hypothetical food webs as a function (a) of the number of taxa and (b) of the average number of major connections (flows >5% of inputs) per taxon. Hypothetical webs were either unstructured in design with uniformly ○ or lognormally ○ distributed transfer coefficients, or structured in design with uniformly ● or lognormally ○ distributed transfer coefficients. Empirical webs are denoted by ■ (see also Plate 4).

had the lowest (4 ± 1 bits). The SIs of empirical and hypothetical webs were equivalent among webs of comparable size, while structured webs had lower flow diversities for a given size than unstructured webs (Figure 7.7, Table 7.1). For a web with m taxa and m^2 flow paths, the SI reaches an upper limit of $-\log_2(1/m^2)$, provided the flows are uniform in size. Alternatively, if the flows among taxa are dominated by n major flows per taxon, then the SI of structured webs reduces to $-\log_2(1/nm)$, which gives a limit for SI of about 12.1 for a large web of size $m = 1500$ and $n = 3$. The latter calculation is consistent with the results obtained from the structured webs (Figure 7.7). Note that the mean number of major connections per taxon for this web type was about 3 (Table 7.1). Hence, the imposition of a realistic structure seems to be selecting for webs with a flow diversity that is dominated by a few major flows among taxa, whereas the unstructured webs are more fully connected and have greater uniformity of flow (Figure 7.5).

Full development capacity

The total potential order or complexity (kcal bits m^{-2} per year) is termed full developmental capacity (Cd) and is the product of TST and the flow diversity (SI) (Ulanowicz 1986; Ulanowicz and Norden 1990). Based on the above generalizations about the limits

to flow diversity and total system throughput, we can place limits on Cd. For fully connected, unstructured webs of uniform flow and $m^2 - m$ flow paths, $Cdmax = -TST \cdot \log_2[1/(m^2 - m)] = -\alpha m\log_2[1/(m^2 - m)]$, where α is a proportionality constant, equivalent to 15 and 21 kcal m^{-2} per year per taxon in unstructured uniform and unstructured lognormal webs, respectively. For an arbitrary range of sizes starting with $m = 100$, this gives $20 \cdot 10^3$ to $28 \cdot 10^3$ kcal bits m^{-2} per year and $166 \cdot 10^3$ to $232 \cdot 10^3$ kcal bits m^{-2} per year for $m = 600$, depending on α, and these values are consistent with the Cd observed for the unstructured webs without realistic trophic structure (Figure 7.8; Plate 5).

For webs with realistic structure or webs with limited TST, a fully connected web also would have $Cd = -TST \cdot \log_2[1/(m^2 - m)]$, which for the hypothetical webs here, were $TST \approx 4GPP$, we estimate that Cd_{max} would be about $84 \cdot 10^3$ kcal bits m^{-2} per year when $m = 1,500$. This is near the upper range of Cd observed among the hypothetical webs with realistic structure (Figure 7.8). When a web is dominated by a few flows per taxon, a more realistic example, the Cd_{max} is approximated by $-TST \cdot \log_2(1/3m)$, which gives $Cd_{max} \approx 4 \cdot GPP \cdot \log_2(1/3m)$ or $35 \cdot 10^3$ kcal bits m^{-2} per year. This value is in the middle of the observed Cd for structured webs (Figure 7.8) and close to the mean Cd of the structured, lognormal webs (Table 7.1).

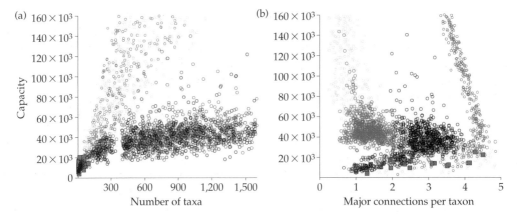

Figure 7.8 Full development capacity of different types of hypothetical food-webs as a function (a) of the number of taxa and (b) of the average number of major connections (flows >5% of total inputs) per taxon. Hypothetical webs were either unstructured in design with uniformly ○ or lognormally ○ distributed transfer coefficients, or structured in design with uniformly ● or lognormally ○ distributed transfer coefficients. Empirical webs are denoted by ■ (see also Plate 5).

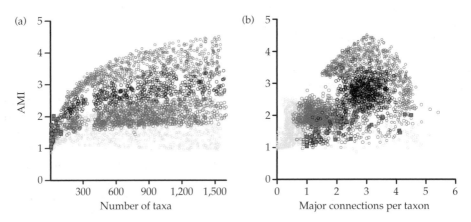

Figure 7.9 AMI content of different types of hypothetical food webs as (a) a function of the number of taxa and (b) of the average number of major connections (>5% of input flows) per taxon. Hypothetical webs were either unstructured in design with uniformly ○ or lognormally ○ distributed transfer coefficients, or structured in design with uniformly ● or lognormally ○ distributed transfer coefficients. Empirical webs are denoted by ■ (see also Plate 6).

Average mutual information

The AMI of empirical and hypothetical webs were consistent among webs of comparable size (Figure 7.9(a)). Among the hypothetical webs, the highest AMI was found among large, unstructured lognormal webs, while the lowest AMI was found among unstructured, uniform webs. Unstructured webs had higher total number of connections than structured webs, and unstructured lognormal webs had the greatest average number of major connections per taxon, but the unstructured uniform webs had the lowest average number of major connections per taxon (Figure 7.9; Plate 6, Table 7.1). Among structured webs, the lognormal variety tended to have higher AMI than the uniform variety (Figure 7.9) and, while the total connectivity was equivalent (Table 7.1), the lognormal webs had the greater number of major connections per taxon (Figure 7.9).

For a food web of size m with nm uniform flows of size TST/nm, the joint probability of flow will be $p(Tij) = (TST/nm)/TST = 1/(nm)$; the row and column sums of the joint probabilities, $p(T_i)$ and $p(T_j)$ respectively, are each equivalent to n/nm. Making these substitutions into equation 7.3 and solving gives the maximum theoretical $AMI = \log_2(m/n)$. Thus, AMI will tend to rise with network size as our results demonstrate (Figure 7.9(a)). A large food web of $m = 1,500$ taxa would have a maximum AMI of about 10.5 bits when there is $n = 1$ uniform connection per taxon. AMI will decline as the connectedness of the food web increases. For example, for a web of size m taxa and $n = m$ connections of uniform flow per taxon, the AMI reduces to $AMI = \log_2(m/m) = 0$. Thus, AMI rises with network size and declines as the uniform number of connections (total and per taxon) rises (as the direction of flow becomes less certain).

The relationship between AMI and SI is complex. For networks of n uniform flows per taxon, $AMI = \log_2(nm)$, and $SI = -\log_2(1/nm)$. SI may be rewritten as $SI = [\log_2(nm) - \log_2(1)] = AMI$. Thus, AMI is equivalent to SI when flows are uniform and highly predictable. SI and AMI were positively correlated in both empirical and hypothetical webs (Figure 7.10; Plate 7). However, the flows in the empirical and hypothetical webs were neither uniform nor highly predictable, and the AMI was less than the SI (Figure 7.10). For a given flow diversity, the range of possible AMIs increased as flow diversity increased. Thus, as flow diversity increases, the potential AMI also increases. As expected, the hypothetical webs constructed of uniformly distributed transfer coefficients had lower AMI:SI than their lognormal counterparts (Figure 7.10) due to greater uniformity (= greater unpredictability of direction) of flow among the former web types. Empirical webs clustered at the low end of the SI and AMI distributions, but were otherwise consistent with the results of the hypothetical networks.

Ascendency

Interestingly, and importantly, structured webs with realistic trophic structure had greater As/Cd than unstructured webs, within each class of transfer coefficient distribution (Table 7.1, Figure 7.11; Plate 8). Thus, the imposition of a realistic trophic structure, which constrained the flow distribution, raised the As/Cd. Average As/Cd of empirical webs was considerably greater than the average As/Cd of the hypothetical webs (Table 7.1), but at the low end of the size range, empirical webs and hypothetical webs had similar As/Cd (Figure 7.11). Thus, the small size of both empirical and hypothetical webs appears to constrain the flow distribution and raise As/Cd. The As/Cd of hypothetical webs declined rapidly with increasing web size toward a relatively stable average value (depending on web design), with considerable variability around the mean (Table 7.1, Figure 7.11). For a given size of web, the As/Cd of lognormal webs was greater than that of uniform webs (Figure 7.11), which is a consequence of the greater inequality of flow among the lognormal webs.

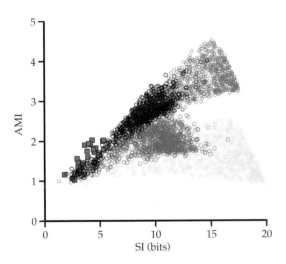

Figure 7.10 AMI as a function of flow diversity (SI) in empirical (▪) and hypothetical webs. Hypothetical webs were either unstructured in design with uniformly ▫ or lognormally ○ distributed transfer coefficients, or structured in design with uniformly ● or lognormally ○ distributed transfer coefficients (see also Plate 7).

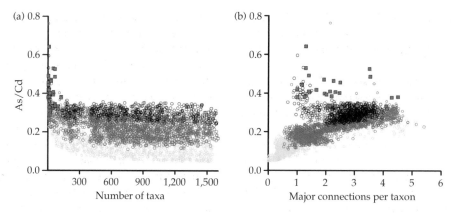

Figure 7.11 Ascendency:capacity as a function of (a) web size and (b) of the average number of major connections (flows >5% of inputs) per taxon (right). Hypothetical webs were either unstructured in design with uniformly ○ or lognormally ○ distributed transfer coefficients, or structured in design with uniformly ● or lognormally ○ distributed transfer coefficients. Empirical webs are denoted by ■ (see also Plate 8).

Discussion

The rules of organization of the random webs had significant effects on the various information indices. These rules of organization defined both the size of the networks (i.e. number of taxa and TST) and function (i.e. number and distribution of connections). Some of the responses of these information metrics to the rules were quite predictable, at least qualitatively, while others were not. It was shown that that flow diversity (SI) increased with network size (Figure 7.7), and with uniformity of flow and connectedness. Average mutual information also increased with network size (Figure 7.9), but the distribution of flows had greater impact on AMI than did network size. The SI had an upper bound depending on the size of the network. For example, if flows are uniformly distributed, then a fully connected web of size m with $m^2 - m$ connections, excluding losses and inputs, will have joint flow probabilities that are determined by the species diversity m. The flows $T_{i,j}$ are given by $TST/(m^2 - m)$, and the joint probabilities (from equation 7.4) by $1/(m^2 - m)$. From equation 7.6, the flow diversity would be equal to $-\log_2[1/(m^2 - m)]$, which gives an SI for a large web ($m = 1{,}500$) of about 21 bits, and this is consistent with the results for the unstructured webs with uniformly distributed transfer coefficients (Figure 7.7). Unevenness in flow reduces SI, and unstructured lognormal webs

had lower SI than unstructured uniform webs (Figure 7.7, Table 7.1).

Average mutual information is a measure of the average constraint or probability of flow from one taxon to the next (Ulanowicz 1997). It is not easily decomposed like flow diversity (SI) and has a behavior that is far more complex. AMI has a minimum value of zero when a network is maximally connected with uniform flows, or when there is an equal probability of flow moving from one taxon to any other taxon and the uncertainty in the direction of flow is greatest. Conversely, AMI is maximized when a web is minimally connected and flows are uniform. In other words, there is no uncertainty about direction of movement.

Zorach and Ulanowicz (2003) provided an extension of AMI by relating it to the number of roles in a network, where a role is a unique flow pattern within a food web that is common to one or more species. The number of roles is a measure of the diversity of functions that are occurring within a network. The empirical food webs analyzed here had about 2–5 roles per food web, and occupied a graphical area termed the window of vitality (Zorich and Ulanowicz 2003), similar to the clustering we observed between AMI and the density of major connections (Figure 7.9(B)). This is an idea similar to the network motifs or recurring circuit elements described by Milo et al. (2004), and seems to be a powerful way of characterizing and classifying webs. However, the application of these metrics to real webs suffers from the same problems of aggregation as the information theoretic metrics.

Interestingly, unstructured uniform and structured uniform networks gave the lowest range of AMI (Table 7.1) and, therefore, the lowest number of roles. Conversely, the lognormal networks gave the greatest range of AMI and highest number of roles. Intuitively, a greater number of roles is consistent with the greater variation in flow strength afforded by the lognormal distribution, and on a mechanistic level this is most likely a consequence of the differences in the distributions of transfer coefficients. However, the actual flow distributions expressed as a percent of total inflow hardly differed among the different web types (Figure 7.5).

Ascendency (As) is given by the product of AMI times TST and is a measure of the absolute order in a system times the power generated, also described as the performance of the system (Ulanowicz 1986, 1997). The quotient As/Cd is a measure of the degree of organization of the system. We had hypothesized that As/Cd would be higher for structured webs than for unstructured webs and this was confirmed by our hypothetical data. Recall that Cd (full capacity) is a measure of the total potential order or complexity, and As is the realized complexity. All web types had a moderate range of As/Cd for webs of constant size, and for large webs the As/Cd appeared to be insensitive to web size, which is consistent with the calculations made above. Thus, As/Cd was most sensitive to flow distribution and insensitive to size, except among the smallest size classes where size apparently constrains the flow distributions and raises As/Cd. This raises the unsettling possibility that the aggregation of species that is characteristic of the relatively small empirical webs artificially raises As/Cd. This may be associated with the general inability to define minor flows in highly aggregated, small food webs. Even larger empirical webs lack resolution of flows of less than a few percent, and no current empirical web has species-level resolution for smaller organisms (Christian and Luczkovich 1999).

We can make some generalizations about the behavior of As/Cd by making the simplifying assumption that there are nm uniform flows in a web of size m. With this assumption, As/Cd reduces to $-\log_2(m/n)/\log(1/nm)$. For one uniform connection per taxon ($n = 1$), the quotient As/Cd $= 1$ and is independent of size m. A network of $n = 1$ is fully specialized. At the opposite extreme ($n = m$), the quotient As/Cd $= 0$, and the network is maximally disorganized. Among hypothetical and empirical webs, the average As/Cd were less than 0.5 (Table 7.1), which affirms a large diversity of flows. Like the hypothetical webs, data on interaction strengths in natural food webs indicate that interaction strengths are characterized by many weak interactions and a few strong interactions, which is thought to increase stability by dampening oscillations between consumers and resources (McCann et al. 1998).

Do empirical food webs reflect real webs? This has been a major source of concern in ecology for the past quarter century (Pimm 1982; Paine 1988; Martinez 1991; Cohen et al. 1993; others). We contend that networks of fully articulated food webs fall within the universe of our random networks. The structured webs (both uniform and lognormal) are a subset that should more closely contain organized, real food webs. Of the metrics considered, the highly aggregated empirical food webs have characteristics that fall within at least the smaller examples of the reference universe of structured webs. Reasonable similarities occur for total system throughput (normalized for inputs), number of major connections per taxon, SI, and Cd. However, AMI and As/Cd tend to be higher for empirical webs than their hypothetical counterparts of similar size. This may be a matter of resolution of flows. Major flows (>5%) are dominant in empirical webs because (1) the number of interaction possibilities is low and (2) the ability to identify rare diet items is low (Cohen et al. 1993). Major flows in these webs may constitute all flow, whereas major flows in the hypothetical webs do not necessarily constitute all flows. These minor flows affect the AMI and thus As and As/Cd. Thus, the true significance of these metrics may not be realized within our current means of characterizing food webs.

CHAPTER 8

Size-based analyses of aquatic food webs

Simon Jennings

Introduction

Size-based analyses of aquatic food webs, where body size rather than species identity is the principle descriptor of an individual's role in the food web, provide insights into food-web structure and function that complement and extend those from species-based analyses. A strength of the focus on body size is that body size underpins predator–prey interactions and dictates how the biological properties of individuals change with size. Thus size-based food-web analyses provide an approach for integrating community and ecosystem ecology with energetic and metabolic theory.

In aquatic ecosystems, the principle primary producers are small unicellular algae, and these strongly support size-structured food chains where most predators are larger than their prey (Sheldon et al. 1972). Individuals of most species begin life as larvae feeding at the base of food chains, but can end life as large terminal predators. Lifetime weight increases of five orders of magnitude and more are typical of many species (Cushing 1975), and a fast growing species may begin life as a prey item for other species only to become the main predator on the same group of species within one year (Boyle and Bolettzky 1996). The significance of size-based predation and the large scope for growth in aquatic animals means that body size is often a better indicator of trophic position than species identity. Thus there are compelling reasons to adopt size rather than species-based analysis of aquatic food webs. In such analyses, a small individual of a large species is treated as functionally equivalent to a large individual of a small species in the same body-mass class (Kerr and Dickie 2001) although this is necessarily a simplification of a more complex reality (Pimm 1991; Hall and Raffaelli 1993).

The systematic study of abundance–body-mass relationships (size spectra) in aquatic communities began when Sheldon and colleagues began to look at the size distributions of phytoplankton in the ocean using a modified Coulter Counter (Sheldon et al. 1972). They observed the remarkable regularity and surprisingly shallow gradient of the relationship between abundance and body size. Subsequently, this approach was extended over the entire size spectrum from plankton to whales, and the links between size-based analyses of food-web structure and size-based analyses of the biological properties of organisms were soon apparent (Kerr 1974). This work was a significant step in understanding food-web processes, since previous studies of abundance–body-mass relationships in both terrestrial and aquatic ecosystems had focused primarily on taxonomically or trophically defined subsets of the entire food web.

As more size spectra were compiled, it became increasingly clear that their similar slopes (Figure 8.1) reflected similar processes of biological organization. These processes appeared to be common to many aquatic ecosystems, even though the physical and biological characteristics of the ecosystems, such as primary production, higher level production, mean temperature, seasonality, and depth, were very different (Boudreau and Dickie 1992). Clearly, an understanding of the processes that led to characteristic size spectra would help to identify general laws that govern biological organization. The applied potential of

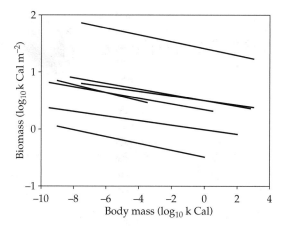

Figure 8.1 Relationships between abundance and body mass (size spectra) for seven aquatic ecosystems. Slopes are comparable but differences in intercept reflect differences in primary production, and hence production at body mass, in each ecosystem. Redrawn with modifications from Boudreau and Dickie (1992).

size-based food-web analysis was also quickly recognized and, in addition to analyses of the impacts of human exploitation on food webs (Dickie 1976), Sheldon and Kerr (1972) boldly used these methods to estimate the biomass of monsters in Loch Ness! The methods developed were classic examples of the macroecological approach (Brown 1995; Gaston and Blackburn 2000).

While the study of abundance–body-mass relationships has contributed to aquatic and terrestrial ecology, studies in the two environments have taken rather different paths. In aquatic ecosystems, most research on abundance–body-mass relationships has been closely tied to energetic analysis of ecosystem function, with an emphasis on the size structure, biomass, and production of all individuals in the food web (Dickie et al. 1987). Conversely, in terrestrial ecosystems, there has been much more focus on abundance and body mass in taxonomically defined subsets of the whole food web, which either derive energy from the same source (e.g. plants) or do not (e.g. birds). In the latter case, the links to patterns of energy flow in the ecosystem are rarely defined (Gaston and Blackburn 2000). Only recently have there been clear attempts to integrate aquatic and terrestrial approaches, and to consider the links between abundance–body-mass relationships in

both subsets and the whole food web (e.g. Brown and Gillooly 2003).

The significance of body size

Since body size determines both the biological properties of individuals and predator–prey interactions, size-based food-web analysis has led to the integration of community and ecosystem ecology with energetic and metabolic theory.

Body size and biological properties

Individual body mass in aquatic communities spans 20 orders of magnitude, from bacteria supporting the microbial loop to whales that filter several tonnes of krill each day. Remarkably, all these organisms obey consistent scaling laws, which dictate how biological features change with size (Brown and West 2000). Species with smaller adult body size have higher metabolic rates (Peters 1983), higher intrinsic rates of increase (Denney et al. 2002; Fenchel 1974), faster growth (Brey 1990, 1999), greater annual reproductive output (Gunderson and Dygert 1988; Charnov 1993), higher natural mortality, and shorter lifespan (Beverton and Holt 1959; Pauly 1980). Conversely, larger adult body size is correlated with lower metabolic rate, lower intrinsic rates of increase, slower growth, lower annual reproductive output, lower natural mortality, and greater longevity (Peters 1983; Charnov 1993).

Scaling relationships between biological properties and body size can be used to parameterize abundance–body-mass relationships, such that they provide estimates of community metabolism, production, turnover time, and other ecosystem properties (Kerr 1974; Schwinghamer et al. 1986; Boudreau and Dickie 1989; Jennings and Blanchard 2004).

Body size and trophic level

Aquatic food chains are strongly size-based and most predators are larger than their prey (Sheldon et al. 1972). While species with a similar maximum body size can evolve to feed at different trophic levels, the whole food web is characterized by a near-continuous rise in mean trophic level with body size (Fry and Quinones 1994; Jennings et al. 2002a). Thus small species that feed at high trophic levels are never abundant relative to small species feeding at lower trophic levels. The trophic continua across body-size classes show that fixed (integer) trophic levels do not appropriately describe the structure of aquatic food webs (France et al. 1998), and I treat trophic level as a continuous measure of trophic position that can refer to any point on the continuum.

In aquatic ecosystems, the commonly adopted procedure of assigning a trophic level to a species is misleading without qualification. While phytoplankton and some herbivores do remain at similar trophic levels throughout their life history, the trophic level of most aquatic species increases with body size. So, comparisons among species are most usefully made at an identifiable and comparable stage in the life history (e.g. size at maturity), or, if detailed life history data are lacking, at a fixed proportion of maximum body size. The latter approach is consistent with the observation that key life history transitions occur at a relatively constant proportion of maximum body size (e.g. Charnov 1993).

The relative contributions of intra and interspecific differences in trophic level to the trophic structure of a marine food web were described by Jennings et al. (2001, 2002a). They used nitrogen stable isotope analysis to estimate trophic level (Box 8.1) and demonstrated that the interspecific relationships between maximum body mass, and trophic level at a fixed proportion of that body mass, were weak or nonsignificant (e.g. Figure 8.2(a)). The results were confirmed with

---

Box 8.1 Nitrogen stable isotope analysis

An impediment to the description of links between body size and trophic structure was created by the unreliability or unsuitability of methods used to estimate trophic level. Conventional diet analysis did not always help since species and individuals in the size spectrum switched diet frequently, digested prey at different rates, and contained unidentifiable gut contents (Polunin and Pinnegar 2002). Diet analysis was also very labor intensive when applied to the range of species and size classes in the size spectrum and estimates of trophic level are needed for prey items, which necessitates study of prey diet. An appealing alternative to diet analysis was nitrogen stable isotope analysis. This provides estimates of trophic level, because the abundance of δ^{15}N in the tissues of consumers is typically enriched by 3.4‰ relative to their prey. The abundance of δ^{15}N reflects the composition of assimilated diet and integrates differences in assimilated diet over time (Post 2002b).

To estimate the mean trophic level of animals in a size class, the tissue of all animals in the size class can be homogenized and the homogenate prepared for analysis (Jennings et al. 2001). The analysis would give an abundance-weighted δ^{15}N for the animals that were present. With knowledge of δ^{15}N fractionation (k), and the δ^{15}N of reference material at the base of the food chain ($\delta^{15}N_{ref}$),

trophic level of individual or size class i can be estimated as

$$TL_i = \left(\frac{(\delta^{15}N_i - \delta^{15}N_{ref})}{k} \right) + TL_b,$$

where $\delta^{15}N_i$ is the mean δ^{15}N of individual or size class i and TL_b is the trophic level assigned to the reference material (e.g. 1 = phytoplankton, 2 = bivalves or zooplankton that feed solely on phytoplankton)

Even when the δ^{15}N of a reference material is not known, estimates of fractionation can usefully be used to assign relative trophic levels to size classes in the spectrum and to estimate the predator–prey mass ratio (PPMR).

While nitrogen stable isotope analysis does overcome many of the disadvantages associated with diet analysis, the method is also subject to a range of biases that must be considered in interpretation. There is considerable variation in fractionation around the reported mean of 3.4‰ and fractionation is influenced by many factors including rates of assimilation and excretion (Olive et al. 2003). While 3.4‰ is likely to be broadly robust when applied to analyses of multiple trophic pathways and many species (Post 2002b), as in the assessment of trophic level in the size spectrum, some variation is expected. In the absence of procedures to correct fractionation estimates, it is appropriate to run sensitivity analyses to assess the potential effects of errors in the assumed fractionation.

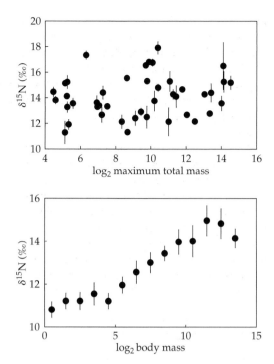

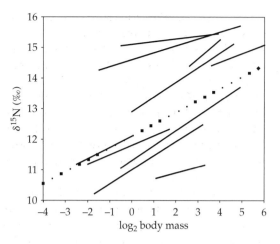

Figure 8.3 The relationship between body mass and relative trophic level (expressed as δ^{15}N) for the central North Sea fish and invertebrate community (broken line) and relationships between body mass and relative trophic level for the 10 most abundant individual species in this community (solid lines). From Jennings et al. (2002*a*).

Figure 8.2 (a) Relationship between relative trophic level (expressed as δ^{15}N) and maximum body mass of North Sea fishes and (b) the relationship between trophic level and body mass for the whole fish community by body-mass class. From Jennings et al. (2001).

a phylogenetically based comparative analysis (Harvey and Pagel 1991). Conversely, when all animals in the communities were divided into body-mass classes without accounting for species identity, trophic level rose almost continuously with body mass (e.g. Figure 8.2(b)).

Additional analyses of intra and interspecific relationships between body mass and trophic level in fishes and invertebrates have shown that increases in trophic level across the size spectrum were predominantly a consequence of intraspecific increases in trophic level with body mass rather than larger species (species with greater maximum body mass) feeding at higher trophic levels. Thus the increase in trophic level of individual species with body size makes an important contribution to the increase in the trophic level of the community with size (Figure 8.3). Analyses of this type have yet to be done for complete food webs rather than a defined body size "window" in the size

spectrum, but "within" the size window considered, most of the animals contributing to the spectrum were included.

One potential weakness with size-based analyses is that the observed linear relationship between body size and trophic level will break down at higher size classes. Thus the very largest animals in aquatic ecosystems are often filter feeding sharks and whales, which typically "feed down the food chain" on smaller and more productive size classes of prey. This is probably because the mobility of these large animals allows them to track waves of zooplankton production, providing a near continuous food supply, and because feeding down the food chain will give access to much higher levels of prey production.

Predator–prey body-size ratios and transfer efficiency

The structure of size-based food webs is determined by predator–prey interactions and the efficiency with which energy is passed from prey to predators. Thus the predator-prey mass ratio (PPMR) and trophic transfer efficiency (TE) are fundamental attributes of size-based food webs and measurements of PPMR and TE are needed to support size-based analysis (Gaedke 1993).

Diet analyses for species that dominate community biomass and production have provided estimates of PPMR for some species and species groups in some ecosystems (Ursin 1973; Rice et al. 1991) but they do not provide mean estimates of PPMR for all animals in a size "window" in the spectrum. Mean estimates of PPMR can be made using size-based nitrogen stable isotope analysis, provided that individuals can be sampled in proportion to their abundance within size classes of all species.

Estimating the abundance of all individuals in food webs is a significant challenge. While methodologies for quantifying phytoplankton and zooplankton abundance are generally well tested and established, many fishes and larger invertebrates are difficult to sample quantitatively. The paucity of vulnerability and catchability estimates needed to estimate their relative and absolute abundance is a major impediment to progressing large-scale food-web studies. Moreover, many different sampling techniques and high levels of replication are needed. Thus a recent attempt to estimate the abundance and trophic level of all animals in part of a marine size spectrum required three types of trawl nets plus acoustic, grab, and corer surveys (Jennings et al. 2002*b*).

If sampling difficulties can be overcome, individuals can be divided into body-mass categories (typically $\log_2$) and biomass weighted mean δ^{15}N is determined (Box 8.1). Mean δ^{15}N can also be determined mathematically from δ^{15}N versus size relationships for individual species or species groups and estimates of their abundance. The mean PPMR is calculated from the slope of the relationship between mean δ^{15}N (y) and $\log_2$ body mass (x) (e.g. Figure 8.2(b), where PPMR $= 2^{(k/\text{slope})}$) and k is the assumed mean fractionation of δ^{15}N per trophic level, usually 3.4 ‰ (Jennings et al. 2002*b*).

Consistent with Dickie's (1976) analysis, the decline in abundance with body mass is a function of the inefficient transfer of energy from prey to predators. This inefficiency is most simply described by TE. TE describes the proportion of prey production that is converted to predator production (TE $= P_{n+1}/P_n$, where P_{n+1} is predator production and P_n is prey production). Ware (2000) reviewed TE in marine food webs. It was typically higher at lower trophic levels, with a mean of

0.13 for energy transfer from phytoplankton to zooplankton or benthic animals and 0.10 for zooplankton or benthic animals to fish. In freshwater food webs, TE is typically 0.15 from primary to secondary consumers (Blazka et al. 1980).

The dynamic size spectrum

For analytical purposes it is convenient to view size spectra as stable; most field sampling programs are designed to account for seasonal variability in biomass and production, and to estimate annual means. Seasonal variations in the size spectrum are particularly pronounced at high latitudes and in small size classes with fast turnover times (turnover time $= 1/P:B$, and $P:B$ scales with body mass as $P:B = 2M^{-0.25}$ where $P:B$ is the production to biomass ratio and M is body mass; Ware 2000). In large size classes, turnover time is slow and there is greater temporal stability in biomass and production.

The dynamics of the size spectrum, although not the focus of this chapter, have important ecological implications. For example, seasonal production cycles lead to the propagation of waves of production up the size spectrum. These waves flatten and broaden as they reach larger individuals with higher turnover times. Pope et al. (1994) produced a mathematical model of the propagating wave. To survive, animals needed to surf (track) the wave, reproducing and growing at a rate that allowed them to feed on the wave of abundant food while avoiding a coevolving wave of larger predators.

The slopes of size spectra

Explorations of the processes governing the slopes of size spectra have included detailed process-based models of predator–prey interactions and more simplistic models based on fundamental ecological principles. All models are underpinned by the recognition that the scaling of metabolism with body size determines the energy requirements of animals in different size classes.

Predator–prey interactions in the size spectrum

Following observations of the relative constancy of slopes of abundance–body-size relationships in several ecosystems (Sheldon et al. 1972; Boudreau and Dickie 1992), theoreticians began to explore how these patterns might arise (Kerr 1974; Dickie 1976). Dickie's (1976) theoretical development was particularly significant and described how the ratio of production and biomass at successive trophic levels, and hence the slope of the size spectrum (in which body mass is an index of trophic level), might be determined.

Production (P) at trophic position (n) is a function of consumption (C) and the efficiency of use of food intake (K),

$$P_n = K_n C_n, \tag{8.1}$$

and consumption by a predator at trophic position $n + 1$ is

$$C_{n+1} = F_{n+1} B_n, \tag{8.2}$$

where F is rate of predation by the predator and B is prey biomass. Thus

$$\frac{C_{n+1}}{C_n} = F_{n+1} K_n \frac{B_n}{P_n}, \tag{8.3}$$

where C_{n+1}/C_n is the ratio of food intake at successive trophic positions; the "ecological efficiency" of Slobodkin (1960).

In a stable and seasonally averaged size spectrum, the rates of production and total mortality must balance, so

$$\frac{P_n}{B_n} = F_{n+1} + U_n, \tag{8.4}$$

where U is mortality due to factors other than predation.

Dickie (1976) showed that equations (8.3) and (8.4) could be used to predict ratios of production and biomass at successive trophic positions, and hence the slope of the size spectrum. Thus,

$$\frac{P_n}{P_{n-1}} = \frac{K_n C_n}{K_{n-1} C_{n-1}} = K_n \frac{F_n}{F_n + U_{n-1}}, \tag{8.5}$$

and

$$\frac{B_n}{B_{n-1}} = \frac{P_n}{P_{n-1}} \frac{F_n + U_{n-1}}{F_{n+1} + U_n}. \tag{8.6}$$

Thus,

$$\frac{B_n}{B_{n-1}} = F_n K_n \frac{B_n}{P_n} \tag{8.7}$$

and

$$\frac{B_n}{B_{n-1}} = \frac{F_n}{F_{n+1}} \frac{C_{n+1}}{C_n}. \tag{8.8}$$

Thus the ratio of the biomass at two successive trophic positions is the ecological efficiency corrected by the ratio of the predation rates, and will be independent of the mean body size of individuals at trophic levels. Further, since predation rates will depend on metabolism, biomass ratios will reflect the scaling of metabolism (R) with M, which we know to be $R \propto M^{+0.75}$ (Peters 1983). Dickie's (1976) analysis also implied that relative predation rates must increase rapidly as trophic level decreases, and thus utilization of production at low trophic levels will be higher.

Subsequent theoretical work focused on the secondary structure of the size spectrum, with an emphasis on whether domes that corresponded with trophic groups were superimposed on the primary scaling that Dickie (1976) had described (Thiebaux and Dickie 1992, 1993; Benoit and Rochet 2004). Empirical evidence for such domes has been found in some freshwater lakes (Sprules and Stockwell 1995), but they may not be a universal feature of aquatic size spectra in communities when animals with a greater range of morphologies and life histories are present.

Energy equivalence rule

While aquatic ecologists were modeling the detailed structure of size spectra (Dickie et al. 1987), terrestrial ecologists were compiling empirical estimates of the slopes of abundance–body-mass relationships for subsets of food webs (review: Gaston and Blackburn 2000). For communities that shared a common energy source, such as plants using sunlight, numerical abundance (N) typically scaled with body mass (M) as $M^{-0.75}$. Since metabolic rate scaled with mass as $M^{0.75}$ (Peters 1983), the rate of energy use in these communities was expected to be independent of body size (Damuth 1981), a prediction now known as the energetic equivalence hypothesis (Nee et al. 1991). Energetic equivalence, for example, has since been shown to predict the scaling of numerical abundance and body mass in phytoplankton and plant communities (Belgrano et al. 2002; Li 2002).

Extending energy equivalence to size-based food webs

While available energy (E) scales as M^0 when individuals share a common energy source, size-based food webs are characterized by larger predators eating smaller prey. Since the transfer of energy from prey to predators is inefficient, the total energy available to individuals must scale $<M^0$.

The rate at which available energy decreases with increasing mass in a size-based food web will depend on the mean PPMR and the TE (Cyr 2000). In an innovative commentary on a paper by Cohen et al. (2003), Brown and Gillooly (2003) proposed that the slope of an abundance–body-mass relationship for a small freshwater lake could be predicted using Cyr's (2000) approach for estimating the change in E with size. Brown and Gillooly (2003) posited that the community Cohen et al. (2003) had studied represented three trophic levels that used different sources of energy, with phytoplankton using sunlight, zooplankton eating phytoplankton, and fish eating zooplankton. For each of these trophic levels, the energetic equivalence hypothesis would predict a consistent scaling of abundance and body mass (Figure 8.4(a)) as illustrated conceptually by Brown and Gillooly (2003). Thus the scaling among trophic levels (Figure 8.4(b)) would be determined by the change in E with size and hence PPMR and TE. A tentative calculation, based on an assumed difference in body size of 10,000 between trophic levels (PPMR) and an estimated TE of 0.1, provided realistic predictions of slope.

One feature of Brown and Gillooly's (2003) analysis was that each trophic level (phytoplankton, zooplankton, or fish) was assumed to extend over a range of body sizes, an approach that was inconsistent with evidence for a trophic continuum. Jennings and Mackinson (2003) formalized and tested the Cyr (2000) and Brown and Gillooly (2003) approach in a size-structured marine food web where trophic level was shown to increase continuously with body size. Here, the within trophic level scaling described by Brown and Gillooly (2003) was assumed to apply to an infinitely small increment in body mass. The expected scaling of $\log_{10} E$ with $\log_{10} M$ was calculated as $\log_{10}$ TE/$\log_{10}$ PPMR. For the calculated scaling of E and M, the scaling of B and M was calculated as $M^{(\log_{10}\text{TE}/\log_{10}\text{PPMR})} \times M^{0.25}$. The analysis of size abundance data for the marine food web showed that the scaling of abundance with M was close to linear and B scaled as $M^{-0.2}$. The mean PPMR did not vary consistently with M and was 106:1. The predicted scaling of B with M in the food web was $M^{-0.24}$, with an assumed TE of 0.10.

The insights of Cyr (2000) and Brown and Gillooly (2003) help to explain the remarkable consistency in the observed slopes of size spectra. This is because mean PPMR and TE place significant constraints on the slope of abundance–body-mass relationships, and these parameters are also remarkably consistent in different ecosystems. Thus mean predator–prey mass ratios, as estimated using a variety of methods, are typically 10^2–10^3:1 and TE is typically 0.1–0.2. (Cushing 1975; Ware 2000; Jennings et al. 2002b). For these

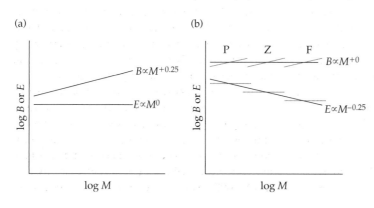

(a)

(b)

Figure 8.4 Relationships between biomass abundance (B) and rate of energy use (E) as a function of body mass (a) within and (b) among trophic levels (here P: phytoplankton, Z: zooplankton, and F: fish). Predicted scalings across trophic levels assume TE $= 0.10$ and PPMR $= 10,000 : 1$. Redrawn with modifications from Brown and Gillooly (2003).

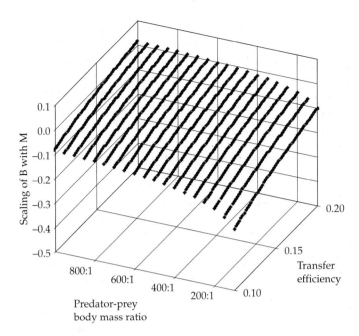

Figure 8.5 Sensitivity of the slope of the biomass abundance (*B*)–body-mass (*M*) relationship (size spectrum) to changes in PPMR and TE.

combinations of parameters, slopes of the size spectrum for the whole food web are expected to range from 0 to −0.2 (Figure 8.5).

Food-chain length

Maximum food-chain length is an important food-web property and affects community structure, ecosystem processes, and contaminant concentrations in animal tissue (Post 2002a). Maximum food-chain length has been correlated with resource availability, ecosystem size, environmental stability, and colonization history. Some of these correlations may result from environmental effects on predator–prey mass ratios.

Size-based food-web analysis provides a method for understanding relationships between predator–prey mass ratios and food-chain length. In a community of given size and species composition, changes in PPMR must influence food-chain length (Figure 8.6). Thus in the community with PPMR = a (Figure 8.6), the trophic level of the individual with the maximum body size (M_{max}) TL_a represents maximum food-chain length and is higher than the trophic level of the same sized individual in the community with PPMR = b (TL_b). Real communities, however, differ from the idealized communities in Figure 8.6 in two ways.

First, in species and size composition. Second, in maximum food-chain length, because the focus on average chain length in idealized communities may obscure the presence of long but rare food chains that support individuals of intermediate body size.

Jennings and Warr (2003) investigated empirical relationships between maximum food-chain length, PPMR, primary production, and environmental stability in marine food webs with a natural history of community assembly. Their analyses provided empirical evidence that smaller mean PPMR was characteristic of more stable environments (stability measured as annual temperature variability) and that food chains were longer when mean PPMR was small. Their results also demonstrated that the heaviest fish predators at each site rarely fed at the highest trophic level and the longest food chains supported fish predators with intermediate body size. However, both maximum and maximum mean food-chain lengths were correlated.

The results suggested that environmental factors favoring smaller mean PPMR allowed individuals feeding at high trophic levels to persist. Jennings and Warr (2003) proposed that relationships between environmental variables and food-chain length were explained by the effects of environment on PPMR. If maximum food-chain length can

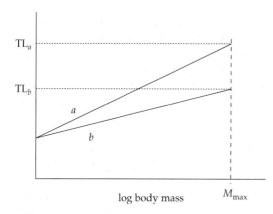

Figure 8.6 Relationships between PPMR and maximum food-chain length in communities of the same size and species composition. When PPMR is relatively small (a) the maximum food-chain length is long (TL$_a$). When PPMR is relatively large (b), the maximum food-chain length will fall (TL$_b$).

depend on PPMR, then feeding behavior, as mediated by the interactions between individuals and their environment, will have a fundamental effect on maximum food-chain length. The life history, morphology, and motility of a predator places ultimate constraints on the prey it can eat, but within these constraints, prey size selection will depend on the environment, competition, prey availability, prey processing costs, prey population stability, and prey production (Stephens and Krebs 1986; Cohen et al. 1993).

It was not clear from these analyses whether the significant relationship between the measure of environmental variability and PPMR demonstrated causality, as annual temperature variation may be a surrogate for many other types of physical and biological variability. For example, at deeper sites with lower annual temperature variation, tide and wave effects are smaller, the water is clearer, and there are more complex seabed habitats, so predators may be able to track and feed on larger, mobile, and patchily distributed prey more easily. Conversely, in turbid shallow sites subject to high levels of physical disturbance, tracking and catching large active prey in the water column may be more difficult and more fishes, of all body sizes, may feed on smaller but more evenly distributed and more productive benthic animals on the seafloor.

Since abundance–body-mass scaling will partly depend on PPMR and PPMR is smaller in more stable environments, steeper abundance–body-mass relationships may be found in more stable environments if there is not a compensatory relationship between PPMR and TE. These relationships have yet to be investigated.

Abundance–body-size relationships in subsets of the food web

While terrestrial macroecologists have often focused on abundance–body-mass relationships for taxonomically or ecologically defined groups of animals (Gaston and Blackburn 2000) such research has not been a consistent focus of aquatic science. Nevertheless, there are studies of groups of individuals which share, or do not share, a common energy source.

Oceanic phytoplankton are an example of a group that share an energy source, since they all use sunlight to photosynthesize. Large-scale and long-term research has shown that the slope of the abundance–body-mass relationship in phytoplankton communities is consistent with that predicted by the energetic equivalence hypothesis, $N \propto M^{-0.75}$ (Li 2002). Thus the sunlight energy used by all phytoplankton cells in a size class is the same as in any other size class (Li 2002). Interestingly, the universality of energetic equivalence is such that it predicts the abundance–body-mass relationship when marine phytoplankton and terrestrial plants are included in the same analysis (Belgrano et al. 2002).

Benthic infaunal communities do not share the same energy source, but they are effectively sustained by energy reaching the seafloor as phytoplankton or detritus and are often treated as a functional unit. Unlike the complete food web, the size structure of infaunal communities does not primarily reflect size-based predation, since many of the largest animals (e.g. bivalve molluscs, burrowing echinoderms) feed at lower trophic levels than some of the smaller animals (e.g. predatory polychaete worms).

Benthic infaunal communities exhibit remarkably consistent abundance–body-mass relationships (Schwinghamer 1981; Schwinghamer et al. 1986;

Duplisea 2000) and Dinmore and Jennings (2004) investigated whether these slopes could be predicted from the energy available to different size classes of animals. Energy availability (E) at size was calculated from the scaling of E with M, based on estimated trophic level at size (from nitrogen stable isotope analysis) and TE. In the infaunal community they studied, trophic level (δ^{15}N) decreased with increasing M and the slope of the relationship between δ^{15}N and M was -1.07. Thus $E \propto M^{0.13}$, $B \propto M^{0.53}$, and $N \propto M^{-0.47}$ when TE was 0.125 and fractionation was 3.4‰ δ^{15}N per trophic level. The predicted scalings of B and N with M were 0.48 and -0.54, respectively and did not differ significantly from the predicted scalings.

This analysis suggested that theory most usefully applied to the prediction of abundance–body-mass relationships in complete food webs may have a role in predicting abundance–body-mass relationships in subsets of the food web. The analysis of the benthic infaunal community is probably robust because it is a relatively defined functional unit, but the approach would be less likely to work if applied to taxonomically rather than functionally defined subsets. This is because taxonomically defined subsets typically receive energy inputs from many sources that can rarely be well quantified.

Practical applications of size-based food-web analysis

Size-based food-web analyses provide a method for assessing the large-scale direct and indirect effects of human activities on marine ecosystems (e.g. climate change, fishing). The potential applications of size-based food-web analysis to management problems were well recognized by the first scientists to work in this field, and a classic paper by Sheldon and Kerr (1972) used knowledge of the regularity of structure in the size spectrum to predict the population density of monsters in Loch Ness. They estimated that 10–20 monsters of approximately 1500 kg were present. This ensured that Loch Ness continued to attract wealthy monster hunters and sceptical tourists, with happy economic consequences for local landlords and

hoteliers. With hindsight, it is clear that Sheldon and Kerr (1972) had actually underestimated monster abundance, since sufficiently monstrous animals may well have fed down the food chain on smaller and more productive size classes of prey, or fed opportunistically on scantily clad terrestrial energy inputs to the Loch!

Species-size–abundance data underpin population, community, and ecosystem analyses and have been collected during monitoring and research programs in many aquatic environments over many years. Examples are the plankton, benthos, and fish surveys conducted to assess the impacts of climate, pollution, and fisheries on aquatic ecosystems. These data can be used for retrospective size-based food-web analysis, when dietary data for species-based food-web analysis are not available (Murawski and Idoine 1992). Assessments of the effects of fisheries exploitation on aquatic food webs have frequently been conducted using size-based approaches (Duplisea and Kerr 1995; Rice and Gislason 1996; Gislason and Rice 1998).

Since the slope of the unexploited size spectrum can be predicted from PPMR and TE, and since there is little evidence that either parameter is affected by exploitation to the same extent as biomass (Ware 2000; Jennings and Warr 2003), the slope of the unexploited size spectrum provides a useful baseline or reference level when assessing the relative impact of human activities on the marine ecosystem. Jennings and Blanchard (2004), for example, used estimates of TE and empirical measurements of PPMR in a marine ecosystem to predict the slope of the unexploited size spectrum and compared this with the slope as predicted from contemporary data (Figure 8.7). Slopes of the size spectra were predicted using the approach previously described, and intercepts were predicted from primary production (PP), where production at any higher trophic level (P_{TL}) declines as

$$P_{TL} = PP \times TE^{TL-1}, \tag{8.9}$$

where TE is the estimated transfer efficiency.

By comparing the unexploited theoretical and observed size spectra, Jennings and Blanchard (2004) predicted that the current biomass of large

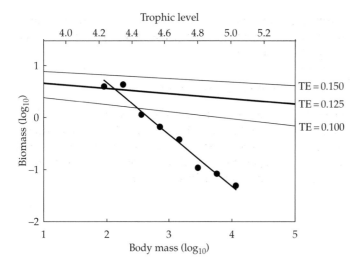

Figure 8.7 Comparison between predicted slopes of unexploited size spectra when TE = 0.100, 0.125, or 0.150 and the slope of the size spectrum in the fished North Sea in 2001. Circles indicate biomass at body-mass for 2001. Trophic levels were assigned to body-mass classes using nitrogen stable isotope analysis. From Jennings and Blanchard (2004).

fishes weighing 4–16 and 16–66 kg, respectively was 97.4% and 99.2% lower than in the absence of fisheries exploitation. Given the scaling relationships between body size and biological properties, the size spectrum could also be parameterized to estimate the effects of fishing on turnover time. The mean turnover time of the exploited community was almost twice as fast as that of the unexploited community (falls from 3.5 to 1.9 years)

Even in data poor systems, the relative constancy of TE and PPMR means that there are a limited number of solutions for the slope of the size spectrum (Figure 8.5). This makes it relatively easy to make crude (order of magnitude) estimates of changes in the relative abundance of small and large animals following exploitation.

Conclusions

Size-based food-web analyses provide insights into food-web structure and function that complement and extend those from species-based approaches. One key strength of size-based analyses is that they provide a transparent approach for integrating community and ecosystem ecology with energetic and metabolic theory. Thus they describe how the environment affects PPMR and how, in turn, predator–prey body-mass ratios and transfer efficiency might determine abundance–body-mass relationships. Although variability in PPMR and

TE among ecosystems appears to be quite limited, and this places constraints on the slopes of size spectra, a better understanding of the links between PPMR and TE would significantly improve understanding of links between aquatic food-web structure and function.

Recent attempts to link size and species-based analyses of food webs are an important step (Jennings et al. 2002a; Cohen et al. 2003), since understanding of the relationships between structure and diversity in aquatic food webs is still poor. Key to this process is the development of analytical methods that account for the changing trophic role of species with body size while preserving the significance of species identity. Are there consistent patterns of change in trophic position with body size, for example, and how do these patterns of change differ in rare or common species and relate to life histories and the environment?

Macroecological approaches in aquatic and terrestrial ecology have followed rather different developmental paths, perhaps reflecting the dominant characteristics of aquatic and terrestrial ecosystems and largely separate research funding mechanisms (Beddington and Basson 1994; Chase 2000). The renewed focus on the common biological properties of individuals and their role in controlling community and ecosystem structure (Brown and West 2000; Brown and Gillooly 2003) should encourage wider examination and testing of macroecological principles and help to identify cross-system generalizations (Belgrano et al. 2002).

Despite the early recognition that size-based food-web analysis could help to address environmental management issues (Dickie 1976; Pope et al. 1988; Boudreau and Dickie 1989; Murawski and Idoine 1992), the application of size-based food-web analysis to management issues is still largely overlooked. This is unfortunate, since the application of a size-based approach helps to place the management of populations (exploited fishes, for example) in an ecosystem context and provides a basis for describing the wider impacts of human activities.

Acknowledgments

I am very grateful to Andy Belgrano for the invitation to write this chapter, to Steve Mackinson for comments on the manuscript and to the UK Department of Environment, Food and Rural Affairs for funding (M0731).

CHAPTER 9

Food-web theory in marine ecosystems

Jason S. Link, William T. Stockhausen, and
Elizabeth T. Methratta

Introduction

Food webs provide the basis for describing community and ecosystem structure. Food webs also
provide the framework for integrating population
dynamics, community structure, species interactions, community stability, biodiversity, and
ecosystem productivity. Food-web interactions
ultimately determine the fate and flux of every
population in an ecosystem, particularly upper
trophic levels of fiscal importance (May 1973;
Pimm 1982). Additionally, food webs often provide a context for the practical management of
living resources (Crowder et al. 1996; Winemiller
and Polis 1996). As such, food webs have high
heuristic value for ecological theory and have been
the subject of considerable interest in general
ecology.

Earliest food webs—first era

The development of ecological theory concerning
the patterns exhibited by food webs and the underlying processes they reflect can loosely be divided
into four temporal eras. Some of the earliest food-
web studies were marine examples (Petersen 1918
quoted in Pomeroy 2001; Hardy 1924). The work by
Lotka (1925) and Volterra (1926) to understand the
relative roles of predation and fishing in European
seas represented some of the first attempts to
characterize food-web interactions. Their work
explored the importance of species interactions
after the cessation and restart of fisheries brought
about by WWI. The emphasis on interactions
became an integral part of ecology, and ultimately

morphed into focusing on binary pairs of species
rather than the entire food web.

However, multispecies food webs were still
being considered in those and following decades.
Clarke (1946) developed a food-web for Georges
Bank that encapsulated the thinking that ecosystems were like machines, with gear-like mechanisms that represent how energy flowed through a
food web. The concurrent trophic-dynamic concept of Lindeman (1942) similarly developed from
food-web observations in a north-temperate lake.
Together, the seminal work of Lindeman and
Clark provided the foundations for much of the
subsequent research that explored the energetics
and structure of food webs (e.g. Odum 1957, 1960).
The importance of species interactions and their
essential role in regulating the flow of energy in an
ecosystem was clearly established by the end of
this era.

Complexity is prominent—second era

In the middle of the twentieth century, the
emphasis on energy flow of food webs in ecology
continued (in many ways developing into the field
of biogeochemistry) but the prominent "food-web
theory" of that era was largely encapsulated by
Elton's (1958) and Paine's (1966) exploration of complexity and stability in ecological communities.
The hypothesis that complexity engendered stability in ecological communities enjoyed considerable
support and approached the status of a mathematical theorem (MacArthur 1955; May 1973) prior
to the theoretical attacks of May (1972, 1973)

98

and others. As the field of food-web ecology progressed, the study of large marine systems was de-emphasized. The conclusion from this era was that complexity was an important, if not the central, factor for food webs.

Complexity and stability—third era

Subsequent to this era, May (1972, 1973), built upon computer simulation results by Gardner and Ashby (1970) and developed a criterion for community stability related to connectance, whereby a system is stable if $i(SC)^{1/2} < 1$ (where i is mean interaction strength, S is the number of species, and C is the connectance of a food web, principally derived from S and the number of links between species or species interactions, L), based on analysis of randomly constructed webs. Under the assumption that mean interaction strength i is independent of food-web size, May's formula predicts that connectance should decline hyperbolically as the number of species increases, with intriguing consequences for stability.

This prediction ushered in the next era of food-web theory and led to numerous comparisons of empirically based, topological food-web catalogs (Briand and Cohen 1984, 1987; Cohen and Briand 1984; Cohen 1989; Schoener 1989; Sugihara et al. 1989; Cohen et al. 1990; Pimm et al. 1991; Havens 1992). Additional topological web metrics (e.g. Pimm 1982; Cohen et al. 1990; Bersier et al. 2002) were introduced in these and associated studies to identify other food-web patterns. By the end of the 1980s, food-web theorists had developed a set of empirical relations based on these food-web catalogs, some of which included >100 webs.

Cohen (1989) summarized several of these empirical relations as five "laws": (1) excluding cannibalism, cycles are rare, (2) food chains are short, (3) the proportions of top, intermediate, and basal species (%T, %I, %B) are independent of food-web scale (the "species scaling law"), (4) the proportions of link types (%T-I, %T-B, %I-I, %I-B) are independent of food-web scale (the "link scaling law"), and (5) linkage density ($L_D = L/S$) is independent of food-web scale (the "link-species scaling law"). The last of these laws supports May's stability criterion as a constraint on trophic

structure, as it implies that connectance declines hyperbolically with increased species richness. Cohen (1989) also presented a model of community organization, the "cascade" model, that gave "remarkable quantitative agreement" between his empirical laws and the model's predictions based on a single parameter—the expected linkage density.

During this era and extending into the present (ca. the past 25 years), in many respects analysis of topological food-web catalogs has dominated the search for pattern and process in food webs. Few of the food webs studied were marine ecosystems, and those few examples were estuarine, coastal, or intertidal and not representative of the vast majority of the world's oceans. It is interesting that marine food webs have been underrepresented in these collections even though the first food-web diagrams were constructed for marine systems (Petersen 1918 quoted in Pomeroy 2001; Hardy 1924).

Networks and current synthesis—fourth era

At the beginning of the fourth (and current) era in the development of food-web theory, the edifice of empirical evidence supporting prior generalizations based on topological food-web catalogs was seriously challenged on a number of grounds. Chief among these criticisms was that the scaling—spatial, temporal, or taxonomic—used to construct the catalogued food webs, was too limited in scope to adequately capture all the S and L of a given food web (Polis 1991; Hall and Raffaelli 1991, 1993; Cohen et al. 1993; Martinez 1993; Goldwasser and Roughgarden 1997; Solow and Beet 1998, Martinez et al. 1999). Other related concerns were that the criteria for aggregation were inconsistent, ontological changes in diet were absent, cannibalism was ignored, and inconsistencies in sampling effort, spatio-temporal resolution and spatio-temporal aggregation giving rise to sampling artifacts could not be assessed (Paine 1988; Winemiller 1990; Hall and Raffaelli 1991; Kenny and Loehle 1991; Martinez 1991; Polis 1991).

To be fair, in part these criticisms reflected that few, if any, of the collected food webs had originally been developed with the intent of testing food-web theory (Warren 1994). Additionally, more recent and exhaustive food webs disagreed with many of the previous theoretical predictions. For example, there is much more omnivory, cannibalism, and cycling than was previously expected (Polis and Strong 1996; McCann et al. 1998; Closs et al. 1999). The new webs also exhibited higher species diversity and topological complexity than typical in older webs (e.g. Winemiller 1990; Hall and Raffaelli 1991; Martinez 1991; Polis 1991; Goldwasser and Roughgarden 1993; Polis and Strong 1996; Reagan et al. 1996). The average number of links per species (i.e. linkage density, L_D) and food-chain lengths were greater in newer food webs. It was also shown that many food-web properties were sensitive to the criteria used to aggregate species, as well as the level of aggregation, thus casting previous scale invariance laws (species, link, and link-species scaling laws) into question (Winemiller 1990; Martinez 1991).

Martinez (1991, 1992) also challenged the hyperbolic scaling relation for connectance (Pimm 1982; Cohen and Newman 1988) and hypothesized "constant connectance," whereby directed connectance is scale invariant. The empirical and theoretical framework outlined by Cohen (1989) to extend May's (1972, 1973) work underwent rigorous critique in the past decade, severely questioning the determinants of food-web stability.

During the past 10 years there has also been a call to move beyond just "S and L" and topological food webs, examining explicitly the flow of energy and rates of species interactions in food webs (bioenergetic and biodemographic, respectively Winemiller and Polis 1996). In particular, the importance of interaction strength relative to food-web stability is still being explored (Haydon 1994; de Ruiter et al. 1995; Raffaelli and Hall 1996; McCann et al. 1998; Closs et al. 1999). Much of the shift in effort for food-web studies is reflected in the recent emphasis on network structure and theory (Dunne et al. 2002a,b, 2004; Krause et al. 2003). In some respects this shift in effort has provided extensions or advances to prior "cascade," "niche," and scaling models (e.g. Williams and Martinez 2000; Dunne et al. 2004). Yet two points remain despite current efforts; (1) much of the fundamental debate about complexity, stability,

and structure needs to be further resolved, despite enhanced caveats to account for increased food web complexity, and (2) the vast majority of food web and network analyses neglect marine ecosystems.

The marine environment

Our premise is that all types of ecosystems are unique, but food-web theory has been developed (certainly tested further) principally from freshwater and terrestrial food webs, largely omitting marine food webs. Additionally, not all marine food webs are equal, with differences in the expected number of species, dominant processes, physical forcing factors, stability, production, structure, interaction strength, etc., observed among the various types of marine ecosystems (see the section below for a categorization of marine ecosystems). Steele (1985) has noted some of the obvious differences between marine and other ecosystems. Here we expand upon those considerations as they relate particularly to food webs.

Scale

As difficult as it is to construct and study freshwater and terrestrial food webs, it is much worse in marine ecosystems. Observations and data are much more costly to obtain in marine systems. The typical sampling endeavor in marine ecosystems is akin to trying to estimate population parameters of small mammals in a field or forest via towing a butterfly net from a low-flying aircraft. Suffice it to say that sampling marine ecosystems is difficult, and as a direct result S and L are going to be sorely under sampled. Food-web theories which rely on those parameters are consequently difficult to test in marine food webs. This parallels the realization reached >10 years ago for freshwater and terrestrial food webs but is exacerbated by the challenge of sampling in the marine environment.

The spatio-temporal scales at which marine ecosystems function are multiple and large (Steele 1985; Mann and Lazier 1991). The scales over which biological interactions occur are probably much larger for marine ecosystems compared to nonmarine systems. Often individuals of the same population can be found dispersed over several thousand square kilometer. Many species have daily movements on the scale of hundreds of kilometer. Currents, fronts, and similar physical phenomena also occur at broad spatial scales and can transport organisms at greater rates and magnitudes than in other systems (Carr et al. 2003). Often species migrate or aggregate seasonally for significant biological events (i.e. spawning, foraging, etc.). Local and regional scale phenomena are also amalgamated with broader oceanographic regimes, such as warming or a shift in the PDO or NAO, with the result that species distributions, ranges, and productivity can shift. Cousins (1996) suggested that food webs be defined by ambits of top predators or the dominant processes that affect them. Applied to marine systems this principle yields food webs with enormous spatial extents. The sum result of all these scaling considerations is that there are likely to be a lot of species interactions (L) that go undetected. Additionally, it is likely that L, and hence linkage density (L_D) and connectivity (C), are probably much higher than typically reported for marine systems and food webs in general, and may even be much higher than predicted by food-web theory.

Marine species

Differences in foraging behavior and the higher degree of ominivory of marine versus terrestrial species have been hypothesized to explain the higher L_D observed for marine food webs (Cohen 1994). For instance, insects are often dominant terrestrial herbivores and specialize on one or a few tree species in an entire lifetime. This strategy reduces the overall number of links for that species and its food web. A single marine copepod, by contrast, is likely to encounter multiple species of phytoplankton, microzooplankton, and micronekton over the course of its life history and drive the L_D much higher for marine food webs (Cohen 1994). Similarly, Link (2002) has noted that most marine fish species are opportunistic generalists, with few trophic specialists. Again the implication is that L, L_D, and C will be greater in these highly but loosely connected food webs.

The ontogenetic shifts in the size (either biomass or length) of marine organisms routinely span 3 to 4 orders of magnitude, and in some instances span 5 to 6 orders of magnitude. This may be rivaled only by a few large tree and mammal species in terrestrial ecosystems. Changes in body size over the course of an organism's life history typically correspond to changes in diet and functional trophic group for many marine species, a process referred to as "metaphoetesis" (Cohen et al. 1993). The result is that one species may actually function like three or four different species across its life history. Thus, ontogenetic shifts in diet may have a significant influence on food-web topology and energy flows.

The largest and most long lived species in the ocean are fundamentally different than those in other ecosystems (Steele 1991). These species which correspond to the apex and top trophic levels in the ocean are also the most heavily exploited species (Pauly et al. 1998; Jackson et al. 2001). The opposite is true in terrestrial ecosystems, where autotrophs tend to be the larger, longer-lived organisms and the most widely harvested for human use (Steele 1991). The resolution of existing food webs reflects this disparity, particularly for lower trophic levels. Basal trophic levels are highly unresolved in most marine food webs, with finer resolution often impractical for webs describing metazoan interactions. As a result, producers are often represented oversimply, if at all, in these food webs.

The taxonomic groups represented in the ocean are numerous. The way in which species are treated has also been found to affect the amount of information derived from a food web, indicating that the number of compartments (i.e. trophic species; the largest set of organisms with identical sets of predators and prey defines a group known as a trophic species; see Cohen and Briand 1984; Cohen 1989) may influence both structural and functional aspects of the food web (Schoener 1989; Polis 1991; Hall and Raffaelli 1991, 1993; Martinez 1993; Cohen 1994; Solow and Beet 1998; Martinez et al. 1999; Abarca-Arenas and Ulanowicz 2002). Lumping and splitting of taxonomic and functional groups and the criteria by which those groups are linked have important ramifications for food-web structure, metrics, and theory. In particular, reduced food-web resolution is associated with lower S, L, L_D, C, and mean chain lengths (Schoener 1989; Polis 1991; Hall and Raffaelli 1991, 1993; Martinez 1993; Cohen 1994; Solow and Beet 1998; Martinez et al. 1999; Abarca-Arenas and Ulanowicz 2002). Aggregation of marine species is common, but in some cases different life history stages are separated into different trophic species to better resolve a food web. (e.g. Christensen and Pauly 1992; Gomes 1993). In some cases, the criteria used to assign a species to a trophic species are based on clear trophic distinctions across life history (Pimm and Rice 1987) whereas in other cases, trophic species are more difficult to define. While taxonomy provides one approach to the aggregation or splitting of species in a food web, the "trophic species" concept (Cohen and Briand 1984; Cohen 1989) is generally a widely accepted convention that has been shown to reduce methodological biases (Cohen et al. 1990; Pimm et al. 1991; Martinez 1994), but is not without its critics (Schoener 1989; Polis 1991). While constructing food webs, caution should be taken when aggregating or splitting species, but it is likely that gross aggregation will continue for most marine food webs.

Additionally, there are marine taxa for which we know little else other than that they exist. This does not even account for those species that we have not even discovered or identified. For example, the discovery of the microbial loop highlights a potentially important but overlooked set of taxa. In marine ecosystems, microbes play important roles in decomposition of detrital material and renewal of resources for primary producers (Schlesinger 1997). Furthermore, evidence suggests that microbes may be as important as zooplankton in consuming organic carbon in some marine ecosystems (Cho and Azam 1988) and that the shunting of carbon into the microbial loop may represent a carbon sink in marine systems (Azam et al. 1983). The additional role of phagotrophic protozoa is under-known but suspected to be an important link between microbes and the zooplankton consumed by higher trophic-level metazoans (Sherr et al. 1986). Despite their important functional roles, detailed microbial

interactions are typically left out of metazoan food webs across ecosystems. Some evidence suggests that the aggregation of microbial groups in food webs may reduce the informational value provided by food webs and that extinctions of highly connected species at basal trophic levels may be even more destabilizing to an ecosystem than the loss of top, highly connected species (Abarca-Arenas and Ulanowicz 2002). Further studies examining the linkages between microbes and metazoans will provide a more comprehensive understanding of food-web structure and function.

Uncommon species, or species with low abundances, may play significant roles in ecosystem functioning (e.g. Lyons and Schwartz 2001) but by definition these species are hard to find. A reconsideration of inconspicuous but potentially significant interactions such as parasitic, symbiotic, or host/pathogen relationships may also be warranted for marine food webs. Keystone species are defined by their ability to influence community structure in ways that are disproportionate to their own abundance (Power and Mills 1995). In some instances, sampling of shelf and deep-sea systems has recovered numerous species represented by only a single individual (Etter and Mullineaux 2001) but the overall contribution of these singleton species to food-web dynamics is difficult to assess. Species which are difficult to collect such as gelatinous zooplankton may also be functionally significant. Due to their low nutritive value, gelatinous zooplankton such as coelenterates, ctenophores, and salps are generally considered to be an unimportant food source for most economically important fish species. However, gelatinous prey are thought to be consumed by over 100 fish species and are the main food source for some predators including pelagic turtles, moonfish, and stromateoid fishes (Verity and Smetacek 1996). They are also the dominant predators of many zooplankton and fish larvae (Purcell 1986). Moreover, the majority of trophic cascades identified in pelagic marine ecosystems are initiated by gelatinous zooplankton (Verity and Smetacek 1996), indicating that these species are probably important determinants of energy flow and ecosystem function. Overlooked or understudied organisms are more likely the norm rather than the exception in marine ecosystems.

Obiter dicta

There are three major points from the discussion about the uniqueness of marine ecosystems. First, many of the S and L (and hence the food web metrics and theory derived from them) are probably undetected, and thus underestimated, for marine ecosystems. Second, as marine food webs become more exhaustive, S, L and those parameters derived from them are expected to be much higher than typically reported for most food webs and probably much higher than generally suspected from food-web theory. Finally, by necessity (enforced via logistical constraints), there will continue to be some degree of aggregation in marine food webs for the foreseeable future. A more rigorous look at how taxonomic or trophic resolution in the construction of marine food webs affects topological metrics, network metrics, and energy flows is needed.

A compilation of marine food webs

We assembled case studies of marine food webs published in the past 25 years (Table 9.1). We limited our inclusion of examples to those food webs that are more representative of larger ocean ecosystems (i.e. not an estuary, not a small embayment, not a rocky intertidal zone, etc.). The list is by no means exhaustive, but the inclusiveness and large number (107) of cases likely make this list representative of the food-web work executed in marine systems.

Metrics and categorization

We categorized the type of food web into one of seven ecosystems (bounded seas (6%), coastal (20%), continental shelf (40%), coral reef (11%), mid ocean gyre (4%), seamount (6%), or upwelling (14%)) along a latitudinal gradient (equatorial (11%), subtropical (31%), temperate (31%), or boreal (27%)). Clearly the more difficult-to-reach ecosystems (sea mounts or mid ocean gyres) were the least studied, reflective of the high cost of data

collection at those locales. However, there was less of a latitudinal gradient than one may have expected or that has been observed in other disciplines (e.g. limnology, fisheries science).

We distinguished three types of food webs (Winemiller and Polis 1996) based upon their information content: (1) topological or descriptive webs, (2) flow or bioenergetic webs, and (3) interaction, functional, or biodemographic webs. Topological webs are qualitative in nature; only the presence/absence of interactions between groups is indicated. Bioenergetic and interaction food webs are quantitative; the relative strengths of trophic interactions between groups are indicated. Bioenergetic webs quantify the transport through consumption of energy/matter among groups, while interaction webs depict the strength of links between groups in terms of their influence on the dynamics of community composition and structure. Using this distinction, we noted which type of web each of the food-web case studies were, but in some instances we also noted that the study could have fallen into more than one category. The vast majority of marine food webs were categorized as bioenergetic (84%), while seven were interaction food webs (10%) (Table 9.1). The remaining food webs were topological (22%) and periodically cooccurred with one of the other two categories. Because most of the marine food webs were highly aggregated, even the topological webs, we did not attempt to calculate or compare common food-web macrodescriptors for these marine food webs.

However, we did note the number of species (or approximate number if not given, if species were aggregated, or if multiple webs were constructed). Given the caveats described in the section above, we recognize that it is not inconceivable for marine food webs to have an S on the order of several hundreds, with L on the order of several thousands or tens of thousands. Yet given the sampling constraints of working in marine ecosystems, the highest number of S observed was on the order of 100 (Table 9.1). Most marine food webs with S of 40 or greater were, relatively speaking, well studied. Still, despite the caveats described above and the known limitations of food-web catalogs, the average S for the marine for food webs examined

Table 9.1 Compilation of marine food webs

Food web	Type of system	Latitude	Source	Topological	Flow/ Energetic	Interaction/ functional	# Spp.	Apex TLs	Upper TLs	Middle TLs	Lower TLs	Basal TLs	Comprehensiveness of spp. coverage
Baltic Sea	Bounded seas	Temperate	Baird et al. 1991	x	x	—	15	Low	Low	Med	Med	Low	Moderate
Baltic Sea	Bounded seas	Temperate	Harvey et al. 2003	—	x	—	15	Med	High	Low	Low	Low	Moderate
Baltic Sea	Bounded seas	Temperate	Sandberg et al. 2000	—	x	—	10–15	n.a.	Low	Low	Low	Low	Limited
North Sea	Bounded seas	Temperate	Christensen 1995	—	x	—	29	High	High	Med	Med	Low	Reasonable
North Sea	Bounded seas	Temperate	Greenstreet et al. 1997	—	x	—	9	Low	Low	Low	Low	n.a.	Limited
North Sea	Bounded seas	Temperate	Andersen and Ursin 1977	?	—	x	20–25	Med	Hi	Med	Low	Low	Moderate
California	Coastal	Temperate	Clarke et al. 1967	x	x	—	24	Hi	Hi	Med	Low	n.a.	Reasonable
Monterey Bay, California	Coastal	Temperate	Olivieri et al. 1993	—	x	—	16	Low	Low	Low	Low	Low	Limited
North Central Chile	Coastal	Temperate	Ortiz and Wolff 2002	—	x	—	23	n.a.	High	High	Med	Med	Moderate
Port Phillip Bay, Australia	Coastal	Temperate	Fulton and Smith 2002	—	x	—	34	Med	Med	Med	Low	Med	Moderate
Suruga Bay, Japan	Coastal	Temperate	Hogetsu 1979	—	x	—	16	n.a.	High	High	High	High	Moderate
Borneo	Coastal	Subtropical	Pauly and Christensen 1993	—	x	—	13	Low	Low	Low	Low	Low	Moderate
Campeche Bank, Mexico	Coastal	Subtropical	Vega-Cendejas et al. 1993	—	x	—	18	n.a.	High	High	Med	Low	Moderate
Central Gulf of California	Coastal	Subtropical	Arreguin-Sanchez and Calderon-Aguilera 2002	—	x	—	26	Low	Med	Med	Low	Low	Moderate
Gulf of Thailand soft bottom community	Coastal	Subtropical	Pauly and Christensen 1993	—	x	—	15	Low	Low	Low	Med	Low	Limited
Hong Kong territorial water	Coastal	Subtropical	Cheung et al. 2002	—	x	—	33	Med	Med	Med	Low	Low	Reasonable
Malaysia	Coastal	Subtropical	Pauly and Christensen 1993	—	x	—	13	Low	Low	Low	Low	Low	Limited

Location	Environment	Climate	Reference				N						Quality
Philippine coast	Coastal	Subtropical	Pauly and Christensen 1993	—	x	—	17	High	Med	Low	Low	Low	Moderate
San Miguel Bay, Philippines	Coastal	Subtropical	Bundy and Pauly 2000	—	x	—	16 (200+)	n.a.	Med	Med	Low	Low	Moderate
South China Sea, coastal benthos	Coastal	Subtropical	Pauly and Christensen 1993	—	x	—	13	Low	Low	Low	Low	Low	Limited
South China Sea, shallow water	Coastal	Subtropical	Pauly and Christensen 1993	—	x	—	15	Low	Low	Low	Med	Low	Moderate
South China Sea, Vietnam and China Coasts	Coastal	Subtropical	Pauly and Christensen 1993	—	x	—	13	Low	Med	Low	Low	Low	Moderate
SW Gulf of Mexico	Coastal	Subtropical	Arregun-Sanchez et al. 1993	—	x	—	24	Med	High	Med	Low	Low	Moderate
Tamiahua, Mexico	Coastal	Subtropical	Abarca-Arenas and Valero Pacheco 1993	—	x	—	13	Med	Med	Low	Low	Low	Moderate
Northeast Greenland, Arctic	Coastal	Boreal	Rysgaard et al. 1999	?	—	—	15–20	n.a.	n.a.	n.a.	Med	Med	Limited
Northern British Columbia	Coastal	Boreal	Beattie and Vasconcellos, 2002	—	x	—	53	Med	High	Med	Low	Low	Reasonable
Prince William Sound	Coastal	Boreal	Okey and Pauly, 1999	—	x	—	50	Med	Med	Low	Low	Low	Reasonable
Bay of Biscay 1970 and 1998	Continental Shelf	Temperate	Ainsworth et al. 2001	—	x	—	38	Med	Med	Low	Low	Low	Moderate
Mid-Atlantic Bight US	Continental Shelf	Temperate	Okey 2001	—	x	—	55	Med	High	High	Low	Low	Reasonable
SE Australia	Continental Shelf	Temperate	Bulman et al. 2001	x	—	—	100+	Med	Med	Med	Med	Low	Reasonable
SE Australia	Continental Shelf	Temperate	Bulman et al. 2001	x	—	—	100+	Med	Med	Med	Med	Low	Reasonable
Southeastern United States	Continental Shelf	Temperate	Okey and Pugilese 2001	—	x	—	42	High	High	Med	Med	Low	Reasonable
US NW Atlantic	Continental Shelf	Temperate	Link 2002	x	—	—	81	Med	High	High	Med	Low	Reasonable
Gulf of Thailand	Continental Shelf	Subtropical	Christensen 1998	—	x	—	26	Med	Med	Low	Low	Low	Reasonable
Minnan-TaiwancHighentan, South China Sea	Continental Shelf	Subtropical	Qiyong et al. 1981	x	—	—	38	n.a.	Med	High	High	Low	Moderate
South China Sea	Continental Shelf	Subtropical	Silvestre et al. 1993	—	x	—	13	Low	Low	Low	Low	Low	Limited

Table 9.1 (*Continued*)

Food web	Type of system	Latitude	Source	Topological	Flow/Energetic	Interaction/functional	# Spp.	Apex TLs	Upper TLs	Middle TLs	Lower TLs	Basal TLs	Comprehensiveness of spp. coverage
								Degree of spp. resolution					
South China Sea, Borneo	Continental Shelf	Subtropical	Pauly and Christensen 1993	—	x	—	13	Low	Low	Low	Med	Low	Moderate
South China Sea, deep shelf	Continental Shelf	Subtropical	Pauly and Christensen 1993	—	x	—	13	Low	Low	Med	Low	Low	Limited
South China Sea, deeper shelf	Continental Shelf	Subtropical	Pauly and Christensen 1993	—	x	—	13	Low	Low	Low	Low	Low	Limited
South China Sea, Gulf of Thailand	Continental Shelf	Subtropical	Pauly and Christensen 1993	—	x	—	15	Low	Low	Low	Low	Low	Limited
South China Sea, Gulf of Thailand	Continental Shelf	Subtropical	Pauly and Christensen 1993	—	x	—	15	Low	Low	Low	Med	Low	Moderate
South China Sea, NW Phillipines	Continental Shelf	Subtropical	Pauly and Christensen 1993	—	x	—	17	Med	Med	Med	Med	Low	Moderate
South China Sea, pelagia	Continental Shelf	Subtropical	Pauly and Christensen 1993	—	x	—	10	Low	Low	Low	Low	Low	Limited
South China Sea, SW SCS	Continental Shelf	Subtropical	Pauly and Christensen 1993	—	x	—	13	Low	Low	Low	Med	Low	Moderate
Southwestern Gulf of Mexico	Continental Shelf	Subtropical	Manickchand-Heileman et al. 1998a,b	x	x	—	19	Med	High	Med	Low	Low	Moderate
US Gulf of Mexico	Continental Shelf	Subtropical	Browder 1993	—	x	—	15	Med	Med	Low	Low	Low	Moderate
Venezuela Shelf	Continental Shelf	Subtropical	Mendoza 1993	—	x	—	16	Med	Med	Low	Low	Low	Moderate
Yucatan, Mexico	Continental Shelf	Subtropical	Arregun-Sanchez et al. 1993	—	x	—	21	Med	Med	Med	Low	Low	Moderate
Strait of Bali	Continental Shelf	equatorial	Buchary et al. 2002	—	x	—	14	Med	Med	Low	Low	Low	Moderate
Antarctic	Continental Shelf	Boreal	Constable et al. 2000, Constable 2001	—	—	x	15–25	Med	High	Med	Low	n.a.	Moderate
Antarctic	Continental Shelf	Boreal	Ainley et al. 1991	x	—	—	20–25	n.a.	High	Med	Med	Low	Moderate
Arctic Ocean	Continental Shelf	Boreal	Tittlemier et al. 2002	?	x	—	9–10	n.a.	Low	Med	Low	n.a.	Moderate

Location			Reference											
Barents Sea	Continental Shelf	Boreal	Hop et al. 2002	?	x	—	10–15	Med	Med	Med	Low	Low	Limited
Barents Sea	Continental Shelf	Boreal	Bogstad et al. 1997, Tjelmeland and Bogstad 1998	—	—	x	5–15	Med	High	Med	Low	n.a.	Moderate
Barents Sea	Continental Shelf	Boreal	Borga et al. 2001	?	x	—	9–15	n.a.	Med	Med	Med	n.a.	Moderate
Eastern Bering Sea	Continental Shelf	Boreal	Trites et al. 1999	—	x	—	25	Med	Low	Low	Low	Low	Moderate
Eastern Bering Sea	Continental Shelf	Boreal	Aydin et al. 2002	—	x	—	38	Med	Low	Low	Low	Low	Reasonable
Hi Arctic Ocean	Continental Shelf	Boreal	Hobson and Welch 1992	?	x	—	40–45	Med	High	High	High	Med	Reasonable
Lancaster Sound 1980s	Continental Shelf	Boreal	Mohammed 2001	—	x	—	31	High	Med	Med	Med	Low	Moderate
Lancaster Sound, Northwest Territories, Canada	Continental Shelf	Boreal	Atwell et al. 1998, Welch et al. 1992	x	?	—	27	Med	Low	High	High	n.a.	Reasonable
Newfoundland Southern and Northeastern Grand Bank	Continental Shelf	Boreal	Gomes 1993	x	—	x	26	Low	Med	Med	Med	Low	Moderate
Newfoundland-Labrador	Continental Shelf	Boreal	Bundy 2001, Bundy et al. 2000	—	x	—	20–25	High	High	High	Med	n.a.	Moderate
Northern Polynya, Arctic Ocean	Continental Shelf	Boreal	Fisk et al. 2001, Hobson et al. 1995	?	x	—	35–40	Med	High	High	High	Med	Reasonable
Norwegian and Barents Seas	Continental Shelf	Boreal	Dommasnes et al. 2001	—	x	—	30	Med	High	Med	Med	Low	Reasonable
Prince Edward Island, Southern Ocean	Continental Shelf	Boreal	Kaehler et al. 2000	x	?	—	38	Med	High	High	Med	Low	Reasonable
Southern Ocean	Continental Shelf	Boreal	Murphy 1995	—	—	x	4–20	Low	High	High	Med	n.a.	Moderate
Southern Ocean, Antarctica	Continental Shelf	Boreal	Reid and Croxall 2001	—	x	—	4–7	Low	Med	Med	Low	n.a.	Limited
West Antarctica Peninsula	Continental Shelf	Boreal	Barrera-Oro 2002	x	x	—	25–35	Med	Med	Low	Low	Low	Moderate
West Greenland	Continental Shelf	Boreal	Pederson and Zeller 2001, Pederson 1994	—	x	—	22	Med	High	Low	Low	Low	Moderate
Western Bering Sea	Continental Shelf	Boreal	Aydin et al. 2002	—	x	—	36	Med	Med	Low	Low	Low	Moderate

Table 9.1 (*Continued*)

Food web	Type of system	Latitude	Source	Topological	Flow/Energetic	Interaction/functional	# Spp.	Apex TLs	Upper TLs	Middle TLs	Lower TLs	Basal TLs	Comprehensiveness of spp. coverage
								Degree of spp. resolution					
Bolinao, NW Philippines	Coral reef	Subtropical	Alinõ et al. 1993	—	x	—	25	Med	Med	Low	Low	Low	Moderate
Florida Keys	Coral reef	Subtropical	Venier and Pauly 1997	—	x	—	20	Med	Low	Med	Low	Low	Moderate
Northern Great Barrier Reef, Australia	Coral reef	Subtropical	Gribble 2001	—	x	—	24	Med	Med	Low	Low	Low	Reasonable
South China Sea, coral reef	Coral reef	Subtropical	Pauly and Christensen 1993	—	x	—	13	Med	Med	Low	Med	Low	Limited
South China Sea, reef flats	Coral reef	Subtropical	Pauly and Christensen 1993	—	x	—	26	Low	Low	Med	Med	Low	Moderate
Caribbean	Coral reef	Equatorial	Opitz 1993	—	x	—	11	Low	Low	Low	Low	Low	Limited
Caribbean	Coral reef	Equatorial	Opitz 1993	—	x	—	20	Low	Low	Med	Low	Low	Moderate
Caribbean	Coral reef	Equatorial	Opitz 1993	—	x	—	50	High	Med	Med	Med	Low	Reasonable
French Frigate Shoals	Coral reef	Equatorial	Polovina 1984	—	x	—	12	Low	Low	Low	Low	Low	Limited
Moorea Island, French Polynesia	Coral reef	Equatorial	Arias-Gonzalez et al. 1997	—	x	—	43–46	n.a.	Med	Med	Med	Med	Reasonable
Moorea Island, French Polynesia	Coral reef	Equatorial	Arias-Gonzalez et al. 1997	—	x	—	43–46	n.a.	Med	Med	Med	Med	Reasonable
Takapoto Atoll lagoon	Coral reef	Equatorial	Niquil et al. 1999	—	x	—	7–10	n.a.	n.a.	n.a.	Low	Low	Limited
Central North Pacific	Mid ocean gyre	Equatorial	Kitchell et al. 1999, 2002	—	x	—	27	High	High	Low	Low	Low	Reasonable
Eastern Tropical Pacific	Mid ocean gyre	Equatorial	Olson and Watters, 2003; Essington et al. 2002	—	x	—	36	Med	High	Low	Low	Low	Reasonable
Pacific Warm Pool Pelagic Ecosystem	Mid ocean gyre	Equatorial	Godinot and Allain 2003	—	x	—	20	High	Med	Low	Low	Low	Moderate
Western and Central Pacific Ocean, warm pool pelagic	Mid ocean gyre	Equatorial	Cox et al. 2002	—	x	—	25	High	High	Med	Low	Low	Reasonable
Azores Archipelago 1997	Seamount	Temperate	Guenette and Morato 2001	—	x	—	43	Med	Med	Low	Low	Low	Moderate
Faroe Islands (NE Atlantic)	Seamount	Boreal	Zeller and Freire 2001	—	x	—	20	Med	High	Med	Med	Low	Moderate
Iceland	Seamount	Boreal	Bjornsson 1998	—	x	x	3	n.a.	Med	Low	n.a.	n.a.	Limited
Icelandic Ecosystem 1950	Seamount	Boreal	Buchary 2001	—	x	—	24	Med	High	Med	Med	Low	Reasonable
Icelandic Ecosystem 1997	Seamount	Boreal	Mendy and Buchary 2001	—	x	—	25	Med	High	Med	Med	Low	Reasonable

Ecosystem	Type	Zone	Reference			n							
New Zealand Southern Plateau	Seamount	Boreal	Bradford-Grieve and Hanchet 2002; Bradford-Grieve et al. 2003	—	x	18	Med	Low	Low	Low	Low	Low	Moderate
Benguela Current	Upwelling	Temperate	Baird et al. 1991	x	x	16	Low	Low	Med	Med	Low	Low	Moderate
Benguela Current	Upwelling	Temperate	Yodzis 1998, 2000, Field et al. 1991	x	—	29	Med	Med	Med	Med	Low	Low	Reasonable
Humboldt Current, Peru	Upwelling	Temperate	Jarre et al. 1991	x	x	16	Med	High	Med	Low	Low	Low	Moderate
Humboldt Current, Peru	Upwelling	Temperate	Jarre-Teichmann and Pauly 1993	—	x	20	High	High	Low	Low	Low	Low	Moderate
Morocco Atlantic Coast mid-1980s	Upwelling	Temperate	Stanford et al. 2001	—	x	38	Med	Low	Low	Low	Low	Low	Moderate
North Central Chile	Upwelling	Temperate	Wolff 1994	—	x	17	Low	Low	Low	Low	Low	Low	Limited
Northern Benguela	Upwelling	Temperate	Heymans et al. 2004	—	x	17	n.a.	Med	Low	Low	Low	Low	Moderate
Northern Benguela	Upwelling	Temperate	Heymans and Baird 2000	—	x	24	Med	Med	Med	Low	Low	Low	Reasonable
Northern Benguela	Upwelling	Temperate	Shannon and Jarre-Teichman 1999	—	x	24	Med	Med	Med	Low	Low	Low	Reasonable
Oregon/Wash. Coast 1981	Upwelling	Temperate	Brodeur and Pearcy 1992	x	—	22	n.a.	High	High	Low	Low	Low	Limited
Oregon/Wash. Coast 1982	Upwelling	Temperate	Brodeur and Pearcy 1992	x	—	21	n.a.	High	High	Low	Low	Low	Limited
Oregon/Wash. Coast 1983	Upwelling	Temperate	Brodeur and Pearcy 1992	x	—	24	n.a.	High	High	Low	Low	Low	Limited
Oregon/Wash. Coast 1984	Upwelling	Temperate	Brodeur and Pearcy 1992	x	—	20	n.a.	High	High	Low	Low	Low	Limited
Peruvian Current	Upwelling	Temperate	Baird et al. 1991	x	x	15	Low	Low	Med	Med	Med	Low	Moderate
Southern Benguela	Upwelling	Temperate	Jarre-Teichman et al. 1998	x	—	19	Med	Med	Med	Low	Low	Low	Moderate

was 20–25. This is a clear area for improvement in future studies.

We also categorized the degree of species resolution across the various groups using five categories loosely corresponding to a trophic level or an associated group of trophic levels: apex, upper, mid, lower, basal. We used a qualitative rank of low, medium, or high to characterize how well all the species in each group was represented and also how much aggregation of species occurred. With very few exceptions (<1%), most published marine food webs did not address the basal trophic levels with any degree of detail (Table 9.1). The other trophic levels in 35–50% of marine food webs had species with at least a medium or greater resolution. Surprisingly, >14% of marine food webs had apex trophic levels that were not considered. When comparing these results to previous web catalogs (Briand and Cohen 1984, 1987; Cohen and Briand 1984; Cohen 1989; Schoener 1989; Sugihara et al. 1989; Cohen et al. 1990; Pimm et al. 1991; Polis 1991; Havens 1992) this pattern confirms the continued trend of poor treatment for lower trophic levels and highlights the dearth of information for many of the apex consumers in marine food webs.

Finally, we noted the comprehensiveness of species coverage relative to the overall number (or suspected number) of the species in the system. We used a subjecitve rank, limited, moderate, or reasonable to provide an overall sense of extensiveness for each of the food webs. By no means are we stating that any of the food webs with limited coverage are of poor quality, rather that they are not likely to fully capture S and L nor contribute directly to associated theories based upon those metrics. Many of the limited food webs (22%) were from organochlorine biomagnification or isotope studies that encapsulated only portions of a food web. Most of the other limited food webs were from studies that emphasized a particular taxonomic group that omitted or greatly aggregated other species. We included these in our compilation, as they do provide some information, but their utility for testing some food-web theories individually may be limited. That >75% of the marine food webs examined are of moderate or reasonable coverage is encouraging. However, by no means are we stating that even a marine food web with an S of 50–100 is exhaustive nor necessarily complete.

Obiter dicta

There are a few general observations from this compilation of marine food webs. First is that most of the species emphasized, most of the species studied with any degree of resolution, and most of the stated goals for assembling these food webs are related to marine fisheries.

Major criticisms of aquatic and marine food webs are that (1) they are highly biased toward fish species, and (2) basal taxa are extremely aggregated in these webs. These criticisms and observations are in part artifacts of the expense and challenges of sampling marine ecosystems, but also reflect the national priorities experienced by many marine research scientists and organizations. A large proportion of researchers who have access to the necessary large-scale marine sampling equipment also have concurrent professional obligations to conduct stock assessments, determine fisheries impacts on economically valuable fish populations, and generally provide fishery management advice. Thus, the focus on data collection tends to be geared toward commercially targeted species. Consequently most of our knowledge on marine species interactions is centered on these groups. As suggested by Raffaelli (2000), more collaborations among scientists from a broader array of marine and ecological disciplines may ameliorate these problems.

The second observation is that other studies, such as those exploring organochlorine or pesticide biomagnification, isotope signatures, or focused studies on select marine organisms (e.g. sea birds, benthos, etc.) can provide useful and ancillary information for marine food webs. These studies need to be recognized as incomplete in a classical food-web sense, but should not be summarily ignored either. In particular, combining multiple and interdisciplinary studies from the same ecosystem may provide the basis for constructing a more complete food web of that system. The implication is that the literature from other disciplines (e.g. environmental contaminants, ornithology, marine pollution, etc.)

may be an unmined source of valuable marine food-web data.

The final observation from this compilation is that, given the pros and cons of Ecopath/Ecosim (E/E, Christensen and Pauly 1992; Walters et al. 1997; discussed in Hollowed et al. 2000 and Whipple et al. 2000), this modeling framework has been a valuable tool for cataloging marine food webs. With some exceptions, most marine food webs have been published since the early 1990s, and the vast majority of these are bioenergetic flow models using E/E. However, unlike other network analyses or topological food webs, many of the food-web macrodescriptors are lacking or not published with E/E models, yet they need to be presented to further explore some of the basic tenets of food-web theory as applied to marine systems. Additionally, most marine food webs, including but not limited to those constructed under an E/E framework, exhibit a high degree of aggregation. We need to broaden our understanding and better our resolution of marine food webs in order to test food-web theory for ecosystems that cover >70% of the planet's surface.

Food-web theory and marine ecosystems

Clearly, there is much work to be done in marine food-web research. More exhaustive and more highly resolved food webs from marine ecosystems will broaden our understanding of food webs in general, let us compare marine food webs to their nonmarine counterparts, and will test how these webs support or contradict food-web theory. Recognizing the current bounds on knowledge about marine food webs, we are not precluded from making a few limited observations about marine food webs and how further exploration and inclusion of them may alter food-web theory.

Marine food web verification of theory

Marine food webs are likely similar to all other food webs with the share of species primarily clustered at intermediate trophic levels. Like aquatic food webs, they are also not likely to integrate lower trophic levels with upper trophic

levels beyond a gross aggregation of basal species. Because of this, the predator to prey ratio will probably continue to be consistent across most food webs as well. Finally, it is likely that marine food webs will continue emphasizing commercially valuable species due to the increased interest in collecting data on those organisms, similar to other ecosystems.

Estimating interaction strengths or functional value in marine food webs still remains a major challenge. We suspect that in simpler boreal (relatively lower S and L) ecosystems that the interaction strengths i will be higher than in more complex temperate ecosystems. Conversely, i in highly complex but highly specialized food webs (e.g. equatorial coral reefs) will continue to be some of the highest interaction strengths measured (*sensu* Emery 1978; Carr et al. 2002). Marine food webs are not unique, however, in that functional or interaction webs are the least represented type of food webs.

In many respects, the topologies of marine food webs already confirm more recent and exhaustive food-web studies (e.g. Winemiller 1990; Hall and Raffaelli 1991; Martinez 1991; Polis 1991; Goldwasser and Roughgarden 1993; Reagan et al. 1996). For example, if one examines the level of C in most marine ecosystems in relation to S, these food webs are distinct outliers in the decreasing hyperbolic curve of C versus S (e.g. Link 2002). In fact, many of the marine food webs do not fall on the curve at all. These observations have clear implications for food-web stability and may ultimately mean that marine food webs support the constant connectance hypothesis (Martinez 1992), but need to extend it in terms of magnitude of S.

Marine food-web disparities with theory

However, in some respects marine food webs are not likely to fit empirically based theory that was developed from land or freshwater. Using the example above, observations of S, C and lack of a hyperbolic decline in the relationship between the two does not necessarily mean that marine food webs concur with the recent niche model (Williams and Martinez 2000; Dunne et al. 2004). The reasoning is threefold. First, these and prior theories were developed with S and L that were an order of magnitude or more lower than what actually occurs in marine food webs. Second, nonlinearities and the high probability of alternate steady states are common in marine systems (Steele 1985). Steele (1985) and Link (2002) discuss how stability may neither be an observable phenomenon in marine systems nor a useful construct given the high degree of perturbation and dynamics experienced by marine ecosystems. Third, we strongly suspect that we have not yet incorporated or even collected all the salient information to best elucidate the situation (Raffaelli 2000).

For those few marine food webs that report topological metrics, these metrics consistently stand apart from those of their terrestrial and freshwater counterparts. Marine food webs will continue to have higher L_D (Bengtsson 1994; Cohen 1994; Link 2002), higher average chain lengths (Bengtsson 1994; Cohen 1994), higher maximum chain lengths (Cohen 1994; Schoener 1989), and higher C values (Bengtsson 1994; Link 2002). Cannibalism, omnivory, and cycles will also continue to be prominent features of most marine food webs. Marine food webs will probably continue to challenge the scale invariance laws. Additionally, network metrics from marine systems will also probably continue to confirm the amazing complexity found in food webs. In particular, indices of energy re-cycling within marine food webs will remain high (e.g. Manickchand-Heileman et al. 1998b, Niquil et al. 1999; Abarca-Arenas and Ulanowicz 2002).

Most of the data presented for marine food webs do not fully cover the entire size range, distribution, and trophic functioning across the entire life history of most marine species. Most of the data presented for marine food webs severely aggregate under-known or unstudied organisms, if they do not omit them entirely. Most of the data presented for marine food webs represents only a small snapshot of the number of species and species interactions in an ecosystem. In fact, many of the marine food-web macrodescriptors (S, L, L_D, C) and related network metrics which we had originally planned to examine and compare with other systems were unfortunately unavailable.

Thus, it is premature to state whether food-web-theory will fully encapsulate all the observations of marine food-web structure and functioning (vis-á-vis Dunne et al. 2004). The most immediate and useful test might be to compare large lake systems to high latitude marine systems with careful controls for the number of species (e.g. focus on less speciose arcto-boreal marine systems in comparisons with temperate and boreal lake systems where there may be more parity in S). Such a comparison might allow useful contrasts while controlling for similarities and differences in life history patterns of the species involved.

Conclusions

The implications from studying marine food webs and associated theory are widespread for living marine resource management. Food webs provide a context for the management of living resources (Crowder et al. 1996; Winemiller and Polis 1996) Changes in marine food webs can potentially alter all populations, including those that support or are economically valuable species. Changes in food-web theory may also alter how we view marine food webs, fundamentally altering the feasibility of our expectations for particular management objectives.

If theories continue to explore the issue of food-web stability and structure, it is likely that marine food webs will expand this discussion, yet perhaps without providing conclusive evidence one way or another. However, Link (2002) has argued that assessing the stability of marine food webs may be a moot point given the ongoing harvesting pressure component populations have experienced over the past several decades (Pauly et al. 1998; Jackson et al. 2001). Assuming that marine ecosystems exhibit at least Lyapunov stability, then two points stand out. One is that with the high degree of interactions in most marine food webs, the resilience is going to be very high for the entire system. That is, to return to a historical equilibrium will take a long time. Second, it is likely that marine food webs may be perturbed beyond historical equilibria and shifted to new stable states (Steele 1985; Link 2002).

Additionally, the sheer complexity of marine food webs makes them difficult to predict. How the populations of marine food webs will collectively fluctuate from one steady state to another remains a major, if not the key management challenge for global marine resource managers. Given the complexity of marine food webs, the high degree of omnivory, and the generalist nature of most fish, it is unclear if predicting unambiguous trade-offs in biomass allocation among species in marine ecosystems is entirely practical.

However, some extant approaches are making valuable contributions to better deal with marine food-web dynamics and to better manage living marine resources. Many of the more recent network approaches and E/E (Christensen and Pauly 1992) explicitly explore trade-offs in energy flows and biomass among groups of species in a food web. Other multispecies and multivariate models (Hollowed et al. 2000; Whipple et al. 2000) have also shown promise toward this end. Although the requisite decision criteria remain to be fully developed, the ability to predict the "climate if not the weather" is a promising intersection between food-web theory and resource management. Translating many of the food-web macro-descriptors and network metrics into a decision criteria format remains a key area of research. The entire ecosystem-based fisheries management

(NMFS 1999) approach is premised on a solid understanding of marine food webs.

The search for unifying concepts in ecology has far-reaching significance. Food webs are complex, and we submit that marine food webs are probably some of the worst cases of food-web complexity. We also submit that the unique biological and physical properties (as discussed above) displayed in marine ecosystems distinguish them from non-marine food webs, perhaps even fundamentally very different than other types of food webs. Conversely, this also means that the potential to make significant contributions to food-web theory via the further examination of marine food webs may be quite high.

Acknowledgments

We thank all members of the Food-Web Dynamics Program, past and present, for their dedicated effort at collecting, auditing, and maintaining a food habits database of unprecedented scale and scope, from which much of our local empirical observations are taken. We also thank the editors of this book for inviting us to contribute this chapter. Finally, we thank J. Dunne and M. Fogarty who providing constructive comments on earlier versions of the manuscript which notably improved its content.

PART III

Stability and diversity in food webs

The following chapters link food-web stability and species diversity in aquatic systems to underlying trophic dynamics across different habitats, spatio-temporal scales, and levels of organization. Previous studies of patterns of diversity have usually focused on single trophic levels. Food webs provide a framework for integrating diverse types of diversity–stability studies, for example, studies that focus on guilds of aggregated species, studies that look at diversity within trophic levels, and studies that focus on the role of indirect interactions.

In addition to diversity, the structure of food webs can also fundamentally influence community dynamics and stability. For example, structure is shown to mediate the relationship between external climate forcing and species dynamics in large, pelagic, marine ecosystems. In general, a fundamental insight emerging from these chapters is that it is necessary to combine analyses of the network structure of complex food webs with dynamical approaches if we want to make progress on how to implement the use of dynamical food-web models in a conservation and management context.

CHAPTER 10

Modeling food-web dynamics: complexity–stability implications

Jennifer A. Dunne, Ulrich Brose, Richard J. Williams, and Neo D. Martinez

Introduction

Understanding the structure and dynamics of ecological networks is critical for understanding the persistence and stability of ecosystems. Determining the interplay among network structure, network dynamics, and various aspects of stability such as persistence, robustness, and resilience in complex "real-world" networks is one of the greatest current challenges in the natural and social sciences, and it represents an exciting and dramatically expanding area of cross-disciplinary inquiry (Strogatz 2001). Within ecology, food-web research represents a long tradition of both empirical and theoretical network analysis (e.g. Elton 1927; Lindeman 1942; MacArthur 1955; Paine 1966; May 1973). All aspects of ecological network research have increasing relevance (McCann 2000) for a world facing biodiversity and habitat loss, invasive species, climate change, and other anthropogenic factors that are resulting in the drastic reorganization of many ecosystems, and in some cases may lead to the collapse of ecosystems and the vital, underappreciated services they provide human society (Daily 1997).

Most theoretical studies of trophic dynamics have focused narrowly on predator–prey or parasite–host interactions, and have thus ignored network structure. In natural ecosystems such interaction dyads are embedded in diverse, complex food webs, where many additional taxa and their direct and indirect effects can play important roles for both the stability of focal species as well as the stability of the broader community. Moving beyond the one- or two-species population dynamics modeling paradigm, other research has expanded the focus to include 3–8 species, exploring dynamics in slightly more complex systems. However, these interaction modules still present a drastic simplification of the diversity and structure of natural ecosystems. Other approaches have focused on higher diversity ecological networks, but have either left out dynamics altogether (e.g. network topology and carbon-budget studies) or ignored network structure in order to conduct analytically tractable dynamical analyses. The nature of modeling requires judicious simplifications (Yodzis and Innes 1992), but such choices can leave empirical ecologists suspicious and theoretical ecologists perplexed, both wondering how to bridge the gap between simple mathematical models and complex natural systems. At worst, theoreticians and empiricists end up ignoring or deriding each other's work, dismissing it as irrelevant for being too abstract or too particular, never the twain shall meet.

In this chapter, we focus on theoretical aspects of the broader food-web research agenda, particularly the background and various approaches used for modeling food-web dynamics in abstract systems with more than two taxa. Much of this type of modeling has oriented itself around the classic (May 1972) and enduring (McCann 2000) complexity–stability debate, especially those aspects which relate to the theoretical and associated empirical food-web research into the relationships between ecosystem complexity, often characterized as number of links and/or number of species in a community, and various aspects of ecosystem stability. "Stability" is a catchall term that has been variously defined to reflect aspects of population and/or system equilibrium, persistence, resilience, resistance, and robustness (see McCann 2000 for some broad, but not exhaustive, stability definitions). In some cases stability refers to the outcome of internal dynamics while in other cases it reflects the response of a population or system to a perturbation. The very notion of what is a plausible definition of ecosystem stability has driven some of the fundamental shifts in modeling methodology that will be described in the following sections. We do not address the more empirically and experimentally based "diversity–stability" debate of recent years (e.g. Naeem et al. 1994; Hector et al. 1999; Tilman et al. 2001), which mostly focuses on the relationship between plant species richness and primary productivity.

This chapter discusses how three approaches to dynamical modeling of ecological networks try to strike a balance between simplifying and embodying aspects of the complexity of natural systems to gain a better understanding of aspects of ecosystem stability. In particular, the degree to which the different approaches incorporate diversity, network structure, nonlinear dynamics, and empirically measurable parameters will be recurring themes. We do not address spatial heterogeneity (see Chapter 2 by Melian et al.), metapopulation dynamics, age–class structure, environmental variability, or stoichiometry (see Chapter 1 by Elser and Hessen). Inclusion of those important factors in dynamical food-web models may well alter many of our notions about the interplay between ecosystem complexity and stability. While we do not explicitly review studies focused on food-web structure or topology (e.g. Cohen et al. 1990; Williams and Martinez 2000; Dunne et al. 2002a; Garlaschelli et al. 2003), we do refer to them to the degree that they have influenced dynamical approaches. A fundamental message of this chapter is that the insights from studies on the network structure of complex food webs need to be merged with dynamical approaches if we are to make serious headway regarding issues of the stability of complex networks and the potential for using dynamical models in a conservation and management context.

We focus on one end of a spectrum of modeling (Holling 1966), the use of abstract models to elucidate general qualitative insights about ecosystem structure, dynamics, and stability. We do not discuss more explicitly applied approaches for simulating particular ecosystems, a strategy which has been used for many aquatic systems (e.g. mass-balance, carbon-budget models such as Ecopath/Ecosim: Christensen and Walters 2004). Such approaches are presented elsewhere in this book (Chapter 3 by Christian et al. and Chapter 7 by Morris et al.). These researchers have often referred to their work as "network analysis." However, in our view, network analysis is a very broad term that encompasses all types of research that treat a system as a network of nodes and "edges," or links, regardless of how those nodes and links are defined or their relationships are analyzed (Strogatz 2001). In ecology, there has often been a tension between research focused on abstract models versus that focused on applied simulation models of particular ecosystems. We shall see that as abstract models seek to represent greater ecological complexity, they take on some of the characteristics of applied simulation modeling, hopefully narrowing the gap between theoretical and empirical work by carefully integrating characteristics and goals of the two modeling extremes. In this spirit, we will end by reviewing a recent application of a multispecies nonlinear dynamical food-web model to the issue of how culling a top predator is likely to affect the hake fishery yield in the Benguela ecosystem (Yodzis 1998, 2000, 2001).

Background

In the first half of the twentieth century, many ecologists believed that natural communities develop into stable systems through successional dynamics. Aspects of this belief developed into the notion that complex communities are more stable than simple ones. Odum (1953), MacArthur (1955), Elton (1958), and others cited an array of empirical evidence supporting this hypothesis. Some popular examples included the vulnerability of agricultural monocultures to calamities in contrast to the apparent stability of diverse tropical rainforests, and the higher frequency of invasions in simple

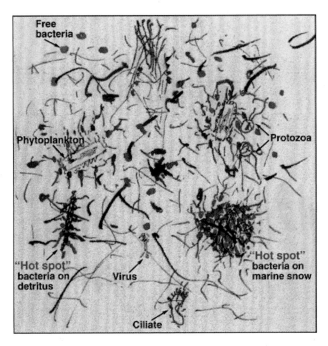

Plate 1 The microbial loop: impressionist version. A bacteria-eye view of the ocean's euphotic layer. Seawater is an organic matter continuum, a gel of tangled polymers with embedded strings, sheets, and bundles of fibrils and particles, including living organisms, as "hotspots." Bacteria (red) acting on marine snow (black) or algae (green) can control sedimentation and primary productivity; diverse microniches (hotspots) can support high bacterial diversity. (Azam, F. 1998. Microbial control of oceanic carbon flux: the plot thickens. *Science* **280**: 694–696.)

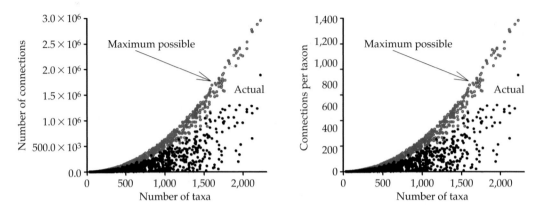

Plate 2 Density of connections of structured, hypothetical food webs as a function of the number of taxa. Shown are the maximum possible number of connections (○), determined by the greatest number of feasible connections for a particular distribution of taxa, and the corresponding actual number of connections (●) for approximately 6,800 networks.

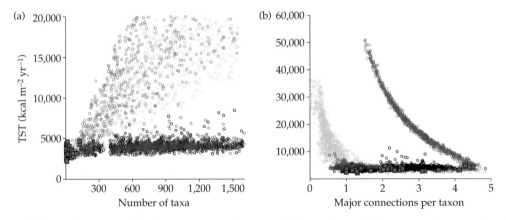

Plate 3 Total System throughput (TST) of different types of hypothetical food-webs as a function of (a) the number of taxa and (b) of the average number of major connections (flows >5% of inputs) per taxon. Hypothetical webs were either unstructured in design with uniformly ○ or lognormally ○ distributed transfer coefficients, or structured in design with uniformly ● or lognormally ○ distributed transfer coefficients. Empirical webs are denoted by ■.

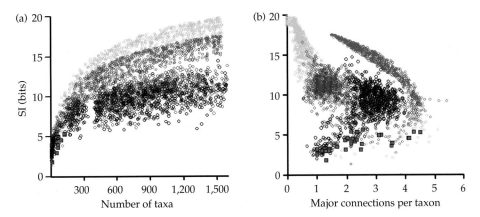

Plate 4 Flow diversity (SI) of different types of hypothetical food webs as a function (a) of the number of taxa and (b) of the average number of major connections (flows >5% of inputs) per taxon. Hypothetical webs were either unstructured in design with uniformly ○ or lognormally ○ distributed transfer coefficients, or structured in design with uniformly ● or lognormally ○ distributed transfer coefficients. Empirical webs are denoted by ■.

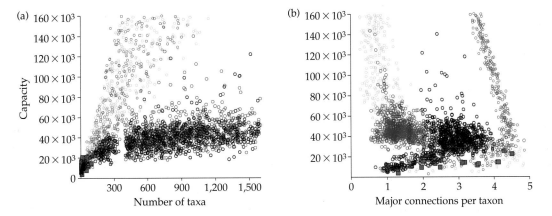

Plate 5 Full development capacity of different types of hypothetical food-webs as a function (a) of the number of taxa and (b) of the average number of major connections (flows >5% of total inputs) per taxon. Hypothetical webs were either unstructured in design with uniformly ○ or lognormally ○ distributed transfer coefficients, or structured in design with uniformly ● or lognormally ○ distributed transfer coefficients. Empirical webs are denoted by ■.

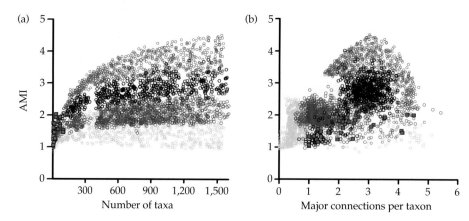

Plate 6 Average Mutual Information (AMI) content of different types of hypothetical food webs as (a) a function of the number of taxa and (b) of the average number of major connections (>5% of input flows) per taxon. Hypothetical webs were either unstructured in design with uniformly ○ or lognormally ○ distributed transfer coefficients, or structured in design with uniformly ● or lognormally ○ distributed transfer coefficients. Empirical webs are denoted by ▪.

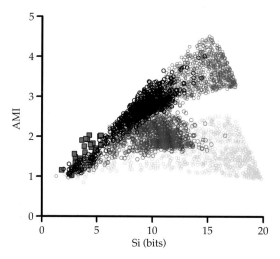

Plate 7 Average Mutual Information (AMI) as a function of flow diversity (SI) in empirical (▪) and hypothetical webs. Hypothetical webs were either unstructured in design with uniformly ○ or lognormally ○ distributed transfer coefficients, or structured in design with uniformly ● or lognormally ○ distributed transfer coefficients.

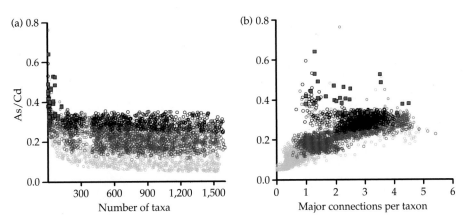

Plate 8 Ascendency: capacity as a function of (a) web size and (b)of the average number of major connections (flows >5% of inputs) per taxon (right). Hypothetical webs were either unstructured in design with uniformly ○ or lognormally ○ distributed transfer coefficients, or structured in design with uniformly ● or lognormally ○ distributed transfer coefficients. Empirical webs are denoted by ■.

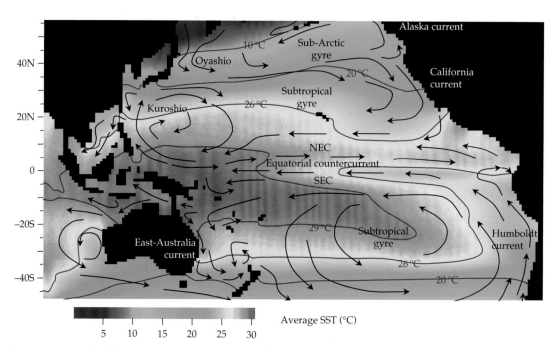

Plate 9 Tropical Pacific (TP). (NEC = North Equatorial Current, SEC = South Equatorial Current.)

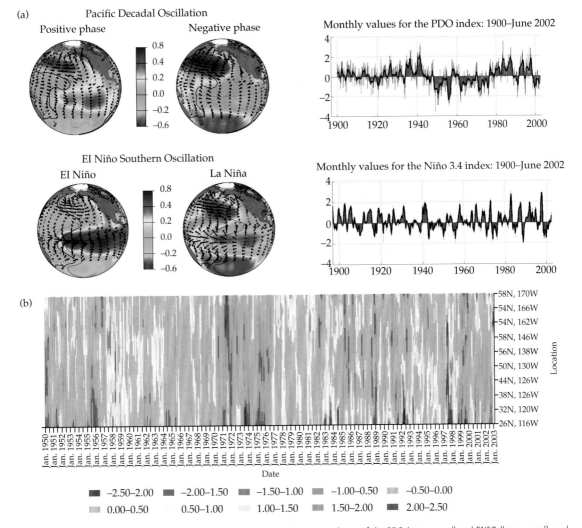

Plate 10 Sea Surface Temperature (SST) anomalies during positive and negative phases of the PDO (upper panel) and ENSO (lower panel), and time series of the climate index. During a positive PDO phase, SST anomalies are negative in the North Central Pacific (blue area) and positive in the Alaska coastal waters (red area) and the prevailing surface currents (shown by the black arrows) are stronger in the pole-ward direction. During a positive phase of the ENSO, SST anomalies are positive in the eastern Tropical Pacific and the eastward component of the surface currents is noticeably reduced. (b) Time series of temperature anomalies in different locations of the North Pacific. The graph shows area of intense warming (yellow and red areas) associated with the ENSO propagating to the North Pacific, a phenomenon termed El Niño North condition (updated from Hollowed et al. 2001).

(a)

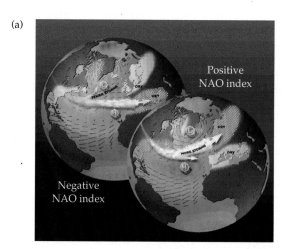

(b)

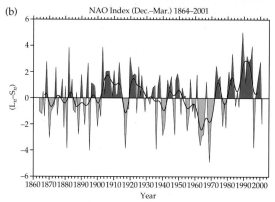

Plate 11 The North Atlantic Oscillation (NAO). (a) During positive (high) phases of the NAO index the prevailing westerly winds are strengthened and moves northwards causing increased precipitation and temperatures over northern Europe and southeastern United State and dry anomalies in the Mediterranean region (red and blue indicate warm and cold anomalies, respectively, and yellow indicates dry conditions). Roughly opposite conditions occur during the negative (low) index phase (graphs courtesy of Dr Martin Visbeck, www.ldeo.columbia.edu/~visbeck). (b) Temporal evolution of the NAO over the last 140 winters (index at www.cgd.ucar.edu/~jhurrell/nao.html). High and low index winters are shown in red and blue, respectively (Hoerling et al. 2001).

island communities compared to more complex mainland communities. It was thought that a community comprised of species with multiple consumers would have fewer invasions and pest outbreaks than communities of species with fewer consumers. This was stated in a general, theoretical way by MacArthur (1955), who hypothesized that "a large number of paths through each species is necessary to reduce the effects of overpopulation of one species." MacArthur concluded that "stability increases as the number of links increases" and that stability is easier to achieve in more diverse assemblages of species, thus linking community stability with both increased trophic links and increased numbers of species. Other types of theoretical considerations emerged to support the positive complexity–stability relationship. For example, Elton (1958) argued that simple predator–prey models reveal their *lack* of stability in the oscillatory behavior they exhibit, although he failed to compare them to multispecies models (May 1973). The notion that "complexity begets stability," which already had great intuitive appeal as well as the weight of history behind it, was thus accorded a gloss of theoretical rigor in the mid-1950s. By the late 1950s it took on the patina of conventional wisdom and at times was elevated to the status of ecological theorem or "formal proof" (Hutchinson 1959). While some subsequent empirical investigations raised questions about the conclusiveness of the relationship (e.g. Hairston et al. 1968), it was not until the early 1970s that the notion that complexity implies stability, as a theoretical generality, was explicitly and rigorously challenged by the analytical work of Robert May (1972, 1973), a physicist by training and ecologist by inclination.

May's 1972 paper and his 1973 book *Stability and Complexity in Model Ecosystems* (reprinted in a 2nd edition in 1974 with an added preface, afterthoughts, and bibliography; most recently reprinted in an 8th edition in 2001 as a *Princeton Landmarks in Biology* volume) were enormously important on several fronts. Methodologically, May introduced many ecologists to a much-expanded repertoire of mathematical tools beyond Lotka–Volterra predator–prey models, and also set the stage for major ecological contributions to the development of theories of deterministic chaos (e.g. May 1974, 1976). Analytically, May used local stability analyses of randomly assembled community matrices to mathematically demonstrate that network stability decreases with complexity. In particular, he found that more diverse systems, compared to less diverse systems, will tend to sharply transition from stable to unstable behavior as the number of species, the connectance (a measure of link richness—the probability that any two species will interact with each other), *or* the average interaction strength increase beyond a critical value. We describe his analyses in more detail in the next section. May (1973) also pointed out that empirical evidence is not conclusive with regard to a particular complexity–stability relationship (e.g. there are a number of highly stable, productive, and simple natural ecosystems such as east-coast *Spartina alterniflora* marshes; complex continental communities have suffered the ravages of pests such as the Gypsy Moth, etc.).

May's analytical results and his conclusion that in "general mathematical models of multispecies communities, complexity tends to beget instability" (p. 74, 2001 edition) turned earlier ecological "intuition" on its head and instigated a dramatic shift and refocusing of theoretical ecology. His results left many empirical ecologists wondering how the astonishing diversity and complexity they observed in natural communities could persist, even though May himself insisted there was no paradox. As May put it, "In short, there is no comfortable theorem assuring that increasing diversity and complexity beget enhanced community stability; rather, as a mathematical generality, the opposite is true. The task, therefore, is to elucidate the devious strategies which make for stability in enduring natural systems" (p. 174, 2001 edition). Most subsequent work related to food webs was devoted to finding network structures, species' strategies, and dynamical characteristics that would allow complex communities to persist. In some cases, the research was undertaken to corroborate May's findings, and other research sought to reanimate what many now felt to be the ghost of positive complexity–stability past. Whatever the motivation, important pieces of this ongoing research agenda include examining the role of omnivory, the number of trophic levels, weak interactions between species, adaptive foraging of consumers, and complex network structure on various measures of ecosystem stability. However, substantially differing methodologies, stability definitions, and a variety of sometimes apparently conflicting results have somewhat muddied the original question of whether complexity begets stability. Instead, the question is becoming more usefully recast along the lines of May's notion of "devious strategies"—what are the general characteristics of complex ecological networks that allow for, promote, and result from the stability and persistence of natural ecosystems.

Local stability analyses of community matrices

In his seminal studies on food-web stability, May (1972, 1973) measured local or neighborhood stability. In this analysis, it is assumed that the community rests at an equilibrium point where all populations have constant values (Figure 10.1(a)). The stability of this equilibrium is tested with small perturbations. If all species return to the equilibrium—monotonically or by damped oscillations—it is stable (Figure 10.1(a)). In contrast, if the population densities evolve away from the equilibrium densities—monotonically or oscillatory— they are unstable (Figure 10.1(b)). Neutral stability represents a third possibility, in which the perturbation neither grows nor decays as the population densities either reach an alternative equilibrium or oscillate with a constant amplitude in limit cycles (Figure 10.1(b)).

In a community of n species, this approach is based on an $n \times n$ Jacobian community matrix of species interaction coefficients that describe the impact of each species i on the growth of each species j at equilibrium population densities. In food webs, these coefficients may be positive (i.e. i is eaten by j), zero (no interaction), or negative (i.e. i eats j). To describe interactions by single coefficients it is necessary to assume linear interactions. The n eigenvalues of the community matrix characterize its temporal behavior (see May 1973 for a detailed description how these eigenvalues are

calculated). For every population, positive real parts of the eigenvalues indicate perturbation growth, negative real parts indicate perturbation decay, and zero real parts indicate neutral stability. Accordingly, if any of the eigenvalues has a positive real part the system will be unstable, that is, at least one of the species does not return to the equilibrium. Stability may be measured either as a binary variable—return to equilibrium or not—or as a metric variable in form of the return time needed by the population densities to settle back to the equilibrium.

May (1972, 1973) used community matrices in which species were randomly linked with random interaction strength to show that the local stability decreases with complexity (measured as connectance), diversity, and average interaction strength among the species. The use of such random community matrices has attracted much criticism. It was shown to be extremely unlikely that any of these random communities even remotely resembles ecosystems with a minimum of ecological realism such as containing at least one primary producer, a limited number of trophic levels, and no consumers eating resources that are two or more trophic levels higher (Lawlor 1978). The nonrandomness of ecosystem structure has been demonstrated in detail by more recent food-web topology studies (e.g. Williams and Martinez 2000; Dunne et al. 2002a, 2004; Morris et al. Chapter 7, this volume). Accordingly, subsequent work added more structural realism to those random community matrices by including empirical patterns of food-web structure and interaction strength distributions.

Varying connectance in 10 species food webs, De Angelis (1975) tested May's (1972) results within a more realistic model environment and found three conditions for positive complexity–stability relationships: (1) the biomass assimilated by consumers is a small fraction (<50%) of the amount of biomass removed from the resource population, (2) the consumers are subject to strong self-dampening (up to 20 times the maximum growth rate of primary producers) which restricts their population growth, and (3) a bias toward bottom-up control of the interactions (i.e. resource biomass has a stronger influence on the interaction strength

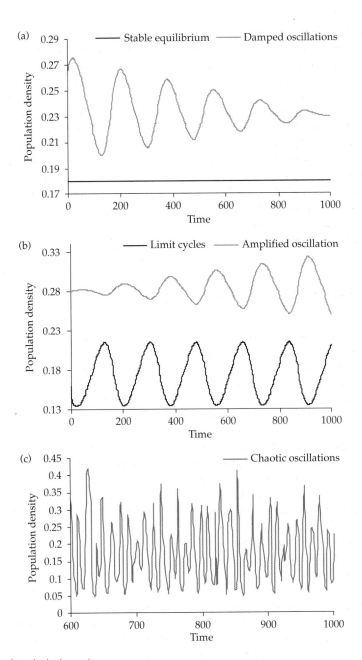

Figure 10.1 Population dynamics in time series.

values than consumer biomass), which also means that consumers have very little influence on the dynamics of their resources. The realism of these conditions, however, is questionable as assimilation efficiencies for carnivores should generally be higher (~85%, Yodzis and Innes 1992) and

such extreme self-dampening or low influence of consumers on resources appear unlikely. Nevertheless, for a long time the results of De Angelis (1975) remained the only mathematical model showing positive complexity–stability relationships.

In a more structural approach, Pimm and Lawton (1977, 1978) used community matrices of four interacting species. They systematically varied food-chain length and introduced omnivory by letting one of the species consume resources of multiple trophic levels. They found that the stability of the community matrices decreases with the number of trophic levels. The results for omnivory, however, varied, as omnivory decreases stability in terms of the fraction of stable models that return to equilibrium but at the same time increases stability of these stable models by decreasing their return time. Nevertheless, they concluded that the number of trophic levels in natural communities is limited by the stability of population dynamics (Pimm and Lawton 1977), and they predicted that omnivory should be a rare phenomenon in natural food webs (Pimm and Lawton 1978). In particular the latter prediction was not found to be consistent with empirical data as more detailed food webs were compiled (e.g. Martinez 1991; Polis 1991). In a further approach to analyze the relationship between food-web structure and dynamics, Yodzis (1981) showed that empirical community matrices corresponding to 40 real food webs are more stable than randomly assembled community matrices. This clearly indicated that natural food webs are structured in a way that stabilizes the internal population dynamics, but the question remained open as to which particular structural properties were important. Interestingly, Moore and Hunt (1988) showed that in those 40 food webs species are blocked into smaller subwebs of high connectance with only a few links connecting the subwebs. Based on analyses of May (1972), they argued that such compartmentalization may increase network stability and thus be responsible for the comparatively high stability of empirical food webs.

Evidence for clustering or compartmentalization in more detailed natural food webs has been suggestive, but scant. Krause et al. (2003), using food-web network data with unweighted links as well as with links weighted by interaction frequency, carbon flow, or interaction strength, identified significant compartments in three of five complex food webs. However, out of 17 versions of the 5 food webs examined, only 6 versions revealed significant compartmentalization. Girvan and Newman (2002) found ecologically meaningful compartments based on unweighted topology for a Chesapeake Bay food web (Baird and Ulanowicz 1989), but had difficulty identifying compartments in other more highly resolved food webs (Girvan personal communication). Topological analysis of 16 diverse, well-resolved food webs showed that their clustering coefficient (Watts and Strogatz 1998) is low in comparison to other types of networks (Dunne et al. 2002a). The low clustering coefficients of food webs is in part attributable to their high connectance (links/species^2) compared to other real-world networks (Dunne et al. 2002a)—in effect, there is such a high concentration of links in most food webs that the possibility of compartmentalization, at least from a purely topological perspective, is drowned out. In addition to being based on unweighted links, which may obscure compartmentalization, the clustering coefficient misses compartments of non-omnivorous species based on a single basal species. In sum, the issue of whether and how food webs are compartmentalized, and the impacts of compartmentalization on stability, remains an open question that requires more research.

Extending the approach of adding empirical realism to local stability analyses, de Ruiter et al. (1995) parameterized the general community matrix model with empirical food-web structures and interaction strengths among the species. In their empirical data, they found a pattern of strong top-down effects of consumers on their resources at lower trophic levels in food webs and strong bottom-up effects of resources on their consumers at higher trophic levels. Adding empirical interaction strength patterns to the community matrices increased their local stability in comparison to matrices with random interaction strength values. Most importantly, these results demonstrated that natural food-web structures as well as the distribution of interaction strengths within those structures contribute to an increased local stability of the corresponding community matrices. Neutel et al. (2002) explained this finding with results that showed that weak interactions are concentrated in long loops. In their analysis, a loop is a pathway of interactions from a certain species through the web back to the same species, without visiting other species more than once. They measured loop weight as the geometric mean of the interaction strengths in the loop and showed that loop weight decreases with loop length. Again, when applied to the community matrix, this empirically documented pattern of interaction strength distributions increased its local stability in comparison to random interaction strength distributions.

Thus, over the last 30 years, studies of the local stability of community matrices have contributed to our understanding of the relationship between complexity and stability. De Angelis (1975) indicated some specific conditions for positive complexity–stability relationships, whereas none of the subsequent studies has inverted the negative complexity–stability relationship demonstrated by May (1972). Nevertheless, other studies consistently suggested that local stability increases when ecological realism is added to the community matrix. In this sense, stability is gained by a limited number of trophic levels (Pimm and Lawton 1977), natural food-web structures (Yodzis 1981; Ulanowicz 2002), and interaction strength distributions (de Ruiter et al. 1995; Neutel et al. 2002). However, to gain mathematical tractability, the community matrix approach, with or without the addition of more plausible ecological structure, makes several assumptions that may restrict the generality of its results. First, community equilibrium is assumed, which excludes chaotic population dynamics that may characterize many natural systems (Figure 10.1(c), Hastings et al. 1993). Second, only small perturbations are tested. Larger perturbations might yield alternative equilibria. Only addressing local or neighborhood stability does not necessarily predict global stability, although they can coincide. Third, linear interaction coefficients are only a rough approximation of species' dynamics that will only hold for arbitrarily small variances in population densities. For any larger population variances, nonlinear species' interactions will be poorly approximated by linear coefficients. This assumption of small population variances, however, is likely to hold once the analyses are restricted to slightly perturbed populations at equilibrium. However, that case is a rather small slice of a much broader ecological pie.

Nonlinear, nonequilibrium population stability in food-web modules

To overcome the limitations of the equilibrium assumption and linear interactions included in community matrix analyses, an alternative approach to analyzing food-web stability emerged in the early 1990s (e.g. Hastings and Powell 1991; Yodzis and Innes 1992; McCann and Yodzis 1994a,b). Giving up those assumptions prohibited the use of community matrices. Instead, these studies defined less analytically tractable, but more ecologically plausible models of nonlinear species' interactions to study the population dynamics in species-poor food-web modules such as three-species food chains. In effect, one aspect of ecological plausibility was traded for another: nonlinear, nonequilibrium dynamics were embraced, but diversity was lost. Most of these nonlinear dynamical studies have used an allometric bioenergetic consumer–resource model of species biomass evolution that developed by Yodzis and Innes (1992). It defines for every species how much biomass is lost due to metabolism or being consumed and how much biomass is gained due to primary production or feeding on other species. In this approach, primary producer growth is based on logistic growth equations, and consumer growth is based on nonlinear functional responses of consumption to resource densities (Skalski and Gilliam 2001). Another aspect of ecological plausibility is added through the basis of most of its parameters on species body size and metabolic categories (i.e. endotherm, ectotherm vertebrate, invertebrate, primary producer). Specifying this bioenergetic model for every species in a food web yields a set of ordinary differential equations (ODEs). The numerical integration of the ODEs results in time series of species' biomass evolution that may be used to characterize population dynamics (e.g. stable equilibria, limit cycles, chaotic cycles, Figure 10.1). Furthermore, if species extinction is defined as falling below a threshold of biomass density (e.g. 10^{-30}) the persistence of the populations in the food web can be studied. This approach led to the identification of persistent chaos in three-species food chains where the population minima are bounded well away from zero (Hastings and Powell 1991; McCann and Yodzis 1994*a*, Figure 10.1(c)). However, on a longer timescale, persistent chaos may transform into transient chaos due to sudden catastrophic population crashes (McCann and Yodzis 1994*b*). Together, these results demonstrate that species and community persistence is not restricted to equilibrium dynamics (Figure 10.1(a) and (b)), with the potential for chaotic population dynamics (Figure 10.1(c)) that may or may not be persistent over short- and long-timescales.

Exploring another aspect of ecological complexity, McCann and Hastings (1997) investigated the impact of omnivory on dynamics by altering the network structure so that the top species of the food chain consumed both the herbivorous and basal species. They found that this type of omnivory stabilizes population dynamics by either eliminating chaotic dynamics or bounding their minima away from zero. In contrast to the earlier community matrix work of Pimm and Lawton (1978), this suggests that omnivory is a stabilizing feature of complex networks. It has to be cautioned, however, that these results were obtained under the assumption that the omnivore has a higher preference for the intermediate species than for the basal species, which rarely appears in natural food webs (Williams and Martinez 2004*a*). Furthermore, the consequences of omnivory in a comparable simple experimental system are highly dependent on nutrient enrichment, since coexistence of both consumers is restricted to intermediate nutrient saturations (Diehl and Feissel 2001).

In an extended study of food-web modules, McCann et al. (1998) showed that weak interactions might generally dampen population oscillations and thus stabilize dynamics in ecological networks. They suggested that such weak interactions among species might be the key for understanding the existence of complex communities, if most interactions in such communities are weak. Post et al. (2000) demonstrated that a set of two three-species food chains linked by the same top consumer can be stabilized when the top consumer can switch its preferences between the two intermediate species. Such preference switching appears likely whenever consumers

migrate between different habitat compartments, such as the pelagic and littoral food webs of lakes (Post et al. 2000). This suggests a potentially important role of spatial patterns on food-web stability that remains to be elucidated. Extending the size of the food-web modules, Fussmann and Heber (2002) demonstrated that the frequency of chaotic dynamics increases with the number of trophic levels, but decreases with other structural properties that cause higher food-web complexity. Interestingly, these results indicate that—dependent on which process dominates in a particular food web—population stability might increase or decrease with food-web complexity.

Overall, the numerical integration of nonlinear population dynamic models of food-web modules has added a great deal to our understanding of dynamical stability of multispecies systems. While not as simple, abstract, and analytically tidy as community matrix analyses, these models appear to capture more faithfully some centrally important aspects of natural ecological communities—aspects which appear very relevant to issues concerning complexity and stability. Most importantly, this modeling approach has clarified that nonequilibrium population dynamics do not prevent species' persistence (Hastings and Powell 1991; McCann and Yodzis 1994*a*). Further results suggest how omnivory (McCann and Hastings 1997; Diehl and Feissel 2001), weak interactions (McCann et al. 1998), and spatial processes (Post et al. 2000) may stabilize food webs. Being restricted to species-poor food-web modules, however, none of these studies can adequately explore or test complexity–stability relationships in diverse networks that reflect more of the complexity of natural systems. While it has been suggested that results from interaction module studies can be generalized to more diverse, complex networks, that notion is based on two potentially problematic assumptions: (1) population dynamics respond similarly to independent variables in food-web modules and complex networks, and (2) population stability coincides with community stability. It has been shown that species' persistence does not coincide with population equilibria (McCann and Yodzis 1994*a*) and that the frequency of chaotic dynamics is influenced by food-web complexity (Fussmann and Heber 2002).

Species persistence in complex food webs

More recently, the numerical integration approach has been extended to complex food-web models (e.g. Williams and Martinez 2004*b,c*). In this approach, bioenergetic models are applied to a diverse assemblage of species in complex trophic networks. The structure of these complex networks is defined by a set of simple topological models: random, cascade, or niche model food webs (Williams and Martinez 2000). The niche model has been found to successfully predict the general structure of complex food webs, while the cascade model (Cohen et al. 1990) predicts some aspects and the random model predicts almost no aspects of food-web structure (Williams and Martinez 2000; Dunne et al. 2004). Dynamics based on the Yodzis and Innes (1992) approach are run on each of the three types of food-web structures, and the ensuing short- and long-term population dynamics are investigated. The explicit integration of complex network structure with nonlinear, non-equilibrium bioenergetic modeling allows for the exploration and testing of complexity–stability relationships in more plausibly complex food webs, while retaining certain aspects of modeling simplicity. The topology models, which are analytically solvable (e.g. Camacho et al. 2002), use two readily estimated parameters, species richness (Brose et al. 2003*a*) and connectance (Martinez et al. 1999), and a few simple rules for assigning links. The dynamics model (Williams and Martinez 2004*b,c*), modified and generalized from Yodzis and Innes (1992), while not analytically tractable, is defined by a narrow set of parameters that can be measured relatively readily (e.g. allometric data), simple rules governing biomass gain and loss, and interaction dynamics that have been studied extensively in empirical settings (e.g. functional response). As described in the previous section, the bioenergetic dynamics model has the additional benefit of having been studied extensively in a simple module setting.

In these integrated structure–dynamics networks, global stability can be measured as persistence, that is, the fraction of species with persistent dynamics (which includes equilibrium, limit cycle, and chaotic dynamics). Using such food-web models, Williams and Martinez (2004*c*) found that persistence decreases linearly for communities of 15–50 species and connectance values between 0.05 and 0.3, thus qualitatively replicating the results of May (1972) in a completely different model environment. Also, they found that food webs with random topology display very low persistence, while adding some network structure in the form of the cascade model increases persistence by an order of magnitude. Use of the niche model structure improves persistence yet again over the cascade model (Williams and Martinez 2004*c*). Given that this coincides with a decrease in empirical adequacy of the structural models, the results strongly suggest that the complex network structure of natural communities provides a higher probability of species persistence. Interestingly, the structure of the persistent food webs that remained after species extinctions was more consistent with empirical data than the original structural models, even in the case of the niche model (Williams and Martinez 2004*c*). These results suggest a tight relationship between food-web structure, dynamics, and stability, since network structure influences the population dynamics that in turn determine persistent food-web structures. In addition to the important role of complex network structure for persistence, Williams and Martinez (2004*b,c*) showed that small variations in the functional response of consumption to resource and/or predator density can have dramatic effects on stabilizing the dynamics of particular species, as well as overall species persistence. One important stabilizing variation in functional response is the slight relaxation of consumption at low resource densities. Prior dynamical models have typically assumed Type II functional responses (Holling 1959) which lack this type of feeding relaxation, or have looked at stabilizing aspects of strong versions of relaxation-type responses (e.g. strong Type III) in model systems with only two species (Murdoch and Oaten 1975; Hassell 1978; Yodzis and Innes 1992). The relaxation of feeding at low resource density is evocative of a variety of well-documented ecological mechanisms (Skalski and Gilliam 2001) including predator interference, prey switching, and refuge seeking. These types of trophic and nontrophic behaviors allow rare or low-biomass resource species to persist in both natural and model ecosystems, increasing overall community persistence.

In another approach to integrating complex structure and dynamics, and in contrast to most prior dynamical studies, Kondoh (2003*a*,*b*, see also Chapter 11 by Kondoh) allowed consumer preferences for resources to adaptively vary in diverse random and cascade model food webs. He found positive complexity–stability relationships in many of these adaptive networks. However, the results are confounded due to inclusion of implausible linear nonsaturating species interactions and internal growth terms that allow consumers to persist without resources. A stability reanalysis of adaptive networks using the Williams–Martinez framework revealed that positive complexity–stability relationships require an arbitrarily high fraction of adaptive consumers (>50%) with a high adaptation rate in simple cascade model food webs, and do not occur in the more empirically well-corroborated niche model food webs (Brose et al. 2003*b*). Together these results indicate that positive complexity–stability relationships due to adaptive foraging are unlikely to be found in natural systems (but see Kondoh 2003*b*).

Integrating speciation events and assembly into dynamical food-web models has shown that—despite a high species-turnover—communities may assemble into persistent diverse networks (Caldarelli et al. 1998; Drossel et al. 2001). However, the food-web network structures that evolve are at best loosely comparable to empirical food webs, and appear less complex in some fundamental ways. For example, the connectances (the number of potential feeding links that are actually realized, links/species^2, as calculated using unweighted links) of eight example webs are very low ($C = 1–6\%$, mean of 3%, Drossel et al. 2001: tables 1 and 2) and connectance decreases with species richness. In comparison, a set of 18 recent empirical food webs have much higher connectances ($C = 3–33\%$, mean of $\sim14\%$, Dunne et al. 2004: table 2) and display no significant relationship between species richness and connectance (see also Martinez 1992). Nevertheless, this evolutionary food-web model gives rise to qualitatively

interesting patterns, such as periods of high community diversity and complexity followed by catastrophic extinction events leading to periods of lower diversity and complexity. The community evolution approach has yet to be analyzed in terms of complexity–stability relationships, and it remains unclear what causes the extinction events (Drossel et al. 2001). While the integration of evolutionary and population dynamics in a food-web assembly model is intriguing, this formulation is hampered by its extraordinarily large number of parameters, rendering the interpretation of results as well as their relevance to natural systems difficult, if not impossible.

In summary, the recent extension of the numerical integration approach to diverse model ecosystems with complex, ecologically plausible network structure offers an approach for exploring conditions that give rise to, and hinder the stability of individual populations within a complex ecosystem, as well as the stability of the whole system (Brose et al. 2003*b*; Williams and Martinez 2004*b*,*c*). Interestingly, many results from the integrated network structure and dynamics models are consistent with prior local stability analyses: a generally negative complexity–stability relationship and an increased stability of empirically consistent food-web structures in comparison to random webs (Williams and Martinez 2004*c*). Integrating adaptive or evolutionary processes in complex community models may lead to further insights regarding how diverse, complex natural ecosystems may persist (Caldarelli et al. 1998; Drossel et al. 2001; Kondoh 2003*a*,*b*; Brose et al. 2003*b*). It remains a disadvantage in comparison to the local stability approach that resource competition among basal species is generally not included in the nonequilibrium models, but it can be readily added. In general, the utility of current and future applications of these more complex dynamical approaches will depend to a great degree on a careful balancing act between implausible abstraction associated with fewer parameters and model clarity versus greater biological plausibility associated with more parameters and the increased danger of model obfuscation.

Application to an aquatic ecosystem

Thus far, this chapter has hopefully helped to introduce readers, many of whom are interested in aquatic systems but perhaps have less of a background in mathematical models, to the world of abstract models of population dynamics in simple to complex multispecies assemblages. The various approaches to modeling have been used to explore the conditions for and limitations upon stability and persistence in idealized ecosystems. However, aspects of the models may also lend themselves to more applied implementation, allowing assessment of stability of natural ecosystems in the face of actual or anticipated perturbations. This issue is of particular, immediate concern in aquatic ecosystems that contain fisheries. While the use of single-species models, which have dominated fisheries research and management, is generally acknowledged as inadequate and is slowly giving way to analysis of simple food-web modules (Yodzis 2001), those modules may still be inadequate for discerning the likely outcome of perturbations that occur in a diverse multispecies context. Yodzis (1998, 2000) has examined this fundamentally important issue for both theory and practice in the context of whether culling (removing or killing) fur seals is likely to have a negative or positive impact on the hake fishery in the Benguela ecosystem.

Within a fisheries context, one common view of Benguela food-web dynamics is that people harvest hake, fur seals eat hake, and therefore if people reduce the fur seal population there will be more hake for people to harvest without a net decrease in the hake stock. However, this is obviously an extraordinary oversimplification of a much more complex ecosystem. Instead, Yodzis (1998) considers a more realistic food web that contains 29 taxa including phytoplankton, zooplankton, invertebrates, many fish taxa, and several other vertebrate taxa (modified from Field et al. 1991). While still a simplification of the diversity and complexity of the system, this 29-taxa web better captures the variety of pathways that can link components of the network. There are over 28 million paths of potential influence from seals to hake in this food-web configuration. While

many of those pathways may have negligible impacts on the hake–seal interaction, it cannot be assumed a priori that scientists can ignore most of those pathways of influence when assessing the interaction between the two species (Yodzis 1998).

Yodzis (1998) utilizes this more realistic food web as the framework for studying multispecies dynamics determined by the nonlinear bioenergetic consumer–resource model discussed previously (Yodzis and Innes 1992). Parameter values for this model were set for the Benguela ecosystem in one of three ways: using field data estimations (e.g. population biomasses, average body mass), through allometric and energetic reasoning, and by allowing unknown parameters to vary randomly. The use of some randomly varying parameters means that probability distributions, rather than particular solutions, were calculated for the outcome of interest: "What is the impact of reducing fur seal biomass on the biomass of hake and other commercial fishes?" In effect, Yodzis is simulating an experimental press-perturbation—an experiment that cannot be done in a scientific or ethical way on the actual ecosystem. This framework and question requires two further simplifications: Yodzis assumes local perturbations (e.g. small culls) and equilibrium dynamics (e.g. constancy in parameter values and dynamics). While we have already seen that there can be problems with these sorts of assumptions, they allow the use of the analytically tractable Jacobian matrix (used by May 1972, 1973) to determine the outcome of small, long-term perturbations in a complex system, which is a reasonable approach to use for this particular question. Thus, the analysis combines elements of all three approaches to multispecies dynamical modeling we have discussed, including local stability analyses of community matrices, nonlinear bioenergetic dynamics, and complex food-web structure. These modeling elements, in conjunction with empirical data on trophic interactions and various characteristics of taxa in the Benguela ecosystem, form a plausible framework in which to simulate a cull of fur seals and its likely effects on commercial fishes in the context of a complex ecosystem.

Using this modeling framework, Yodzis (1998) found that if fur seals are culled, there is a significant probability that two of three commercial fishes (hake, anchovy, and horse mackerel) will have negative responses, and that across all of the commercial stocks it is more likely that there will be a loss than a gain. Perhaps even more importantly, Yodzis' analyses (1998, 2000) underscore the dangers of relying on overly simplistic models to make predictions. One important issue is how sensitive the model is to inclusion of apparently unimportant trophic links—that is, can trophic links that reflect some relatively small percentage of a predator's consumption be ignored without affecting the outcome of the model? The answer is "no" for most trophic links. As the cutoff level (of percentage of diet) for inclusion rises above 10%, results start to vary drastically from outcomes seen on the complete web, thus showing that "taxonomic completeness and resolution, detailed food-web structure, does make an enormous difference in predicting the outcomes of generalized press-perturbations, such as culling fur seals in the Benguela ecosystem" (Yodzis 1998). The interaction between hake and fur seals within the Benguela food web is highly influenced by many other species in the web (Yodzis 2000), and it is very likely that those types of influences are found in many, if not most ecosystems. The presence of these potentially strong diffuse effects in complex food webs renders the use of small modules as a means of simplified community analysis highly suspect for many ecosystems and many types of questions (Yodzis 2000).

Conclusions

We have reviewed three dynamical modeling approaches for studying complexity–stability relationships in abstract food webs, and some of their central characteristics are compared in Table 10.1. This type of modeling has its roots in linear, equilibrium dynamics research. Some studies continue to utilize equilibrium assumptions, which may be appropriate for certain kinds of questions, as discussed in the previous section. However, since the early 1990s alternative approaches have taken advantage of increasing computing power and advances in nonlinear dynamics to explore population dynamics in detail, using nonlinear and nonequilibrium assumptions which are more likely to reflect the types of dynamics that occur in natural systems. Rather than being restricted to the binary stable/unstable outcomes that typify local stability analyses, researchers can now explore a wider range of types of population and whole-system stability. This is especially the case with the extension of nonlinear, nonequilibrium dynamical analysis to food webs with complex, plausible network structure. For example, one new measure of-stability is the percentage of species in a model food web that displays persistent dynamics (Martinez and Williams 2004c), which can be further broken down into the percentages of species that display different types of persistent dynamics (e.g. stable equilibria, limit cycles, chaos) (Figure 10.1).

Table 10.1 Summary of the stability analysis methodologies

	Local stability	Food-web module stability	Complex food-web persistence
Interactions	Linear	Nonlinear	Nonlinear
Equilibrium assumption	Yes	No	No
Small perturbations	Yes	No	No
Chaotic dynamics allowed	No	Yes	Yes
Community size	Small to diverse	Small	Diverse
Population dynamics analyses	Stable/unstable	Yes	Yes
Community stability analyses	Yes	No	Yes
Community persistence	Yes	No	Yes
Competition among primary producers	Yes	No	No
Nutrient dynamics	No	No	No

Some of the promising new directions for this type of research include explorations into evolutionary and adaptive dynamics, nutrient dynamics, spatial heterogeneity, and nontrophic interactions. However, adding new layers of ecological plausibility to models carries with it the risk of generating a parameter space so large and uncontrolled that interpretation of model outcomes is rendered an exercise in creative tale-spinning, and the links to empiricism that may have motivated the addition may be rendered worthless. Any elaborations need to be carefully justified, made as simple as possible, and explored and explained in detail.

Rather than being shackled just to questions of whether species richness and connectance "beget" stability or instability, researchers are expanding their search for the "devious strategies" that promote different aspects of population and ecosystem persistence and robustness. Regardless of the methodology used, there are some emerging, consistent results indicating that natural food webs have nonrandom network structures and interaction strength distributions that promote a high degree of population stability and species persistence. As recently pointed out by McCann (2000), model and empirical evidence (e.g. Berlow 1999) that many weak interactions can buffer a few strong, unstable consumer–resource interactions, thus stabilizing the whole community, is consistent with MacArthur's (1955) hypothesis that having more feeding links acts to dampen large population swings. Nonrandom network structure, regardless of interaction strength, also appears to buffer against population extinction,

and the more realistic the food-web structure, the greater the species persistence. From a purely topological perspective, increasing connectance may buffer against the probability of cascading extinctions (Dunne et al. 2002b) partly because middle and high connectance webs have more uniform link distributions (the distribution of trophic links per species) than low connectance webs, which tend to have power-law distributions (Dunne et al. 2002a). Power-law networks appear especially vulnerable to loss of highly connected nodes (e.g. Albert et al 2000; Dunne et al. 2002b).

It should be clear that there is still an enormous amount of work to be done to elucidate May's "devious strategies," and in the face of widespread biodiversity loss and ecosystem perturbation, there is an increasingly compelling set of reasons to do that research. Food-web research has proven to be a useful vehicle for many theoretical and empirical questions in ecology, and some of the most heroic work will be accomplished by people who are willing to try to integrate the lessons from abstract models with real-world applications (e.g. Yodzis 1998). Beyond ecology, the analyses and results that emerge from the study of complex food webs are likely to reverberate back outwards to influence our understanding of other types of biotic and abiotic networks.

Acknowledgments

Jennifer Dunne acknowledges the support from the following grants: NSF DEB/DBI-0074521, DBI-0234980, and ITR-0326460.

Is biodiversity maintained by food-web complexity?—the adaptive food-web hypothesis

Michio Kondoh

The complexity–stability debate

Food-web complexity and population stability

In nature, a number of species are interconnected by trophic links, resulting in a complex food-web network (Warren 1989; Winemiller 1990; Martinez 1991; Polis 1991; Goldwasser and Roughgarden 1993). In a food web, a species affects, and is affected by, other species not only through direct resource–consumer interactions but also through the indirect interactions mediated by other species (Schoener 1993; Menge 1994; Abrams et al. 1996). Indirect effects between species are quite sensitive to food-web topology and the positions of the focal species within the web (Yodzis 2000). Small differences in food-web topology or interaction strength can lead to large changes in indirect effects (Yodzis 2000).

The high sensitivity of indirect effects to detailed food-web structure implies difficulty in relating population abundance and its variation to inter-specific indirect effects and in finding a consistent pattern in the strength, or even the sign, of the indirect effects in complex food webs. Does a particular species within the web increase or decrease when the level of a certain other species increases or decreases? What is the species-level consequence of the addition or removal of a link to/from the web? All those questions are unlikely to be answered for the difficulty in describing food-web structure with a high degree of accuracy (Yodzis 2000).

Nevertheless, it may still be possible to predict how food-web structure affects the grosser dynamic

properties of the food web. An example of such a property is "population stability" (May 1973; Goodman 1975; Pimm 1991). Theory suggests that there is statistical regularity in how population stability changes with changes in the food-web structure (see Pimm 1991). Several components of population stability have been considered, including resilience (i.e. the rate at which population density returns to equilibrium after a disturbance), feasibility (the system is at equilibrium with all existing populations), asymptotic local stability (the system returns to equilibrium after a slight disturbance), global asymptotic stability (the system returns to equilibrium after any level of disturbance), persistence, and permanence (the system does not lose species). I briefly review these studies in the following section.

Complexity–stability paradox

May (1972) first tackled the issue of how population stability is related to food-web structure by explicitly incorporating the effects of interspecific indirect interactions on population dynamics into a dynamic model. May (1972) performed linear stability analysis for a number of randomly generated community matrices and varied the following parameters: number of species (species richness), the probability that a pair of species is connected (connectance), and the strength of interspecific interactions relative to intraspecific negative interactions (interaction strength). The analysis showed that the local asymptotic stability

THE ADAPTIVE FOOD-WEB HYPOTHESIS

of the population, on average, decreased with increasing food-web complexity. There was a threshold in food-web complexity above which the system was unlikely to be stable. The condition for stability is given by $\alpha(SC)^{1/2} > 1$, where α is the interaction strength, S is species richness (number of species), and C is the connectance. The prediction is that the relationship between food-web complexity and population stability is negative.

The negative complexity–stability relationship, first predicted by May (1972), was supported by a number of subsequent theoretical studies (Gilpin 1975; Pimm and Lawton 1977, 1978; Keitt 1997; Schmitz and Booth 1997; Chen and Cohen 2001b). Increasing the species richness or connectance decreases feasibility (Gilpin 1975). Increasing the length of trophic chains decreases resilience (Pimm and Lawton 1977; but see Sterner et al. 1997). Omnivory, that is, feeding at more than one trophic level, lowers local asymptotic stability (Pimm and Lawton 1978). Increasing connectance or species richness lowers permanence and global asymptotic stability (Pimm and Lawton 1977). Although the stability indices and complexity indices used in these studies vary, they agree with each other in that a more complex food web is less likely to persist, owing to its inherent instability.

The theoretical prediction that increasing complexity destabilizes populations contradicts the intuition that large and complex food webs, which in reality persist in nature, should be stable (Pimm 1991). This is because such webs are unlikely to be observed if complex food webs are unstable. The counterintuitive prediction has catalyzed extensive studies on this issue, resulting in a number of hypotheses to explain the maintenance mechanisms of complex food webs (DeAngelis 1975; Lawlor 1978; Nunney 1980; Yodzis 1981; De Ruiter et al. 1995; McCann et al. 1998; Haydon 2000; Roxburgh and Wilson 2000; Sole and Montoya 2001; Chen and Cohen 2001a; Vos et al. 2001; Neutel et al. 2002; Emmerson and Raffaelli 2004). Some studies have focused on how complex food webs are assembled through the sequential process of "trial (i.e. immigration or speciation) and error (extinctions)" (Post and Pimm 1983; Law and Morton 1996; Wilmers et al. 2002). Other studies have focused on more detailed food-web

architecture, with the expectation that a food web with more realistic architecture should be stable (DeAngelis 1975; Lawlor 1978; Yodzis 1981; De Ruiter et al. 1995; McCann et al. 1998; Roxburgh and Wilson 2000; Neutel et al. 2002; Emmerson and Raffaelli 2004).

The starting point of the latter idea is to recognize that for a given food-web complexity, several possible food-web architectures exist, some of which may be stable, while others are not. An important implication of this heterogeneity is that the theoretical finding that a more complex food web is less likely to be stable does not necessarily mean that all complex food webs are unstable. Indeed, studies that have examined how the incorporation of more detailed food-web structure alters the stability of the food web (DeAngelis 1975; Lawlor 1978; Yodzis 1981; De Ruiter et al. 1995; Roxburgh and Wilson 2000; Neutel et al. 2002; Emmerson and Raffaelli 2004) have shown that a complex food web can be stable. In the following section, I focus on this approach, which explicitly considers the effect on population dynamics of realistic food-web architecture.

Static food-web architecture and the resolution of the complexity–stability paradox

The negative complexity–stability relationship of randomly generated food-web models does not necessarily predict a negative relationship in natural food webs. For one reason, within the whole parameter space of complexity (food-web topology, species-specific parameters, and interaction-specific parameters), only a small fraction is likely to be realized in nature (Lawlor 1978). The complexity–stability relationship, which is obtained from the whole parameter space, may not represent the relationship within the "realistic" parameter region if food-web models with "realistic" parameter sets behave differently from those with "unrealistic" parameter sets (DeAngelis 1975; Lawlor 1978).

Haydon (2000) explicitly showed that variation in stability within random food-web models depends on food-web complexity. He analyzed a number of randomly generated community matrices to examine how the maximum stability changes with changing food-web complexity. The model analysis revealed that while more complex food webs are, on average, less likely to be stable, the maximally stable subset of more complex food webs is more stable than that of less complex food webs (Haydon 2000). This result has two important implications. First, within food-web models with varying complexity, the maximally stable food webs could be more complex. Second, the variance in stability is higher in more complex food webs. These findings indicate the importance of considering only a realistic subset of randomly generated food-web models in evaluating the complexity–stability relationship, especially when the food web is highly complex.

Studies that predict a negative relationship have been criticized, as the food-web models used in these studies often assume an unrealistic food-web structure (Lawlor 1978). In May's (1973) model, for example, interaction strength is randomly assigned, and therefore, the models include biologically implausible communities such as communities without autotrophs or with extremely long food chains. If the stability of food-web models with unrealistic structures largely differs from the stability of food-web models with realistic structures, an analysis of randomly generated food webs can lead to an incorrect prediction that is biased by the unrealistic models. Indeed, some studies that focused on the complexity–stability relationship within realistic parameter regions have concluded that the exclusion of unrealistic food-web models increases the stability of populations or even creates a positive complexity–stability relationship (Lawlor 1978; Yodzis 1981).

Some studies have found that when using food-web models with realistic structures, increasing complexity does not destabilize populations (DeAngelis 1975; McCann et al. 1998). DeAngelis (1975) showed that if the effect of predators on prey is sufficiently larger than that of prey on predators (e.g. low conversion rate), increasing the complexity enhances the asymptotic stability of the equilibrium point. Other studies have proposed a correlation between food-web complexity (species richness, connectance) and interaction strength, which influences the complexity–stability relationship (May 1973; Lawlor 1978; Nunney 1980;

Vos et al. 2001). For example, if a predator has more potential prey species, it is likely that the interaction strength of each trophic link decreases due to the limited ability of a predator in dealing with multiple prey species. In the presence of the negative correlation between connectance and interaction strength, increasing complexity may stabilize population dynamics (May 1973; Lawlor 1978; Nunney 1980; Vos et al. 2001).

Analyses of food-web models with empirically obtained topology and interaction strength (Yodzis 1981; De Ruiter et al. 1995; Roxburgh and Wilson 2000; Neutel et al. 2002; Emmerson and Raffaelli 2004) often suggest that natural food webs are constructed in a way that enhances stability. Neutel et al. (2002) showed that pyramid-like biomass distribution between trophic levels results in an interaction strength that enhances food-web stability. Emmerson and Raffaelli (2004) pointed out that interaction strength derived from the relative body sizes of prey and predator promotes population stability. These studies suggest that there may be natural processes "choosing" the interaction strength that results in food-web stability. The generality of this idea, however, is still questionable, as these hypotheses have been tested for only a few select food webs.

Although the predicted complexity–stability relationship varies between studies, these studies are similar in that they focus on the position of trophic links and interaction strength to explain the maintenance of complex food webs. The message of these studies may be summarized as follows:

1. A complex food web can be stable if realistic interaction strengths and/or topology are considered in constructing the food-web model.
2. The instability of complex systems in previous theoretical studies is due to averaging over many food-web models with varying stability.

Adaptation and complexity–stability relationship

Flexible food-web Structure

Previously predicted complexity–stability relationships were derived mainly from mathematical models that considered dynamic populations and static linkage patterns (May 1972; DeAngelis 1975; Gilpin 1975; Pimm and Lawton 1977, 1978; Lawlor 1978; Yodzis 1981; De Ruiter et al. 1995; Haydon 2000; Neutel et al. 2002). These models explicitly or implicitly assume that the interaction strength is a fixed property of a trophic link, that interaction strength only changes for changing population abundances, or that a trophic link has a fixed position within a web (Paine 1988). These simplified assumptions allow for a direct comparison of food-web models with varying structure and make it possible to identify the effects of interaction strength or food-web topology on population stability. The theoretical contributions of this approach are unquestionable; however, at the same time, these assumptions have set limitations as to how interaction strength influences population stability and complexity–stability relationships.

An important property of trophic links that has been more or less overlooked in previous complexity–stability debates is flexibility. Earlier studies tended to assume static food-web structures. However, in reality, the position and strength of a trophic link are not fixed, but are spatiotemporally variable (e.g. Warren 1989; Winemiller 1990). The linkage pattern changes over time as a trophic link may be activated or inactivated. A trophic link that exists for a specific period during a life cycle (e.g. ontogenic niche shift) is disconnected during other periods. If such flexibility is a general property of a trophic link, then a food-web structure composed of a single set of nodes and links represents only a certain time period and cannot serve to represent population dynamics over the long term.

The effect of food-web flexibility on population stability depends on the timescale in which the food-web structure changes. If the food-web structure changes at a timescale that is sufficiently slower than the timescale of population dynamics, then the food-web structure does not change as population levels change. food-web structure can be approximated as static in evaluating the population dynamics. As Food-web flexibility makes the parameters that characterize a food web change over time, increasing flexibility only enlarges the parameter region where stability should be

evaluated (e.g. Pimm 1991). On the other hand, if the timescale of structural changes is comparable to that of population dynamics, it may influence the population dynamics (e.g. Holling 1959; Abrams 1982, 1984).

Adaptation and food-web structure

A general mechanism through which food-web flexibility arises is adaptation. Adaptation is a distinguishing feature of organisms and occurs at several different biological levels. First, natural selection drives population-level adaptation, or evolution (Stephens and Krebs 1987). A population, which is described as a node in a food web, is in reality a heterogeneous unit comprising individuals with varying genotypes. The relative abundances of each genotype change from generation to generation according to their relative reproductive contribution to the next generation. This results in phenotypic shifts at the population level. Second, adaptation occurs at the individual level. Organisms have the potential to learn through their experience and to modify their behavior accordingly (Hughes 1990). Population-level characteristics, which are given by averaging across individuals within a population, can change over time if each individual continuously adjusts its behavior to the changing environment. If the behavior or morphology that influences the strength of a trophic interaction is dynamic, then interaction strength should also be dynamic.

There are two types of adaptations that can influence the strength of a trophic interaction, that is, adaptive defenses by prey (Fryxell and Lundberg 1997; Abrams 2000; Bolker et al. 2003) and adaptive foraging by predators (Emlen 1966; MacArthur and Pianka 1966). Consider a predator and a prey species connected by a trophic link. From the viewpoint of the prey, the trophic link is a path through which it loses its body mass, opportunities for reproduction, or life. Therefore, prey adapts to reduce the risk of predation through a variety of behavioral or morphological changes (antipredator defenses; Endler 1986; Lima and Dill 1996), which include choosing a habitat with less risk of predation, developing a morphological structure that prevents consumption, producing toxins, and shifting their active time. Such anti-predator defenses weaken the strength of the trophic link. From the viewpoint of the predator, on the other hand, a trophic link is a path through which energy and nutrients are gained. A predator influences the strength of trophic interactions to maximize its energetic gain from the potential resources (Emlen 1966; MacArthur and Pianka 1966). Some resources are extensively searched for or captured to strengthen the trophic link, while others are discarded from the diet, thereby weakening the link.

Behavioral or morphological shifts potentially influence population dynamics. A number of theoretical studies have examined the effects of adaptation on simple systems such as prey–predator systems, one prey–two predator systems, and two prey–one predator systems (Tansky 1978; Teramoto et al. 1979; Holt 1983; Abrams 1984, 1992, 2000; Sih 1984; Ives and Dobson 1987; MacNamara and Houston 1987; Lima 1992; Matsuda et al. 1993, 1994, 1996; Wilson and Yoshimura 1994; Sutherland 1996; Abrams and Matsuda 1996; Holt and Polis 1997; Křivan 1998, 2000; Bolker et al. 2003). These studies have shown that the effects of adaptation on population dynamics are diverse and, to a large extent, context dependent (see Abrams 2000; Bolker et al. 2003). Inclusion of an adaptive diet shift either stabilizes or destabilizes prey–predator interactions and either increases or decreases the amplitude of population cycles. Consequences depend on the parameters affected by the focal trait (i.e. pleiotropic effects), relative speed of adaptation to speed of population dynamics, the shape of the functional response, and the constraints under which a trait changes.

Foraging adaptations and complexity–stability relationships

Among the various foraging strategies for a predator to maximize its energetic gains, one possible strategy is to consume only diets of higher quality or quantity from a set of nutritionally substitutable diets (a foraging switch or foraging shift; Stephens and Krebs 1987). This is a natural consequence if different strategies are required to find or capture different diets, because utilization

of a less-profitable resource lowers the net energy gain per unit effort. Indeed, a number of examples have been discussed in which organisms switch to more valuable or abundant diets as the relative abundance and/or quality of potential diets changes (see Stephens and Krebs 1986).

Analyses of food-web models with relatively simple structures have suggested that a foraging switch has a major effect on population dynamics and community structure (Tansky 1978; Teramoto et al. 1979; Holt 1983; Sih 1984; MacNamara and Houston 1987; Wilson and Yoshimura 1994; Abrams and Matsuda 1996; Holt and Polis 1996; Sutherland 1996; McCann and Hastings 1997; Křivan 2000; Post et al. 2000; Kondoh 2003a,b). For example, in a two prey–one predator system, the adaptive diet choice of the predator not only inhibits rapid growth of the prey but also allows minor prey to rally without predation pressure. This prevents apparent competition (Holt 1977), that is, a negative indirect effect between prey species sharing the same predator, leading to species extinctions (Tansky 1978; Teramoto et al. 1979; Abrams and Matsuda 1996), and enhances the coexistence of intraguild prey and predators (Holt and Polis 1997; McCann and Hastings 1997; Křivan 2000) or competing prey species (Wilson and Yoshimura 1994; McCann and Hastings 1997).

More recent studies have examined how food-web flexibility arising from the adaptive foraging switch by the predator alters the dynamics of more complex multi prey–multi predator systems (Pelletier 2000; Kondoh 2003a,b). These studies have revealed that the stability of food webs comprised of adaptive foragers (i.e. adaptive food webs) responds to increasing complexity in a way that is completely different from that of food webs without adaptive foragers. More specifically, an adaptive foraging switch enhances the persistence of complex food webs (Pelletier 2000; Kondoh 2003a,b) and even generates a positive complexity–stability relationship (adaptive food-web hypothesis; Kondoh 2003a,b).

In the following sections, I present dynamic food-web models after Kondoh (2003a,b) to demonstrate the effects of an adaptive foraging switch on population dynamics and the complexity–stability relationship. The model analysis examines how the relationship changes with changing levels of foraging adaptation. Species richness and connectance are used as indices of food-web complexity. Population stability is measured by population persistence, which is defined as the probability that a randomly chosen species goes extinct within a given time period (Kondoh 2003a).

Model and results

Model description

A food web is comprised of [pN] basal species and [(1 − p)N] nonbasal species, where N is the number of species and p is the fraction of basal species. For simplicity, I consider a case where half of the species are basal species ($p = 1/2$). The basal species persist on their own, while the other species rely energetically on other species (either basal or nonbasal species). Nonbasal species consume at least one resource species, which is randomly chosen from the web. Basal species are never connected to each other by a trophic link. All other pairs are connected with the connection probability C. If a link connects a basal species with a nonbasal species, the latter species always eats the former. The direction of a link between nonbasal species is randomly determined (either prey or predator with the same probability of 0.5). Note that the expected connectance of a food web with species richness N and connection probability C is given by $[(1 − p) + C\{(N − 3)/2 − p(pN − 2)\}]/N$. The behavior of the adaptive food web is given by combining population dynamics and adaptive dynamics (see Appendix for details).

I use the Lotka–Volterra model for population dynamics (Appendix). The per-capita predation rate of predator i on prey j is determined by interaction-specific foraging efficiency (f_{ij}) and the foraging effort of the predator (a_{ij}), $\alpha_{ij} = f_{ij}a_{ij}$. A predator specifies (nonbasal species) has a fixed amount of "foraging effort (finding effort or capturing effort)" ($\sum_{j \in \text{sp.}i\text{'s resource}} a_{ij} = \text{constant} = 1$) and can increase the predation rate on a prey species by allocating its foraging effort (a_{ij}) to that prey species. An increase in foraging effort to one prey species is associated with a decrease in the total foraging effort allocated to all other potential prey species.

I assumed that fraction F of the nonbasal species represents adaptive foragers, which allocate more effort to a prey species that offers higher gain per unit effort (i.e. product of density, X_i, and foraging efficiency, f_{ij}). The dynamic equation for this simple decisionmaking rule should meet the following conditions: (1) more effort is allocated to a prey species that is more abundant (larger X_j) or has a higher predation efficiency (higher f_{ji}), (2) the total predation effort of a prey species ($\sum_{k \in \text{sp.}i\text{'s resource}} a_{ik}$) is kept constant over time, (3) the foraging effort is within the limited range of $0 \leq a_{ij} \leq 1$. For adaptive dynamics I use the equation that is used in my previous studies (Appendix). Each adaptive consumer is assigned with a constant adaptation rate, G_i (Kondoh 2003a). The remaining fraction $(1 − F)$ of species in the web does not have the ability to make an adaptive foraging switch. These predators allocate their foraging effort equally to all potential prey species [$a_{ij} = \text{constant} = 1/(\text{number of potential consumer species of species } i)$].

Dynamic structure of an adaptive food web

The average number of potential prey per predator increases with increasing food-web complexity (potential connectance and species richness). How the prey number changes with changing food-web complexity depends on the ability of the predator to forage adaptively (Figure 11.1; see Kondoh 2003a,b). A nonadaptive forager allocates its foraging effort to all potential prey species. Therefore, the expected number of prey species utilized by a nonadaptive predator (defined by a prey to which a certain level of effort, $a_{ij} > 10^{-13}$, is allocated) is given by $[1 + \{C(N − 2)(N + Np − 1)/2 \times (N − 1)\}]$. This is an increasing function of C and N, suggesting that the prey number per predator increases with increasing food-web complexity. In contrast, an adaptive forager may consume only a fraction of the potential prey species, as prey species of low quality or quantity are discarded from the actual diet. Indeed, in food webs comprised of adaptive foragers, the number of prey species per adaptive predator is skewed to smaller numbers.

A food-web-level consequence of an adaptive foraging switch is a reduction in the number of

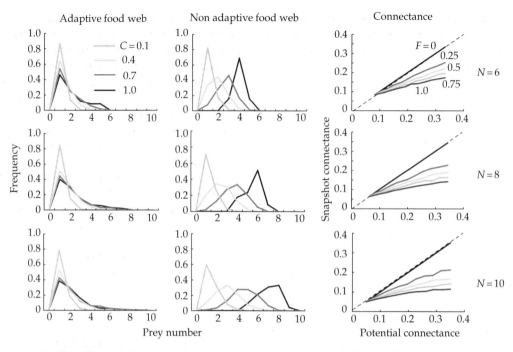

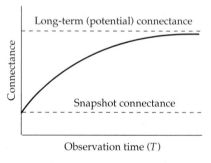

Figure 11.1 The effect of foraging adaptation on food-web structure. Number of prey per predator in a food web with varying food-web complexity ($N = 6, 8, 10$ and $C = 0.0–1.0$) in the presence or absence of adaptive forager. Snapshot connectance in a food web with varying species richness ($N = 6, 8, 10$) and potential connectance ($C = 0.0–1.0$) for varying fraction ($F = 0, 0.25, 0.5, 0.75, 1$) of adaptive forager. $p = 0.5$, $G = 0.25$. The dotted line represents (potential connectance) = (snapshot connectance).

active links and realized connectance, since an adaptive forager tends to use only a small fraction of the potential prey species. We can define "snapshot connectance" as the proportion of active trophic links ($a_{ij} > 10^{-13}$). Snapshot connectance is given as a function of species richness, potential connectance, and a fraction of adaptive foragers (Figure 11.1). It increases with increasing complexity (potential connectance and species richness), reaching an asymptote at high complexity levels. In general, snapshot connectance decreases with an increase in the fraction of adaptive foragers.

In an adaptive food web, a trophic link that is inactive at a given time point may be activated at another point, triggered by population fluctuations or changes in parameters such as foraging efficiency (Kondoh 2003a; Kondoh submitted). The number of active links is therefore described not as a timescale-independent constant but as a cumulative function of observation time (Kondoh submitted; Figure 11.2). This temporally varying

Figure 11.2 The "observed" connectance described as a cumulative function of observation time (T). Snapshot connectance and potential connectance are given as the connectance at $T = 0$ and infinity, respectively.

food-web topology makes connectance timescale dependent (Kondoh submitted). If the number of trophic links is evaluated in a short time period, then "observed" connectance should be identical to snapshot connectance. The number of active links increases with an increased timescale of observation, as a link is more likely to be activated during that time period, and the connectance evaluated from an infinitely long observation (referred to as "long-term connectance" hereafter) should be identical to potential connectance (Kondoh submitted).

Population dynamics in an adaptive food web

The effects of foraging adaptations on the relationship between complexity (potential connectance and species richness) and population stability are evaluated by comparing the relationships for a varying fraction of adaptive foragers (F) and adaptation rate (G). I used population persistence (P_p), the probability that a species randomly chosen from an N-species community does not become extinct within a given time period, as an index of population stability. Population persistence is given by $P_p(C, N) = P_c(C, N)^{1/N}$, where P_c is the probability that no species becomes extinct within the time period and is directly evaluated by measuring the proportion of food-web models where no species are lost within a time period ($t = 10^5$) in an assemblage of 10,000 randomly generated models.

The relationship in the absence of adaptive foraging is examined by setting $F = 0$, $G = 0$ and $a_{ij} = 1/$(number of potential prey species of predator i). The average foraging effort allocated

to a resource decreases with increasing species richness or connection probability as the expected number of resource species increases. The model analysis shows that increasing species richness (N) or connection probability (C) always decreases population persistence (Figure 11.3). A population is more likely to go extinct in a web with higher C or N when parameter values change over time.

In the presence of foraging adaptation, a completely different complexity–stability relationship emerges (Figure 11.3). When the predator species are all adaptive ($F = 1$) and their adaptation rate is sufficiently high ($G = 0.25$), population persistence increases with increasing species richness or connection probability (Kondoh 2003a,b). The positive relationships become less clear with a decreasing fraction of adaptive foragers (F) or decreasing adaptation rate (G), confirming the role of adaptive foragers (Kondoh 2003a,b).

Adaptive food-web hypothesis

Relationship between species richness and population persistence

The effect of changing species richness on population stability depends on the fraction of adaptive foragers and their adaptation rate (Kondoh 2003a,b).

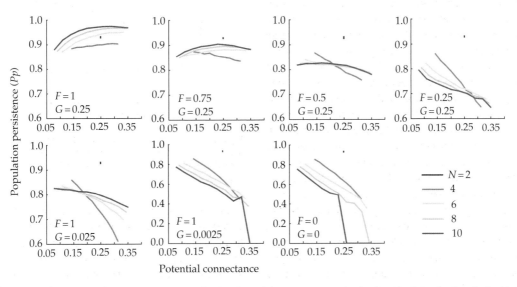

Figure 11.3 The relationship between potential connectance and population persistence in food webs with varying fraction of adaptive foragers ($F = 0$, 0.25, 0.5, 0.75, 1) and adaptation rates ($G = 0.25$, 0.025, 0.0025).

When there are only a few adaptive foragers or when their switching ability is low (i.e. low adaptation rates), population persistence tends to decrease with increasing species richness (Kondoh 2003a,b). This implies that, for a given potential connectance, a species is more likely to become extinct in a larger food web. In contrast, when the fraction of adaptive foragers is sufficiently high and the ability of foraging adaptation is high, the relationship is positive (Kondoh 2003a,b). A population is less likely to go extinct in a larger food web. Foraging adaptation inverts the negative species richness–stability relationship.

The positive relationship that emerges in the presence of adaptation (Kondoh 2003a,b) does not necessarily mean that a larger food web is less likely to lose a species (i.e. higher community persistence of a larger food web). Even if population persistence is independent of species richness, a food web with more species is more susceptible to species losses simply because it has more species. The relationship between species richness and community persistence, therefore, depends on the relative intensities of the opposite effects of increasing species richness on community persistence. Community persistence, that is, the probability that a food web will not lose any species, will be higher in a community with more species only if the stabilizing effect that arises from foraging adaptation is stronger than the negative effect of having more species. Indeed, in the present model, community persistence decreases with increasing species richness even when adaptive foragers are present, suggesting that the negative effect is stronger than the positive effect.

Relationship between connectance and population persistence

In the presence of adaptive foragers, connectance is a timescale-dependent variable, as the topology of an adaptive food web changes over time (Kondoh submitted). This implies that the connectance–stability relationship is also timescale dependent and should, therefore, be separately defined for each timescale (Kondoh submitted). The connectance–stability relationship for each timescale is summarized as follows.

The relationship between potential connectance and population persistence depends on the fraction of adaptive foragers and their adaptation rate (Kondoh 2003a,b). In the absence of adaptation, increasing the potential connectance tends to lower population persistence (Kondoh 2003a,b), implying that a food web comprised of species with a wider resource range is more likely to lose species. In contrast, in the presence of adaptation, the relationship is positive (Kondoh 2003a,b). This suggests that a species is less likely to go extinct when the species in the food web have, on average, a wider range of prey species. As potential connectance represents connectance that is observed by infinitely long observation, the result can be interpreted as a relationship between long-term connectance and population persistence (Kondoh submitted).

For a given fraction of adaptive foragers and a given adaptation rate, snapshot connectance always increases with increasing potential connectance (Kondoh 2003a; Figure 11.1). Since the relationship between potential connectance and snapshot connectance is positive, the sign of the relationship between snapshot connectance and population persistence should be the same as the relationship between potential connectance and population persistence. The relationship is positive in the presence of adaptive foragers (Kondoh submitted), while in the absence of adaptation, the relationship is negative (Kondoh submitted). Note that the positive relationship is stronger than the relationship between potential connectance and population persistence, as snapshot connectance is a saturating function of potential connectance (Kondoh submitted).

The applicability of the adaptive food-web hypothesis to a wider range of food-web models is still an open question, because only a few studies have attempted to reexamine the hypothesis in other food-web models. In my own tests, a similar pattern has been obtained for a variety of food-web models with different assumptions. The models in which the hypothesis holds include models using food-web topology of random (Kondoh 2003a,b), cascade (Kondoh 2003a,b), and niche models (unpublished data; but see Brose et al. 2003; where the adaptive food-web hypothesis does not hold as another stability index is used), models using trophic interactions of Holling's Type I and II (Kondoh 2003*b*; unpublished data) functional responses, and models with different fractions of species with a positive intrinsic growth rate (Kondoh 2003*a*,*b*). Thus, the hypothesis appears to be robust for a wide range of food-web models.

Mechanisms creating the positive relationship

It remains difficult to provide a formal explanation of how a foraging switch turns the previously negative complexity–stability relationship into a positive relationship, as the present analysis is based on numerical calculations. However, it is possible to speculate on possible mechanisms by making a gross approximation of how the extinction process is affected by food-web complexity in an adaptive food web.

Consider a population whose density becomes so low that further reductions in the population level will lead to extinction of the population. In this case, there are only two reasons for the species to become extinct—a low growth rate owing to low resource consumption and a high mortality due to overconsumption by the predator. In the presence of adaptive food choice, increasing complexity can enhance population persistence by either increasing resource availability or preventing overconsumption.

An increase in food-web complexity may enhance resource availability for an adaptive forager. Consider a predator species in a food web with species richness N and connectance C. The expected number of potential prey species of the predator is $[1 + (N - 2)C/2]$, which increases with either increasing C or increasing N. In the presence of adaptive diet choice, most trophic links are inactive and the abundances of a potential prey and a potential predator are less correlated with each other. If the predator is capable of switching to the most efficient resource, a larger number of potential prey will increase the probability that the consumer finds a more valuable or more abundant resource. This increases the growth rate of the predator and thus decreases its extinction probability. The stabilizing effect of having more potential prey has been discussed in previous studies (see Petchey 2000).

Another mechanism that creates a positive relationship in an adaptive food web is related to predation pressure. Without adaptive foraging, prey species that share the same predator are unlikely to coexist as a result of predator-mediated negative indirect effects between prey species (apparent competition; Holt 1977). This implies that if increasing complexity increases the number of prey shared by the same predator, then it is likely to increase the probability of species extinction. In contrast, in the presence of foraging adaptation, increasing complexity may prevent extinction of a minority species caused by predation pressure. The explanation for this is as follows. Consider a prey species, whose abundance is low, and an adaptive predator. In a food web with N species and potential connectance C, the expected number of potential predator species for a prey species is given by $[(N - B - 1)C/2]$. If the predator does not have an alternative resource, it will consume the prey irrespective of its abundance. This implies that a prey population whose abundance is low is still susceptible to predation pressure. Such trophic interactions destabilize prey–predator systems. If there is an alternative resource, a predator is likely to shift to another prey and, therefore, the focal prey species is unlikely to be consumed. Therefore, the probability that a population with low density is not consumed can be grossly approximated by the probability that a predator has an alternative prey species. The probability is given by $[1 - (1 - C/2)^{(N-2)}]^{[(N-1)C/2]}$, which increases with increasing food-web complexity when the complexity is sufficiently high, suggesting that food-web complexity potentially prevents species extinction due to overconsumption.

Further studies are required to understand the mechanism that creates the positive complexity–stability relationship in the adaptive food web, as the present section only provides a few possibilities. There may be other mechanisms. For example, complexity and foraging adaptation may alter the dynamic trajectory of the system and change the probability that the system will approach its limits. In some food-web topologies, a diet switch can stabilize population dynamics, whereas other studies suggest that adaptive dynamics may destabilize population dynamics and increase the extinction probability. It is thus necessary to extend the studies in which simple food webs are used, to more complex systems.

Food-web complexity and an ecosystem service

Although a complexity–stability relationship is a community-level phenomenon, it is also possible to regard this relationship as representative of an interbiological level interaction between a community and a population. A food web is a biological unit that determines the persistence of the species within the web. The population persistence can therefore be viewed as an index that represents the ability of an ecosystem to maintain a population (see Dunne et al. 2002 for a similar argument). High species persistence indicates that a population receives a larger "service" from the community. Applying this idea to the present study, the adaptive food-web hypothesis suggests that in the presence of adaptive foraging, a more complex food web provides a larger community service, supporting the species in the population.

The positive relationship between species richness and population persistence implies that community service is related to maintaining species and species richness in an interesting way. On the one hand, as suggested by the positive complexity–stability relationship, population persistence increases with increasing species richness (i.e. community service is greater in a larger food web). On the other hand, species richness will be higher when community service, which is measured by population persistence, is higher. These two interactions taken together imply positive feedback between species richness and population persistence (Figure 11.4; Kondoh 2003*a*). Species richness is more likely to be maintained in a food web with more species, while high species richness contributes to high population persistence. This implies that a community with high species richness is self-sustaining (Kondoh 2003*a*).

This self-sustainability of a complex food web, however, does not necessarily mean that a more complex food web will be less susceptible to loss of species. This is because positive feedback may operate in reverse. In an adaptive and complex

Figure 11.4 The positive feedback between food-web complexity and population persistence. Species richness enhances population persistence through the complexity–adaptation interactive effect, while increased population persistence maintains the species richness.

food web, a species is less likely to be lost owing to the high population persistence that arises from the interactive effects of an adaptive foraging switch and high food-web complexity. However, once a species is lost, the ecosystem service, which is measured in terms of population persistence, decreases, thereby enhancing further species extinctions. This feedback lowers species richness and population persistence at the same time, resulting in cascading extinctions. A complex and adaptive food web may be more fragile when it is under strong pressure, lowering species richness.

Conclusion

The adaptive food-web hypothesis suggests that adaptation, a distinguishing feature of organisms at individual and population levels, potentially resolves the community-level paradox of how a complex food web persists in nature. An adaptive foraging switch provides a food web with flexibility, which enhances population persistence. Furthermore, this stabilizing effect is stronger in a more complex food web, resulting in a positive complexity–stability relationship.

Changes in interaction strength caused by adaptive responses to temporal population variability are a realistic feature of trophic interactions, and have been well studied in behavioral biology. In this sense, the adaptive food-web hypothesis is in line with previous studies that have asserted that a "realistic" food-web structure is the key to understanding the persistence of complex food webs (Yodzis 1981; De Ruiter et al. 1995; Roxburgh and Wilson 2000; Neutel et al. 2002; Emmerson and Raffaelli 2004). The adaptive food-web hypothesis, however, differs from previous hypotheses in three important ways.

First, the adaptive food-web hypothesis predicts a positive relationship between food-web complexity and population stability, whereas most previous studies have either shown that a complex food web can be as stable as a less complex food web or have not dealt with the complexity–stability relationship (Yodzis 1981; De Ruiter et al. 1995; Roxburgh and Wilson 2000; Neutel et al. 2002; Emmerson and Raffaelli 2004, but see DeAngelis 1975; Nunney, 1980). The positive relationship predicted by the present hypothesis has the potential to drive a paradigm shift in the issue of biodiversity maintenance, as it leads to a completely novel view that complex food webs are self-sustaining and supported by positive feedback between complexity and population stability.

Second, the adaptive food-web hypothesis shows the importance of dynamic food-web structure in biodiversity maintenance. The view that biodiversity is maintained in a food web in which the structure is changing continuously implies that the persistence of species should be considered separately from the persistence of the food web. The persistence of species is supported by the flexible and nonpersistent structure of the food web, whereas a species is less persistent in a nonadaptive food web with a persistent architecture.

Third, the adaptive food-web hypothesis indicates that evolutionary history or individual-level experience plays a central role in determining population dynamics and thus the maintenance of a community. The adaptive foraging switch of a predator in general requires information regarding the relative quantity and/or quality of potential prey species so that it can evaluate which potential prey is energetically more valuable (Stephens and Krebs 1987). Lack of such information would result in incorrect responses by the predator to environmental changes. Therefore, a trophic system in which a predator performs poorly in terms of recognizing prey species or evaluating their relative value is likely to behave differently from

systems in which a predator performs well. More specifically, a trophic system will be less stable with more timid predators, such as invading predators, especially when the system is complex.

The prediction that adaptive foraging alters the complexity–stability relationship suggests that the relationship might be affected by the relative body sizes, mobility, spatial distribution, and generation time of prey and predator, as these factors all potentially influence the predator's capability of diet choice. A larger predator is less likely to be able to choose exclusively the small prey individual even if most profitable; a predator with high mobility is more likely to choose the most profit prey; even if the prey's body size is much smaller than the predator's body size, prey's clustered distribution may allow for predator's adaptive diet choice; generation time should influence the relative timescale of population dynamics to adaptive dynamics. Considering these effects through adaptation, it might be an interesting approach in food-web study to look at how these factors affect population dynamics. Further, it is notable that relative prey–predator body size, mobility, and spatial distribution seem to be very much different between aquatic and terrestrial systems. Predators tend to be larger than prey in aquatic system, while this pattern is less clear in terrestrial systems; terrestrial systems are more likely to form clustered distribution of less mobile organisms. This may imply that the habitat type may affect the complexity–stability relationship of food webs.

The food-web models, with which the adaptive food-web hypothesis is derived, are very much simplified in many ways. A simplifying assumption is the low interspecific heterogeneity within a web—there are only two types of behavior, nonadaptive foraging and adaptive foraging. While a nonadaptive forager does not switch diets at all, adaptive species are able to discriminate every potential resource species and their adaptive switches take place at the same rate. This is however unlikely in real nature. Adaptive rates should be heterogeneous within a web. A consumer with high capability of behavioral flexibility or short generation time would have a higher adaptive rate. An adaptive consumer may well discriminate a certain type of species, while being unable to discriminate another type of species. How does such heterogeneity affect the adaptive food-web hypothesis? Inclusion of the heterogeneity would be a next important step to generalize the present hypothesis.

Acknowledgment

I thank Jennifer Dunne for her valuable comment on an earlier manuscript. This research is partly supported by Japan Society for the Promotion of Science Research Fellowship for Young Scientists and the Grant for the Biodiversity Research of the twenty-first Century COE (A14).

Appendix

The dynamics of biomass X_i of species i $(1, \ldots, N)$ is described by:

$$\frac{dX_i}{dt} = R_i + \sum_{j \in sp.i's\,resources}^{N} e_{ij}F_{ij} - \sum_{j \in sp.i's\,consumers}^{N} F_{j,i}$$

(11.1)

where R_i is the intrinsic population growth rate, the second and third terms are the biomass gain and loss due to trophic interactions, respectively,

and e_{ij} is the conversion efficiency of predator i eating prey j. R_i and F_{ij} are given by:

$$R_i = \begin{cases} X_i(r_i - s_iX_i) & \text{(for basal species)} \\ -m_iX_i & \text{(for nonbasal species)} \end{cases}$$

(11.2)

and

$$F_{ij} = \alpha_{ij}X_iX_j,$$

(11.3)

where r_i is the intrinsic growth rate of the basal species, s_i is the self-regulation term for the basal species, $-m_i$ is the mortality rate of the non basal species, and α_{ij} is foraging efficiency, which is affected by the behavior of either prey j or predator i.

The dynamics of foraging effort is described by:

$$\frac{da_{ij}}{dt} = G_ia_{ij}\left(f_{ij}X_j - \sum_{k \in sp.i's\,resource} a_{ik}f_{ik}X_k \right),$$

(11.4)

where G_i is the adaptation' rate of consumer i. In the analysis, f_{ij} is set to a random value between 0.0 and 1.0; $a_{ij}(0) = 1/$[number of potential prey for predator i]. Each numerical calculation starts with the following initial condition: $X_i = 0.0$–0.1, $r_i = 0.1$, $s_i = 0.1$, $e = 0.15$, $m_i = 0.01$.

Climate forcing, food web structure, and community dynamics in pelagic marine ecosystems

L. Ciannelli, D. Ø. Hjermann, P. Lehodey, G. Ottersen, J. T. Duffy-Anderson, and N. C. Stenseth

Introduction

The study of food webs has historically focused on their internal properties and structures (e.g. diversity, number of trophic links, connectance) (Steele 1974; Pimm 1982; Cohen et al. 1990). A major advance of these investigations has been the recognition that structure and function, within a food web, are related to the dynamic properties of the system (Pimm 1982). Studies that have focused on community dynamics have done so with respect to internal forcing (e.g. competition, predation, interaction strength, and energy transfer; May 1973), and have lead to important advances in community ecology, particularly in the complex field of community stability (Hasting 1988). During the last two decades, there has been increasing recognition that external forcing—either anthropogenic (Parsons 1996; Jackson et al. 2001; Verity et al. 2002) or environmental (McGowan et al. 1998; Stenseth et al. 2002; Chavez et al. 2003)—can profoundly impact entire communities, causing a rearrangement of their internal structure (Pauly et al. 1998; Anderson and Piatt 1999; Steele and Schumacher 2000) and a deviation from their original succession (Odum 1985; Schindler 1985). This phenomenon has mostly been documented in marine ecosystems (e.g. Francis et al. 1998; Parsons and Lear 2001; Choi et al. 2004).

The susceptibility of large marine ecosystems to change makes them ideal to study the effect of external forcing on community dynamics. However, their expansive nature makes them unavailable to the investigational tools of food web dynamics, specifically *in situ* experimental perturbations (Paine 1980; Raffaelli 2000; but see Coale et al. 1996; Boyd et al. 2000). To date, studies on population fluctuations and climate forcing in marine ecosystems have been primarily descriptive in nature, and there have been few attempts to link the external forcing of climate with the internal forcing of food web interactions (e.g. Hunt et al. 2002; Hjermann et al. 2004). From theoretical (May 1973) as well as empirical studies in terrestrial ecology (Stenseth et al. 1997; Lima et al. 2002) we know that the relative strength of ecological interactions among different species can mediate the effect of external forcing. It follows that, different communities, or different stages of the same community, can have diverging responses to a similar external perturbation. In a marine context, such phenomenon was clearly perceived in the Gulf of Alaska, where a relatively small increase (about 2°C) in sea surface temperature (SST) during the mid-1970s co-occurred with a dramatic change of the species composition throughout the region (Anderson and Piatt 1999). However, in 1989 an apparent shift of the Gulf of Alaska to pre-1970s climatic conditions did not result in an analogous return of the community to the pre-1970s state (Mueter and Norcross 2000; Benson and Trites 2002). An even clearer example of uneven community responses following the rise and fall of an external perturbation is the lack of cod recovery from the Coast of Newfoundland and Labrador in spite of the 1992 fishing moratorium (Parsons and Lear 2001).

In this chapter, we review how marine pelagic communities respond to climate forcing. We emphasize the mediating role of food web structure (i.e. trophic interactions) between external climate forcing and species dynamics. This we do by summarizing studies from three different and well-monitored marine pelagic

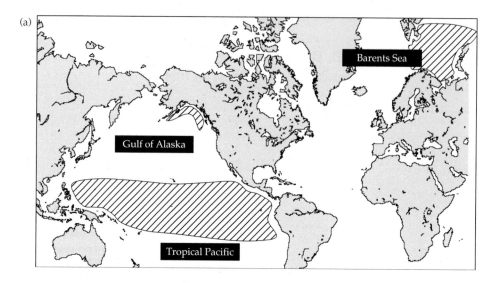

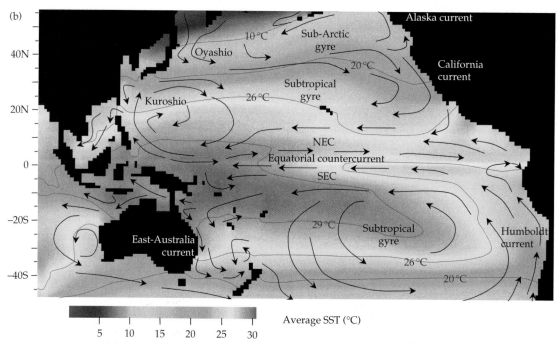

Figure 12.1 (a) Location of the three pelagic ecosystems reviewed in the present study. Also shown are the detailed maps of each system. (b) Tropical Pacific (TP). (NEC = North Equatorial Current, SEC = South Equatorial Current.) See plate 9. (c) Gulf of Alaska (GOA). (ACC = Alaska Coastal Current.) (d) Barents Sea (BS).

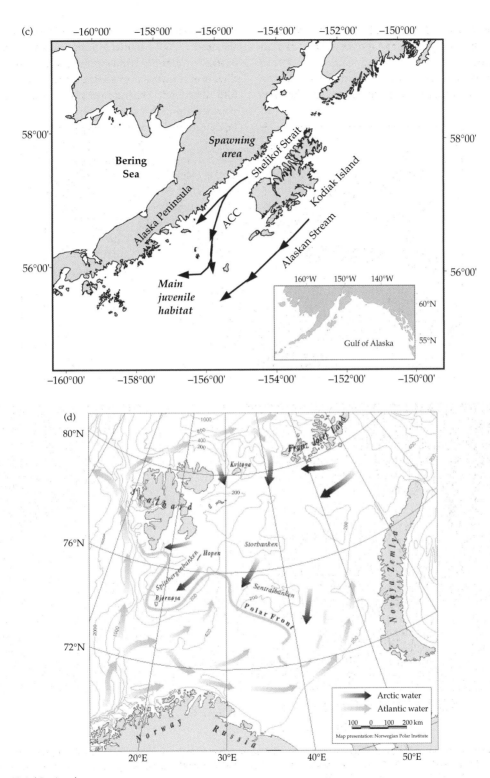

Figure 12.1 (*Continued*)

ecosystems (Figure 12.1(a)): (1) the Tropical Pacific (TP); (2) the western Gulf of Alaska (GOA), and (3) the Barents Sea (BS). These communities are strongly impacted by climate (Anderson and Piatt 1999; Hamre 2003; Lehodey et al. 1997, respectively), but have also fundamental differences in the way they respond to its forcing. In the GOA and (particularly) in the BS systems, food web interactions play a major role in determining the fate of their communities, while in the TP trophic forcing plays a minor role compared to the direct effects of climate. We suggest that such differences are to a large extent the result of dissimilar food web structures among the three pelagic ecosystems.

In this chapter we describe the physics, the climate forcing, and the food web structure of the investigated systems. We then examine their community dynamics in relation with the food web structures and climate forcing. The chapter ends with generalizations on how to link trophic structure and dynamics in large, pelagic, marine ecosystems. We emphasize climate and community processes occurring in the pelagic compartment at the temporal scales perceivable within a period less than a human generation (10–40 years). We recognize that the present review is based on information and data that were not originally meant to be used in community studies, and for this reason it is unbalanced in the level of information provided for each trophic assemblage. Typically, the information is available in greater detail for species that are commercially important. However, to our knowledge, this is the first explicit attempt to link external (climate) and internal (trophic) forcing in the study of community dynamics in large marine ecosystems (but see Hunt et al. 2002; Hjermann et al. 2004), and should be most relevant to advance the knowledge of structure and dynamics also in marine pelagic food webs.

The geography and the physics

Tropical Pacific

The physical oceanography of the TP, roughly between 20°N and 20°S, is strongly dominated by the zonal equatorial current systems (Figure 12.1(b); see Plate 9). Under the influence of the trade winds blowing from east to west, the surface water is transported along the same direction (north and south equatorial currents: NEC and SEC). During transport, surface water is warmed up and creates a warm pool with a thick layer (about 100 m) of water above 29°C on the western side of the oceanic basin. The warm pool plays a key role in the development of El Niño events (McPhaden and Picaut 1990). In the eastern and central Pacific, this dynamic creates an equatorial divergence with an upwelling of deep and relatively cold water (the "cold tongue") and a deepening thermocline from east to west. The general east–west surface water transport is counterbalanced by the north and south equatorial countercurrents (NECC and SECC), the equatorial undercurrent (EUC) and the retroflexion currents that constitute the western boundaries (Kuroshio and east Australia currents) of the northern and southern subtropical gyres. The TP presents a weak seasonality, except in the far western region (South China Sea and archipelagic waters throughout Malaysia, Indonesia, and the Philippines) that is largely under the influence of the seasonally reversing monsoon winds. Conversely, there is strong interannual variability linked to the El Niño Southern Oscillation (ENSO).

Western Gulf of Alaska

The Gulf of Alaska (herein referred to as GOA) includes a large portion of the sub-Arctic Pacific domain, delimited to the north and east by the North American continent, and to the south and west by the 50° latitude and 176° longitude, respectively (Figure 12.1(c)). In the present chapter we focus on the shelf area west of 150° longitude—the most studied and commercially harvested region of the entire GOA. The continental shelf of the GOA is narrow (10–150 km), and frequently interrupted by submerged valleys (e.g. the "Skelikof Sea Valley" between Kodiak and the Semidi Islands) and archipelagos (e.g. Shumagin Islands). The offshore surface (<100 m) circulation of the entire GOA is dominated by the sub-Arctic gyre, a counterclockwise circulation feature of the North Pacific. A pole-ward branch of the sub-Arctic gyre, flowing along the shelf edge, forms the Alaska current/Alaska stream. This current varies in width and speed along its course—from 300 km and 10–20 cm s^{-1} east of 150° latitude to 100 km and up to 100 cm s^{-1} in the GOA region (Reed and Schumacher 1987). The coastal surface circulation pattern of the GOA is dominated by the Alaska coastal current (ACC) flowing southwestward along the Alaska Peninsula. The ACC is formed by pressure gradients, in turn caused by freshwater discharge from the Cook Inlet area. The average speed of the ACC ranges around 10–20 cm s^{-1}, but its flow varies seasonally, with peaks in the fall during the period of highest freshwater discharge (Reed and Schumaker 1987). The ACC and its associated deep-water undercurrents, play an important biological role in the transport of eggs and larvae from spawning to nursery areas of several dominant macronekton species of the GOA (Kendall et al. 1996; Bailey and Picquelle 2002). Royer (1983) (cited in Reed and Shumacher 1987) suggested that the Norwegian coastal current is an analog of the ACC, having similar speed, seasonal variability, and biological role in the transport of cod larvae from the spawning grounds to the juvenile nursery habitats.

Barents Sea

The BS is an open arcto-boreal shelf-sea covering an area of about 1.4 million km^2 (Figure 12.1(d)). It is a shallow sea with an average depth of about 230 m (Zenkevitch 1963). Three main current systems flow into the Barents determining the main water masses: the Norwegian coastal current, the Atlantic current, and the Arctic current system (Loeng 1989). Although located from around 70°N to nearly 80°N, sea temperatures are substantially higher than in other regions at similar latitudes due to inflow of relatively warm Atlantic water masses from the southwest. The activity and properties of the inflowing Atlantic water also strongly influence the year-to-year variability in temperature south of the oceanic Polar front (Loeng 1991; Ingvaldsen et al. 2003), as does regional heat exchange with the atmosphere (Ådlandsvik and Loeng 1991; Loeng et al. 1992). The ice coverage shows pronounced interannual fluctuations. During 1973–75 the annual

maximum coverage was around 680,000 km^2, while in 1969 and 1970 it was as much as 1 million km^2. This implies a change in ice coverage area of more than 30% in only four years (Sakshaug et al. 1992). In any case, due to the inflow of warm water masses from the south, the southwestern part of the BS does not freeze even during the most severe winters.

Climate forcing

Pacific inter-Decadal Oscillation

The GOA and the TP systems are influenced by climate phenomena that dominate throughout the Pacific Ocean. These are the Pacific inter-Decadal Oscillation (hereon referred to as Pacific Decadal Oscillation, PDO; Mantua and Hare 2002), and the ENSO (Stenseth et al. 2003). The PDO is defined as the leading principal component of the monthly SST over the North Pacific region (Mantua et al. 1997). During a "warm" (positive) phase of the PDO, SSTs are higher over the Canadian and Alaskan coasts and northward winds are stronger, while during a cool phase (negative) the pattern is reversed (Figure 12.2; see also, Plate 10). The typical period of the PDO is over 20–30 years, hence the name. It is believed that in the last century there have been three phase changes of the PDO, one in 1925 (cold to warm), one in 1946 (warm to cold), and another in 1976 (cold to warm; Mantua et al. 1997), with a possible recent change in 1999–2000 (warm to cold) (McFarlane et al. 2000; Mantua and Hare 2002). The pattern of variability of the PDO closely reflects that of the North Pacific (or Aleutian Low) index (Trenberth and Hurrell 1994). The relationship is such that cooler than average SSTs occur during periods of lower than average sea level pressure (SLP) over the central North Pacific, and vice versa (Stenseth et al. 2003). It bears note that a recent study by Bond et al. (2004) indicates that the climate of the North Pacific is not fully explained by the PDO index and thus it has no clear periodicity.

El Niño Southern Oscillation

Fluctuations of the TP SST are related to the occurrence of El Niño, during which the equatorial surface waters warm considerably from the International Date Line to the west coast of South America (Figure 12.2). Linked with El Niño events is an inverse variations in SLP at Darwin (Australia) and Tahiti (South Pacific), known as the Southern Oscillation (SO). A simple index of the SO is, therefore, often defined by the normalized Tahiti minus Darwin SLP anomalies, and it has a

period, of about 4–7 years. Although changes in TP SSTs may occur without a high amplitude change of the SO, El Niño and the SO are linked so closely that the term ENSO is used to describe the atmosphere–ocean interactions over the TP. Warm ENSO events are those in which both a negative SO extreme and an El Niño occur together, while the reverse conditions are termed La Niñas

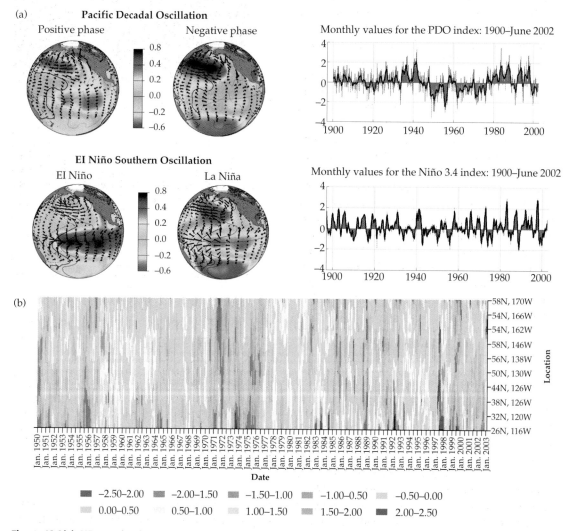

Figure 12.2(a) SST anomalies during positive and negative phases of the PDO (upper panel) and ENSO (lower panel), and time series of the climate index. During a positive PDO phase, SST anomalies are negative in the North Central Pacific (blue area) and positive in the Alaska coastal waters (red area) and the prevailing surface currents (shown by the black arrows) are stronger in the pole-ward direction. During a positive phase of the ENSO, SST anomalies are positive in the eastern Tropical Pacific and the eastward component of the surface currents is noticeably reduced. (b) Time series of temperature anomalies in different locations of the North Pacific. The graph shows area of intense warming (yellow and red areas) associated with the ENSO propagating to the North Pacific, a phenomenon termed El Niño North condition (updated from Hollowed et al. 2001). See Plate 10.

(Philander 1990; Stenseth et al. 2003). Particularly strong El Niño events during the latter half of the twentieth century occurred in 1957–58, 1972–73, 1982–83, and 1997–98.

Typically, the SST pattern of the TP is under the influence of interannual SO-like periodicity (i.e. 4–7 years), while the extra-TP pattern is under the interdecadal influence of the PDO-like periodicity (Zhang et al. 1997). However, El Niño/ La Niña events can propagate northward and affect the North Pacific as well, including the GOA system, a phenomenon known as Niño North (Figure 12.2; Hollowed et al. 2001). During the latter half of twentieth century, there have been five warming events in the GOA associated with the El Niño North: in 1957–58, 1963, 1982–83, 1993, and 1998. The duration of each event was about five months, with about a year lag between a tropical El Niño and the Niño North condition (Figure 12.2). The likelihood of an El Niño event to propagate to the North Pacific is related to the position of the Aleutian Low. Specifically, during a positive phase of the PDO, the increased flow of the Alaska current facilitates the movement of water masses from the transition to the sub-Arctic domain of the North Pacific, in turn increasing the likelihood of an El Niño North event (Hollowed et al. 2001). It has also been reported that the likelihood of El Niño (La Niña) events in the TP is higher during a positive (negative) phase of the PDO (Lehodey et al. 2003).

North Atlantic Oscillation

The BS is influenced by North Atlantic basin scale climate variability, in particular that represented by the North Atlantic Oscillation (NAO) (Figure 12.3; see also Plate 11). The NAO refers to a north–south alternation in atmospheric mass between the subtropical and subpolar North Atlantic. It involves out-of-phase behavior between the climatological low-pressure center near Iceland and the high-pressure center near the Azores, and a common index is defined as the difference in winter SLP between these two locations (Hurrell et al. 2003). A high (or positive) NAO index is characterized by an intense Icelandic Low and a strong Azores High. Variability

(a)

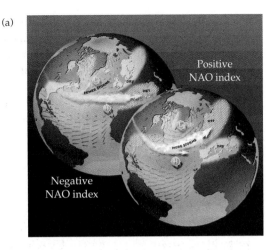

(b)

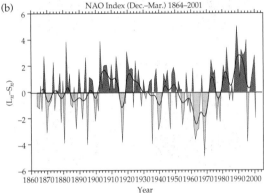

Figure 12.3 The NAO. (a) During positive (high) phases of the NAO index the prevailing westerly winds are strengthened and moves northwards causing increased precipitation and temperatures over northern Europe and southeastern United State and dry anomalies in the Mediterranean region (red and blue indicate warm and cold anomalies, respectively, and yellow indicates dry conditions). Roughly opposite conditions occur during the negative (low) index phase (graphs courtesy of Dr Martin Visbeck, www.ldeo.columbia.edu/~visbeck). (b) Temporal evolution of the NAO over the last 140 winters (index at www.cgd.ucar.edu/~jhurrell/nao.html). High and low index winters are shown in red and blue, respectively (Hoerling et al. 2001). See Plate 11.

in the direction and magnitude of the westerlies is responsible for interannual and decadal fluctuations in wintertime temperatures and the balance of precipitation and evaporation over land on both sides of the Atlantic Ocean (Rogers 1984; Hurrell 1995). The NAO has a broadband spectrum with no significant dominant periodicities (unlike ENSO). More than 75% of the variance of the NAO occurs at shorter than decadal time-scales (D. B. Stephenson, web page at www. met.rdg.ac.uk/cag/NAO/index.html). A weak peak in the power spectrum can, however, be detected at around 8–10 years (Pozo-Vazquez et al. 2000; Hurrell et al. 2003). Over recent decades the NAO winter index has exhibited an upward trend, corresponding to a greater pressure gradient between the subpolar and subtropical North Atlantic. This trend has been associated with over half the winter surface warming in Eurasia over the past 30 years (Gillett et al. 2003).

A positive NAO index will result in at least three (connected) oceanic responses in the BS, reinforcing each other and causing both higher volume flux and higher temperature of the inflowing water (Ingvaldsen et al. 2003). The first response is connected to the direct effect of the increasingly anomalous southerly winds during high NAO. Second, the increase in winter storms penetrating the BS during positive NAO will give higher Atlantic inflow to the BS. The third aspect is connected to the branching of the Norwegian Atlantic Current (NAC) before entering the BS. Blindheim et al. (2000) found that a high NAO index corresponds to a narrowing of the NAC towards the Norwegian coast. This narrowing will result in a reduced heat loss (Furevik 2001), and possibly in a larger portion of the NAC going into the BS, although this has not been documented (Ingvaldsen et al. 2003). It should be noted that the correlation between the NAO and inflow to and temperature in the BS varies strongly with time, being most pronounced in the early half of the twentieth century and over the most recent decades (Dickson et al. 2000; Ottersen and Stenseth 2001).

Food web structure

To facilitate the comparison of the three food webs, we have grouped the pelagic species of each system in five trophic aggregations: *primary producers, zooplankton, micronekton, macronekton, and apex predators*. This grouping is primarily associated with trophic role, rather than trophic level. Macronekton includes all large (>20 cm) pelagic

species that are important consumers of other pelagic resources (e.g. micronekton), but are preyed upon, for the most part, by apex predators. Micronekton consist of small animals (2–20 cm) that can effectively swim. Typically, macronekton, and to a smaller extent, micronekton and apex predators, include commercial fish species (tunas, cod, pollock, herring, and anchovies) and squids. In the following, we summarize available information on food web structure, covering for the most part trophic interactions, and, where relevant (e.g. TP), also differences in spatial distribution among the organisms of the various trophic assemblages.

Tropical Pacific

The TP system has the most diverse species assemblage and most complex food web structure among the three pelagic ecosystems included in this chapter (Figure 12.4). Part of the complexity of the TP food web is due to the existence of various spatial compartments within the large pelagic ecosystem. The existence of these compartments may ultimately control the relationships with (and accessibility to) top predators, and affect the community dynamics as well (Krause et al. 2003). In the vertical gradient, the community can be divided into epipelagic (0–200 m), mesopelagic (200–500 m), and bathypelagic groups (<500 m), the last two groups being subdivided into migrant and non-migrant species. All these groups include organisms of the main taxa: fish, crustacean, and cephalopods. Of course, this is a simplified view of the system as it is difficult to establish clear vertical boundaries, which are influenced by local environmental conditions, as well as by the life stage of species.

In addition to vertical zonation, there is a pronounced east–west gradient of species composition and food web structure in the TP. Typically, there is a general decrease in biomass from the intense upwelling region in the eastern Pacific toward the western warm pool (Vinogradov 1981). While primary productivity in both the western warm pool and the subtropical gyres is generally low, the equatorial upwelling zone is favorable to relatively high primary production and creates a large zonal

Tropical Pacific

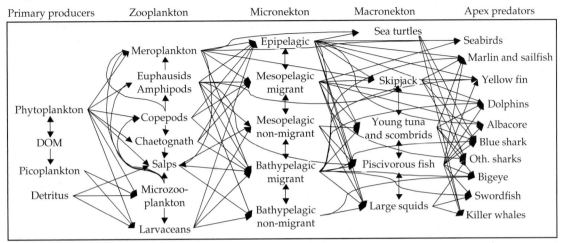

Gulf of Alaska

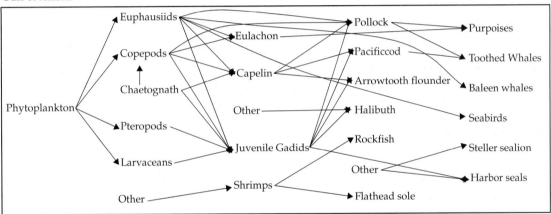

Barents Sea

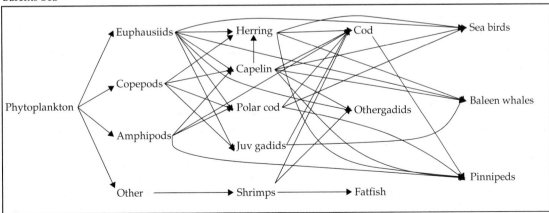

Figure 12.4 Simplified representation of the food web for each studied system. Arrows point from the prey to the predator DOM: Dissolved Organic Matter

band, in the cold tongue area, of rich mesotrophic waters. However, primary productivity rates in this area could be even higher as all nutrients are not used by the phytoplankton. This "high-nutrient, low-chlorophyll" (HNLC) situation is due for a large part to iron limitation (Coale et al. 1996; Behrenfeld and Kolber 1999). Another important difference between the east and west TP appear in the composition of plankton (both phyto and zooplankton). In regions where upwelling is intense (especially the eastern Pacific during La Niña periods), diatoms dominate new and export production, while in equatorial and oligotrophic oceanic regions (warm pool, subtropical gyres) a few pico- and nanoplankton groups (autotrophic bacteria of the microbial loop) dominate the phytoplankton community (Bidigare and Ondrusek 1996; Landry and Kirchman 2002).

Based on size, the zooplankton assemblage can be subdivided into micro- (20–200 μm), meso- (0.2–2.0 mm), and macrozooplankton (2–20 mm). Flagellates and ciliates dominate the microzooplankton group; however, nauplii of copepods are abundant in the eastern equatorial region, in relation with more intense upwelling. Pico- and nanoplankton are consumed by microzooplankton, which remove most of the daily accumulation of biomass (Landry et al. 1995; Figure 12.4). Copepods dominate the mesozooplankton group, as well as the entire zooplankton assemblage of the TP (Le Borgne and Rodier 1997; Roman et al. 2002). Gueredrat (1971) found that 13 species of copepods represented 80% of all copepod species in the equatorial Pacific. However, meso- and macrozooplankton include a very large diversity of other organisms, such as amphipods, euphausiids, chaetognaths, and larval stages (meroplankton) of many of species of molluscs, cnidaria, crustaceans, and fish. Another group that has a key role in the functioning of the pelagic food web but has been poorly studied is what we can name "gelatinous filter feeders." In particular, this group includes appendicularians (a.k.a., larvaceans) and salps. Salps and larvaceans filter feed mainly on phytoplankton and detritus. Though by definition, the zooplankton described above is drifting in the currents, many species undertake diel vertical migrations, mainly stimulated by the light intensity.

Fish, crustaceans (large euphausiids), and cephalopods dominate the micronekton group, with typical sizes in the range of 2–20 cm. These organisms, together with the gelatinous filter feeders, are the main forage species of the top and apex predators (Figure 12.4). Many species of zooplankton and micronekton perform diel vertical migrations between layers of the water column that are over 1,000 m apart. One important benefit of this evolutionary adaptation is likely a decrease of the predation pressure in the upper layer during daytime (e.g. Sekino and Yamamura 1999). The main epipelagic planktivorous fish families are Engraulidae (anchovies), Clupeidae (herrings, sardines), Exocœtidae (flyingfish), and small Carangidae (scads), but an important component also is represented by all juvenile stages of large-size species (Bramidae, Coryphaenidae, Thunnidae). The oceanic anchovy (*Enchrasicholinus punctifer*) is a key species in the epipelagic food web of the warm pool as it grows very quickly (mature after 3–5 months) and can become very abundant after episodic blooms of phytoplankton. Meso- and bathypelagic species include euphausiids, deep shrimps of the Sergestidae, Peneidae, Caridae, and numerous fish families, Myctophidae, Melamphaidae, Chauliodidae, Percichthyidae, and Stomiatidae. The micronekton consume a large spectrum of prey species among which the dominant groups are copepods, euphausiids, amphipods, and fish. More detailed analyses (Legand et al. 1972; Grandperrin 1975) showed that prey composition can differ substantially between micronekton species, especially in relation to predator–prey size relationships: smallest micronekton prey mainly upon copepods, medium size micronekton consumes more euphausiids, and large micronekton are mostly piscivorous.

Tuna dominate the macronekton in the TP food web, although this group also includes large-size cephalopods, and sea turtles. Skipjack tuna (*Katsuwonus pelamis*) is the most abundant and productive species of the TP and constitute the fourth largest fisheries in the world (FAO 2002; ~1.9 million tons per year). Juveniles of other tropical tuna, particularly yellowfin tuna (*Thunnus albacares*) and bigeye tuna (*Thunnus obesus*) are frequently found together with skipjack in the surface layer, especially around drifting logs that aggregate many epipelagic species. With these well-known species, there are many other scombrids (*Auxis* sp., *Euthynnus* spp., *Sarda* spp., *Scomberomorus* spp., *Scomber* spp.), and a large variety of piscivorous fish (Gempylidae, Carangidae, Coryphaenidae, Trichiuridae, Alepisauridae), and juveniles of apex predators (sharks, marlins, swordfish, and sailfish). The largest biomass of the most productive species, skipjack and yellowfin tuna, is in the warm waters of the Western and Central Pacific Ocean (WCPO), but warm currents of the Kuroshio and east Australia extend their distribution to 40°N and 40°S (roughly delineated by the 20°C surface isotherm). Most macronekton species are typically predators of the epipelagic micronekton but many of them take advantage of the vertical migration of meso- and bathypelagic species that are more particularly vulnerable in the upper layer during sunset and sunrise periods.

Apex predators of the TP food web include adult large tuna (yellowfin *T. albacares*, bigeye *Thunnus obesus* albacore *Thunnus alalunga*), broadbill swordfish (*Xiphias gladus*), Indo-Pacific blue marlin (*Makaira mazara*), black marlin (*Makaira indica*), striped marlin (*Tetrapturus audax*), shortbill spearfish (*Tetrapturus angustirostris*), Indo-Pacific sailfish (*Istiophorus platypterus*), pelagic sharks, seabirds, and marine mammals. The diets of apex predator species reflect both the faunal assemblage of the component of the ecosystem that they explore (i.e. epi, meso, bathypelagic) and their aptitude to capture prey at different periods of the day (i.e. daytime, nightime, twilight hours). All large tuna species have highly opportunistic feeding behavior resulting in a very large spectrum of prey from a few millimeters (e.g. euphausiids and amphipods) to several centimeters (shrimps, squids and fish, including their own juveniles) in size. However, it seems that differences in vertical behavior can be also identified through detailed analyses of the prey compositions: bigeye tuna accessing deeper micro- and macronekton species. Swordfish can also inhabit deep layers for longer periods than most apex predators. This difference in vertical distributions is reflected in the diets, swordfish consuming a larger proportion of squids

(e.g. Ommastrephidae, Onychoteuthidae) than the other billfishes. Blue sharks consume cephalopods as a primary component of their diet and various locally abundant pelagic species (Strasburg 1958, Tricas 1979). Whitetip and silky sharks are omnivorous. They feed primarily on a variety of fish including small scombrids, cephalopods, and to a lesser extent, crustaceans. The whitetip sharks consume also a large amount of turtles (Compagno 1984) and occasionally stingrays and sea birds. Thresher, hammerhead, and mako sharks feed on various piscivorous fish including scombrids and alepisaurids, and cephalopods.

Marine mammals encountered in the TP system permanent baleen whales, toothed whales, and dolphins. However, most of baleen whales are not permanent in the tropical pelagic food web as they only migrate to tropical regions for breeding, while they feed in polar waters during the summer. The diets of toothed whales (killer whale, sperm whale, and short-finned pilot whale) are mainly based on squids (the sperm whale often taking prey at considerable depths), and fish. Killer whales are also known to prey on large fish such as tuna and dolphinfish and sometime small cetaceans or turtles. All the dolphins consume mesopelagic fish and squid. Spotted and common dolphins are known also to prey upon epipelagic fish like flying fish, mackerel, and schooling fish (e.g. sardines).

Tropical seabirds feed near or above (flyingfish, flying squids) the water surface, on a large variety of macroplankton and micronekton (mainly fish and squids), including vertically migrating species likely caught during sunset and sunrise periods. They have developed a remarkably efficient foraging strategy associated with the presence of subsurface predators (mostly tuna) that drive prey to the surface to prevent them from escaping to deep waters. Therefore, as stated by Ballance and Pitman (1999), "subsurface predators would support the majority of tropical seabirds and would indirectly determine distribution and abundance patterns, and provide the basis for a complex community with intricate interactions and a predictable structure. This degree of dependence has not been found in non-tropical seabirds." An example of such potential interaction is that a decrease in tuna abundance would not have a positive effect on seabirds despite an expected increase of forage biomass, but instead would have a negative effect as this forage becomes less accessible.

Western Gulf of Alaska

The trophic web of the GOA includes several generalist (e.g. Pacific cod) and opportunistic (e.g. sablefish) feeders (Figure 12.4). In addition, different species exhibit a high diet overlap, such as juvenile pollock, and capelin, or the four dominant macroneckton species (arrowtooth flouder, Pacific halibut, cod, and pollock; Yang and Nelson 2000). As with other systems included in our review, diet patterns can change during the species ontogeny. In general, zooplanktivory decreases in importance with size, while piscivory increases. Also, within zooplanktivorous species/stages, euphausiids replace copepods as the dominant prey of larger fish. In the GOA, as well as in the BS systems, the structure of the food web is also influenced by climate forcing, as shown by noticeable diet changes of many macronekton and apex predators over opposite climate and biological phases.

In contrast to tropical regions, primary production in Arctic and sub-Arctic marine ecosystems varies seasonally with most annual production confined to a relatively short spring bloom. This seasonal pattern is mainly the result of water stratification and increased solar irradiance in the upper water column during the spring. In the GOA system, primary productivity varies considerably also with locations. Parsons (1987) recognized four distinguished ecological regions: (1) the estuary and intertidal domain, 150 $g C m^{-2}$ per year; (2) the fjord domain, 200 $g C m^{-2}$ per year, (3) the shelf domain, 300 $g C m^{-2}$ per year, (4) the open ocean domain, 50 $g C m^{-2}$ per year. These values, particularly for the shelf area, are considerably higher than those observed in the BS and similar to those observed in the east TP. A number of forcing mechanisms can explain the high productivity in the GOA system, including a seasonal weak upwelling (from May to September, Stabeno et al. 2004), strong tidal currents with resulting high tidal mixing, high nutrient discharge from fresh water run off, and the presence of a strong

pycnocline generated by salinity gradients (Sambrotto and Lorenzen 1987).

Euphausiids, copepods, cnidarians, and chaetognaths constitute the bulk of the zooplankton assemblage of the GOA food web. Given the paucity of feeding habits data at very low trophic levels, we assume that the zooplankton species feed mainly on phytoplankton (Figure 12.4). This is a common generalization in marine ecology (e.g. Mann 1993), which, nonetheless, underscores the complex trophic interactions within the phytoplankton and zooplankton assemblage (e.g. microbial loop). However, microbial loop organisms are particularly important in oligotrophic environments, such as the west TP system, and supposedly play a minor role in more productive marine ecosystems, such as the GOA, the BS systems, and the cold tongue area of the TP system.

The most abundant and common micronekton species in the GOA food web are capelin (*Mallotus villosus* Müller), eulachon (*Thalicthys pacificus*), sandlance (*Ammodytes hexapterus*), juvenile gadids (including Pacific cod and walleye pollock), Pacific sandfish (*Trichodon trichodon*), and pandalid shrimps (*Pandalus* spp.). Pacific herring (*Clupea harengus pallasi*) are also important, but their presence is mainly limited to coastal waters and to the northern and eastern part of the GOA. The range of energy density of these micronekton species is very broad (Anthony et al. 2000), as it is also the nutritional transfer to their predators. Eulachon have the highest energy density (7.5 $kJ g^{-1}$ of wet weight), followed by sand lance and herring (6 $kJ g^{-1}$ of wet weight), capelin (5.3 $kJ g^{-1}$ of wet weight), Pacific sandfish (5 $kJ g^{-1}$ of wet weight), and by juvenile cod and pollock (4 $kJ g^{-1}$ of wet weight). Juvenile pollock feed predominantly on copepods (5–20%) and euphausiids (69–81%), the latter becoming more dominant in fish larger than 50–70 mm in standard length (Merati and Brodeur 1996; Brodeur 1998). Other common, but less dominant prey include fish larvae, larvaceans, pteropods, crab larvae, and hyperid amphipods. Capelin diet is similar to that of juvenile pollock, feeding mostly on copepods (5–8%) and euphausiids (72–90%) (Sturdevant 1999). Food habits of other micronekton species are poorly known in this area; however, it is reasonable to assume that most of their diet is also based on zooplankton species.

The GOA shelf supports a rich assemblage of macronekton, which is the target of a large industrial fishery. The majority are demersal species (bottom oriented), such as arrowtooth flounder (*Atherestes stomias*), halibut (*Hippoglossus stenolepis*), walleye pollock (*Theragra chalcogramma*), Pacific cod (*Gadus macrocephalus*), and a variety of rockfishes (*Sebastes* spp.). Walleye pollock currently constitute the second largest fisheries in the world (FAO 2002, second only to the Peruvian anchoveta); however, the bulk of the landings comes from the Bering Sea. In the GOA, arrowtooth flounder (*A. stomias*) presently dominate the macronekton assemblage, with a spawning biomass estimated at over 1 million metric tones.

The macronekton species of the GOA food web can be grouped in piscivorous (arrowtooth flounder, halibut), zooplanktivorous (pollock, atka mackerel, some rockfish), shrimp-feeders (some rockfish, flathead sole), and generalist (sablefish and cod) (Yang and Nelson 2000). Within these subcategories there have been noticeable diet changes over time. For example, in recent years (1996) adult walleye pollock diet was based primarily on euphausiids (41–58%). Other important prey include copepods (18%), juvenile pollock (10%), and shrimps (2%). However, in 1990 shrimps where more dominant in pollock diet (30%) while cannibalism was almost absent (1%) (Yang and Nelson 2000). Adult Pacific cod has also undergone similar diet changes. Recently they feed on benthic shrimps (20–24%), pollock (23%), crabs (tanner crab *Chionoecetes bairdi*, pagurids, 20–24%), and eelpouts (Zoarcidae). In contrast, during the early 1980s they fed primarily on capelin, pandalid shrimps, and juvenile pollock (Yang 2004). Arrowtooth flounder feed predominantly on fish (52–80%), among which pollock is the most common (16–53%), followed by capelin (4–23%) and herring (1–6%). Shrimps are also well represented in the diet of arrowtooth flounders (8–22%), however their importance, together with that of capelin, has also decreased in recent years, while that of pollock has increased (Yang and Nelson 2000). Pacific halibut feed mainly on fish, particularly pollock (31–38%). Other common prey include crabs (26–44%, tanner crab and pagurids) and cephalopods (octopus 3–5%). Flathead sole feed almost exclusively on shrimps (39%) and brittle stars (25%). Atka mackerel is zooplanktivorous (64% copepods, 4% euphausiids), but also feed on large jelly fish (Scyphozoa 19%). Sablefish can equally feed on a variety of prey, including pollock (11–27%), shrimps (5–11%), jellyfish (9–14%), and fishery offal (5–27%). The rockfish of the GOA food web can be grouped among those that feed mainly on shrimps (rougheye, shortspine), euphausiids (Pacific ocean perch, northern, dusky), and squids (shortraker).

Apex predators of the GOA include seabirds, pinnipeds, and cetaceans. Among seabirds the most common are murres (a.k.a. common guillemot) (*Uria aalge*), black-legged kittiwakes *Rissa tridactyla*, a variety of cormorants (double-crested, red-faced, and pelagic), horn and tufted puffins, storm-petrels, murrelets, shearwaters, as well as three species of albatross (laysan, black-footed, and short-tailed). Pinnipedia include Steller sea lions (SSLs, *Eumetopias jubatus*) and harbor seals (*Phoca vitulina*). Cetacea include killer whales (*Orcinus orca*), Dall's porpoise (*Phocoenoides dalli*), harbor porpoise (*Phocoena phocoena*), humpback whale (*Megaptera novaeangliae*), minke whale (*Balenoptera acutorostrata*), and sperm whale (*Physester macrocephalus*) (Angliss and Lodge 2002). Humpback, minke, and sperm whales are transient species, and are present in Alaska waters only during the feeding migration in summer. Apex predators have also undergone drastic diet changes during the last 30 years. For example, a new study (Sinclair and Zeppelin 2003) indicates that in recent times SSL fed on walleye pollock and Atka mackerel, followed by Pacific salmon and Pacific cod. Other common prey items included arrowtooth flounder, Pacific herring, and sand lance. In contrast, prior to the 1970s, walleye pollock and arrowtooth flounder were absent in SSL diet, while capelin was a dominant prey. The food habits of cetaceans are poorly known, but they can reasonably be grouped in piscivorous (porpoises, killer whales, and sperm whales, feeding mainly on micronekton species) and zooplanktivorous (minke and humpback whales). Seabird diets are comprised of squid, euphausiids, capelin, sand lance, and pollock. Piatt and Anderson (1996) demonstrated a change in seabird diets since the last major reversal of the PDO, from one that primarily comprised capelin in the late 1970s to another that contained little to no capelin in the late 1980s.

Barents Sea

The high latitude BS ecosystem is characterized by a relatively simple food web with few dominant species: for example, diatom → krill → capelin → cod, or diatom → copepod nauplii → herring larvae → puffin (Figure 12.4). However, a more detailed inspection of the diet matrix reveals some level of complexity, mostly related with shift in diet preferences among individuals of the same species but different age. The primary production in the Barents Sea is, as an areal average for several years, about 110 gCm^{-2} per year. Phytoplankton blooms that deplete the winter nutrients give rise locally to a "new" productivity of on average 40–50 gCm^{-2} per year, 90 gCm^{-2} per year in the southern Atlantic part, and <40 gCm^{-2} per year north of the oceanic polar front. In the northern part of the BS system, the primary production in the marginal ice zone (polar front) is important for the local food web, although the southern part is more productive (Sakshaug 1997).

The zooplankton community is dominated by arcto-boreal species. The biomass changes inter-annually from about 50–600 $mg\,m^{-3}$, with a long-term mean of about 200 $mg\,m^{-3}$ (Nesterova 1990). Copepods of the genus *Calanus*, particularly *Calanus finmarchicus*, play a uniquely important role (Dalpadado et al. 2002). The biomass is the highest for any zooplankton species, and the mean abundance has been measured to about 50,000, 15,000, and 3,000 thousand individuals per m^2 in Atlantic, Polar Front, and Arctic waters, respectively (Melle and Skjoldal 1998). In the Arctic region of the BS, *Calanus glacialis* replaces *C. finmarchicus* for abundance and dominance (Melle and Skjoldal 1998; Dalpadado et al. 2002). The *Calanus* species are predominantly herbivorous, feeding especially on diatoms (Mauchline 1998). Krill (euphausiids) is another important group of crustaceans in the southern parts of the BS.

Thysanoessa inermis and *Thysanoessa longicaudata* are the dominant species in the western and central BS, while *Thysanoessa raschii* is more common in the shallow eastern waters. Three hyperiid amphipod species are also common; *Themisto abyssorum* and *Themisto libellula* in the western and central BS, and *Themisto compressa* in the Atlantic waters of the southwestern BS. Close to the Polar Front very high abundance of the largest of the *Themisto* species, *T. libellula*, have been recorded (Dalpadado et al. 2002).

The dominant micronekton species of the BS food web are capelin, herring (*C. harengus*), and polar cod (*Boreogadus saida*). The BS stock of capelin is the largest in the world, with a biomass that in some years reaches 6–8 million metric tones. It is also the most abundant pelagic fish of the BS. Capelin plays a key role as an intermediary of energy conversion from zooplankton production to higher trophic levels, annually producing more biomass than the weight of the standing stock. It is the only fish stock capable of utilizing the zooplankton production in the central and northern areas including the marginal ice zone. *C. finmarchicus* is the main prey of juvenile capelin, but the importance of copepods decreases with increasing capelin length. Two species of krill, *T. inermis* and *T. raschii*, and the amphipods *T. abyssorum*, *T. compressa*, and *T. libellula* dominate the diet of adult capelin (Panasenko 1984; Gjøsæter et al. 2002). The krill and amphipod distribution areas overlap with the feeding grounds of capelin, especially in the winter to early summer. A number of investigations demonstrate that capelin can exert a strong top-down control on zooplankton biomass (Dalpadado and Skjoldal 1996; Dalpadado et al. 2001, 2002; Gjøsæter et al. 2002). Polar cod has a similar food spectrum as capelin, and it has been suggested that there is considerable food competition between these species when they overlap (Ushakov and Prozorkevich 2002). When abundant, young herring are also important zooplankton consumers of the BS food web. Stomach samples show that calanoid copepods and appendicularians makes up 87% of the herring diet by weight (Huse and Toresen 1996). Although data on capelin larvae and 0-group (half-year-old fish) abundance suggests that the predation of herring on capelin larvae may be a strong or dominant effect on capelin dynamics (Hamre 1994; Fossum 1996; Gjøsæter and Bogstad 1998), capelin larvae are only found in 3–6% of herring stomachs (Huse and Toresen 2000). However, the latter authors comment that fish larvae are digested fast, and that predation on capelin larvae may occur in short, intensive "feeding frenzies," which may have been missed by the sampling.

The dominant macronekton species of the BS are Atlantic cod (*Gadus morhua*), saithe (*Pollachius virens*), haddock (*Melanogrammus aeglefinus*), Greenland halibut (*Reinhardtius hippoglossoides*), long rough dab (*Hippoglossoides platessoides*), and deepwater redfish (*Sebastes mentella*). The stock of Arcto-Norwegian (or northeast Arctic) cod is currently the world's largest cod stock. As macronekton predators, cod is dominating the ecosystem. They are opportunistic generalists, the spectrum of prey categories found in their diet being very broad. Mehl (1991) provided a list of 140 categories, however, a relatively small number of species or categories contributed more than 1% by weight to the food. These are amphipods, deep sea shrimp (*Pandalus borealis*), herring, capelin, polar cod, haddock, redfishes, and juvenile cod (cannibalism). The trophic interaction between cod and capelin is particularly strong in the BS and, as shown later, occupies a central role in regulating community dynamics. However, cod diet varies considerably during the life cycle, the proportion of fish in the diet increasing with age. The diet of haddock is somewhat similar to that of cod, but they eat less fish and more benthic organisms.

Approximately, 13–16 million seabirds of more than 20 species breed along the coasts of the BS. The most plentiful fish-eaters among Alcidae are Atlantic puffin (*Fratercula arctica*; 2 million breeding pairs) and guillemots, such as Brünnich's guillemot (*Uria lomvia*) and common guillemot (*Uria aalge*), 1.8 million and 130,000 breeding pairs, respectively. Brünnich's guillemot is the most important consumer of fish, of which polar cod (up to 95–100% on Novaya Zemlya) and capelin (up to 70–80% on Spitsbergen) are dominant in the diets. Daily food consumption by a Brünnich's guillemot is 250–300 g, and as much

as 1,300 metric tons per day for the entire population. It is suggested that this estimate represents 63% of the total amount of food consumed by seabirds in the BS (Mehlum and Gabrielsen 1995). Consumption of fish by seabirds is between 10% and 50% of the yearly catch of fish in the fisheries.

In addition to seabirds, there are about 20 species of cetaceans, and seven species of pinnipeds in the apex predator assemblage of the BS. The majority of the cetaceans are present in the BS on a seasonal basis only. Among these, the most common are minke whale (*B. acutorostrata*), white whale (*Delphinapterus leucas*), white-beaked dolphin, and harbor porpoise (*P. phocoena*). Annual food consumption of minke whale has been estimated at approximately 1.8 tons, including about 140,000 tons of capelin, 600,000 tons of herring, 250,000 tons of cod, and 600,000 tons of krill (euphausiids) (Bogstad et al. 2000). Among pinnipeds, the most common is the harp seal (*Phoca groenlandica*) whose abundance in the White Sea (a large inlet to the BS on the northwestern coast of Russia) is 2.2 million. Other pinnipeds are present in lower numbers. These include, ringed seal (*Phoca hispida*), harbor seal (*P. vitulina*), gray seal (*Halichoerus grypus*), and walrus (*Odobenus rosmarus*). Yearly food consumption by harp seal in the BS is estimated at a maximum of 3.5 million tons, including 800,000 tons of capelin, 200,000–300,000 tons of herring, 100,000–200,000 tons of cod, about 500,000 tons of krill, 300,000 tons of amphipods, and up to 600,000 to 800,000 tons of polar cod and other fishes (Bogstad et al. 2000; Nilssen et al. 2000). In spite of the high capelin consumption of harp seals, cod remains the primary consumer of capelin in the BS, with an estimated annual removal (mean 1984–2000) of more than 1.2 million tons (Dolgov 2002).

Community dynamics

In this section we summarize the community changes of the reviewed systems in relation to recent climate forcing. It is possible that some of the community changes that we describe are the result of human harvest, or a synergism between human and environmental factors. At the current state of knowledge it is impossible to quantify the relative contribution of climate and fishing. However, as illustrated below, the synchrony of the biological changes among different components of the food web, and the large ecological scale of these changes point to the fact that climate must have occupied a central role.

Tropical Pacific

Primary production may change drastically during an El Niño event. As the trade winds relax and the warm pool extends to the central Pacific, the upwelling intensity decreases and the cold tongue retreats eastward or can vanish in the case of particularly strong events. The eastward movement of warm water is accompanied by the displacement of the atmospheric convection zone allowing stronger wind stresses in the western region to increase the mixing and upwelling in the surface layer and then to enhance the primary production. Therefore, primary production fluctuates with ENSO in an out-of-phase pattern between the western warm pool and the central-eastern cold-tongue regions.

These changes in large-scale oceanic conditions strongly influence the habitat of tuna. Within the resource-poor warm waters of the western Pacific most of the tuna species are able to thrive, partly because of the high productivity of the adjacent oceanic convergence zone, where warm western Pacific water meets the colder, resource rich waters of the cold tongue. The position of the convergence zone shifts along the east–west gradient and back again in response to ENSO cycles (in some cases 4,000 km in 6 months), and has direct effect on the tuna habitat extension. In addition to the impacts on the displacement of the fish, ENSO appears to also affect the survival of larvae and subsequent recruitment of tuna. The most recent estimates from statistical population models used for tuna stock assessment (MULTIFAN-CL, Hampton and Fournier 2001) pointed to a clear link between tuna recruitment and ENSO-related fluctuations. The results also indicated that not all tuna responded in the same way to climatic cycles. Tropical species (such as skipjack and yellowfin) increased during El Niño events. In contrast, subtropical species

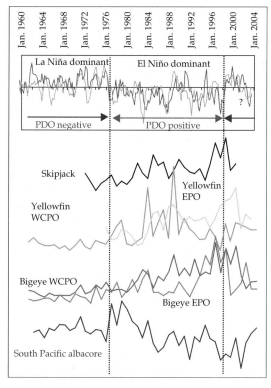

Figure 12.5 PDO, ENSO, and recruitment time series of the main tuna species in the Tropical Pacific (updated from Lehodey et al. 2003). WCPO = Western and Central Pacific Ocean, EPO = Eastern Pacific Ocean. The y axis of the climate indices graph has been rotated. Note that during a positive PDO phase (before 1976), El Niño events become more frequent and tropical tunas recruitment (skipjack, yellowfin, and bigeye) is favored. Conversely, during a negative PDO phase La Niña events are more frequent and the recruitment of subtropical tuna (albacore) is favored. See text for more explanations regarding effect of climate on tuna recruitment.

(i.e. albacore) showed the opposite pattern, with low recruitment following El Niño events and high recruitment following La Niña events (Figure 12.5).

Western Gulf of Alaska

After the mid-1970s regime shift, the GOA has witnessed a dramatic alteration in species composition, essentially shifting from a community dominated by small forage fishes (other than juvenile gadids) and shrimps, to another dominated by large piscivorous fishes, including gadids and flatfish (Anderson and Piatt 1999; Mueter and Norcross 2000). In addition, several species of seabirds and pinnipeds had impressive declines in abundance. An analysis of the GOA community dynamics, for each trophic assemblage identified in the previous sections, follows.

Evidence for decadal-scale variation in primary production, associated with the mode of variability of the PDO, is equivocal. Brodeur et al. (1999) found only weak evidence for long-term changes in phytoplankton production in the northeast Pacific Ocean, though Polovina et al. (1995) suggest that production has increased due to a shallowing of the mixed layer after the 1976–77 PDO reversal. A shoaling mixed layer depth exposes phytoplankton to higher solar irradiance, which in sub-Arctic domains tends to be a limiting factor in early spring and late fall. There is more conclusive evidence for long-term variations in zooplankton abundance in the North Pacific Ocean. Brodeur and Ware (1992) and Brodeur et al. (1999) have demonstrated that zooplankton standing stocks were higher in the 1980s (after the 1976–77 phase change in the PDO), relative to the 1950s and 1960s. Mackas et al. (2001) determined that inter-annual biomass and composition anomalies for zooplankton collected off Vancouver Island, Canada, show striking interdecadal variations (1985–99) that correspond to major climate indices.

During the last 20–30 years, many micronekton species of the GOA community have undergone a sharp decline in abundance, to almost extinction levels (Anderson and Piatt 1999; Anderson 2000). On average, shrimp and capelin biomass decreased after the mid-1980s, while eulachon and sandfish are currently reaching historical high levels (Figure 12.6). The biomass trend of juvenile gadids (cod and pollock) is inferred by the recruitment estimates (Figure 12.7), based on fish stock assessment models (SAFE 2003). While cod recruitment has remained fairly constant, pollock recruitment had a series of strong year-classes in the mid-1970s, followed by a continuous decline up to the 1999 year-class. However, the actual strength of the 1999 year-class will only be available in coming years, as more data accumulates on the abundance of this cohort.

In the macronekton guild, the most remarkable change in biomass has been that of arrowtooth flounder, with a sharp increase in both biomass and recruitment since the early 1970s (Figure 12.7). Arrowtooth flounder now surpass, in total biomass, the once dominant gadid species, such as walleye pollock or Pacific cod. Halibut has also undergone a similar increase in biomass during the available time series. The biomass of adult

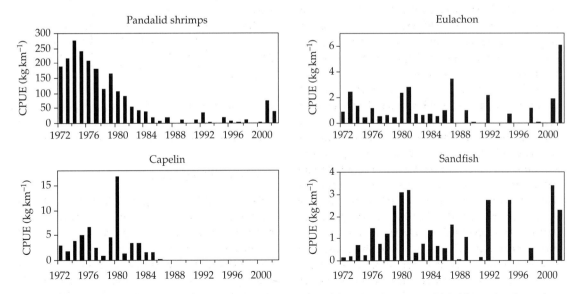

Figure 12.6 Abundance trend of non-gadid micronekton species of the GOA food web. Time series were obtained from survey data and are expressed as catch per unit effort (CPUE kg km^{-1}) (Anderson and Piatt 1999; Anderson 2000).

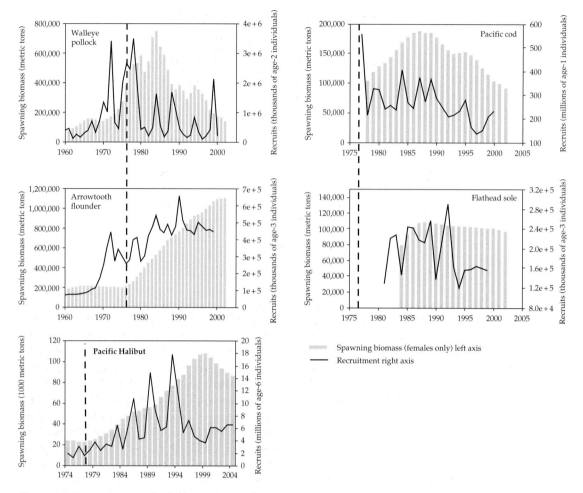

Figure 12.7 Biomass time series for the dominant macronekton species of the GOA food web (SAFE 2003). Pacific halibut time series refers to the legal area 3A, which include most of the north GOA system (halibut data courtesy of Steve Hare, International Pacific Halibut Commission, Seattle WA, USA). The vertical dashed line indicates the 1976 regime shift.

walleye pollock peaked in the early 1980s but has since then declined (Figure 12.7), as a result of several successive poor year classes. Currently, adult pollock (age 3 +) biomass is similar to that of the adult Pacific cod, whose biomass has generally increased since the 1960s, and is recently in slight decline.

Among apex predators, the best-documented and studied case of species decline has been that of the SSLs (Figure 12.8). In 1990, the National Marine Fisheries Service declared SSL a "threatened" species throughout the entire GOA region. Later on, in 1997, the western stock (west of 144° longitude) was declared "endangered" due to continuous decline, while the eastern stock stabilized at low levels and continued to be treated as "threatened". Harbor seals (*P. vitulina*) have also declined in the GOA, compared to counts done in 1970s and 1980s. The exact extent of their decline is unavailable. However, in some regions, particularly near Kodiak Island, it was estimated to be 89% from the 1970s to the 1980s (Angliss and Lodge 2002). Declines in seabird populations have been observed in the GOA, though it should be noted that trends are highly species- and site-specific. For example, black-legged kittiwakes are declining precipitously on Middleton Island (offshore from Prince William Sound), but

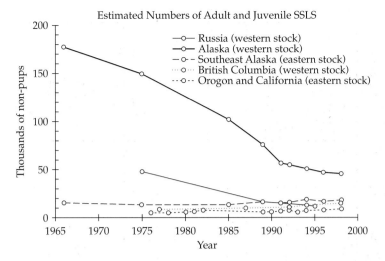

Figure 12.8 SSL population trends for various locations of the North Pacific.

populations are increasing in nearby Prince William Sound, in the central GOA. Likewise, common murres are steadily increasing on Gull Island, but are rapidly declining on nearby Duck Island (both located in the vicinity of Cook Inlet). Cormorants on the other hand, appear to be declining throughout the GOA (Dragoo et al. 2000).

Barents Sea

The bulk of primary production in the BS occurs in two types of areas: close to the ice edge and in the open sea. The spring bloom along the ice edge can occur as early as mid-April when the melt water forms a stable, nutrient-rich top layer. As nutrients are exhausted, but new areas of nutrient-rich water is uncovered by the receding ice, the bloom follows the ice edge in the spring and summer. In cold years, the increased area of ice leads to an earlier and more intense ice edge bloom (Rey et al. 1987; Olsen et al. 2002). Olsen et al. (2002) found that this leads to a tendency for a higher annual primary production in cold years. However, most of it is ungrazed, due to mismatches with primary copepods consumers, for example, capelin and cod.

The zooplankton shows large interannual variations in abundance, species composition, and timing of the development of each species. The capelin can have a significant impact on zooplankton abundance, being able to graze down the locally available zooplankton in a few days (Hassel et al. 1991). Capelin biomass has been fairly stable through the 1970s and early 1980s, but since then has had two major collapses, one in 1985–90 and another in 1994–98 (Figure 12.9). The Norwegian spring-spawning herring has undergone large fluctuations in abundance throughout the twentieth century (Toresen and Østvedt 2000; Figure 12.9). At the turn of the century the spawning stock biomass was around 2 million tons, increasing to more than 15 million tons in 1945. From 1950, the biomass decreased steadily while the landings increased. In the late 1960s, the stock collapsed to a very low level (<0.1 million tons), mainly because of over-exploitation (Dragesund et al. 1980). Strong regulation of the fishery allowed the stock to recover very slowly during the 1970s, and more rapidly after 1983 (due to the strong 1983 year-class). Finally, the stock started to increase again around 1990, reaching about 10 million tons in 1997.

The cod stock declined from 3–4 million tons in the 1950s to less than 1 million tons in the late 1980s (Figure 12.10). Fishing changed the structure of the spawning stock greatly during the same period, by decreasing both the age at maturity (Jørgensen 1990; Law 2000) and the mean age (G. Ottersen, personal communication) of the spawning stock from 10 to around 7 years.

After this crisis in the cod fishery in the late 1980s, the population has picked up; while the spawning stock was only 118,000 tons in 1987, it is now 505,000 tons. The other gadoid stocks of haddock

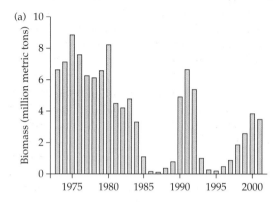

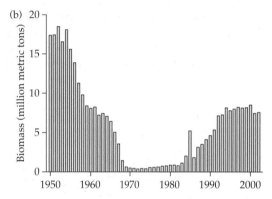

Figure 12.9 Time series of (a) capelin and (b) herring biomass in the BS.

and saithe have decreased by around 50% since the 1960s and 1970s; the portion of large piscivorous individuals has probably decreased even more. The biomass of the long-lived deep-water redfish decreased from around 1 million tons in the 1970s to 0.14 tons in 1986, and is still very small. The same can be said for Greenland halibut and probably long rough dab.

Abundance trends of BS pinnipeds are largely unknown, though an upward trend has been noted for the abundance of harp seal, walrus, and common seal; ringed seal in the western part of the BS may be declining. Among seabirds, the common guillemot has declined dramatically since the 1960s. Already in 1984, before the first collapse of the capelin stock, its abundance had declined by 75% (compared to 1964) because of drowning in fishing gears and perhaps also the collapse in the herring stock around 1970. In 1984–85, the capelin stock collapsed, and in 1986 the large 1983 year-class of herring emigrated from the BS. As a consequence, the largest colonies of common guillemot were further reduced by 85–90% in 1986–88.

Linking climate forcing, food web structure and community dynamics

In the TP, ENSO events are a central forcing variable of the tropical tunas population dynamics (Lehodey et al. 1997, 2003; Lehodey 2001). The links between climate, habitat, and tuna recruitment have been investigated in detail with a

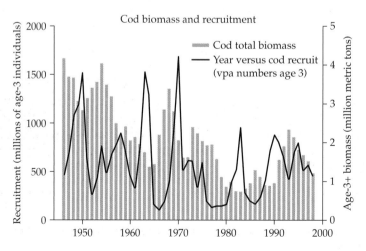

Figure 12.10 Cod biomass (age 3 and older) and recruitment (age 3) in the BS.

spatial population dynamic model (SEPODYM) that describes the population responses of tuna to changes in both feeding and spawning habitats (Lehodey et al. 2003). Results suggest that skipjack and yellowfin recruitment increases in both the central and western Pacific during El Niño events, as the result of four mechanisms: the extension of warm water farther east (ideal spawning habitat is found in warm, 26–30°C water), enhanced food for tuna larvae (due to higher primary production in the west), lower predation of tuna larvae, and retention of the larvae in these favorable areas as a result of ocean currents. The situation is reversed during La Niña events, when westward movement of cold waters reduces recruitment in the central Pacific. When all the favorable conditions occur together, then high peaks of recruitment are observed. This was the case, for example, in the final phase of the powerful 1997–98 El Niño event. In the second half of 1998, the skipjack purse seine catch was concentrated in a small area in the equatorial central Pacific, and contained a high

number of juvenile skipjack between four and eight months of age. Satellite imagery indicated that this same area was the site of a major bloom in phytoplankton some 4–8 months before (Murtugudde et al. 1999). The catch in 1998 was an all-time record; ironically, it led to a drop of 60% in the price of skipjack, which were so abundant they could not all be processed by the canneries.

While the main skipjack and yellowfin spawning grounds in the western and central TP are associated with the warm pool, those of albacore roughly extend through the central Pacific on each side of the equatorial 5°N–5°S band. The out-of-phase primary productivity between western (warm pool) and central (equatorial upwelling) Pacific led the model to predict similar out-of-phase recruitment fluctuations between species associated to one or the other areas of the TP. In addition, the extension of the warm waters in the central Pacific during El Niño events that extends the skipjack spawning grounds may conversely reduce those of the albacore (Figure 12.11). In summary, it appears that

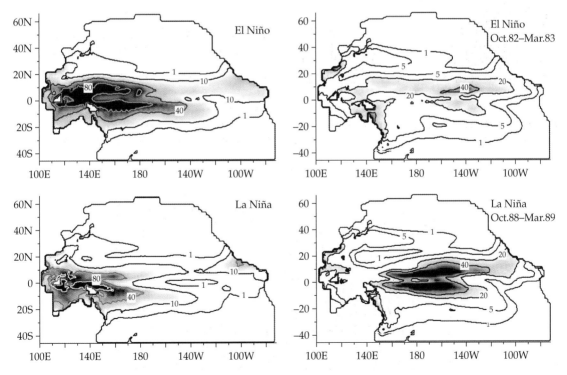

Figure 12.11 Predicted spatial distribution of the larvae and juveniles of skipjack (left) and albacore (right) tuna in the TP during ENSO phases.

tuna recruitment and population fluctuations in the TP would be controlled through physical, bottom-up and "middle-down" (larvae predation by epipelagic micronekton) rather than top-down mechanisms, the intermediate "middle" component including the juvenile and young tuna.

The species dynamics of the GOA community appear to be strongly influenced by both climate forcing and species interactions within the food web. For example, the gradual increase of macronekton during the mid-1980s resulted from a series of strong year-classes that followed the mid-1970s shift of the PDO index. High pollock recruitment may have been the result of a series of favorable conditions, including higher water temperature and lower spring wind stress (Bailey and Macklin 1994) during the larval stages, as well as limited predation and density-dependent mortality during the juvenile stage (Ciannelli et al. 2004). Immediately after the 1976 PDO regime shift, both of these conditions were common in the GOA area. Similarly, flatfish increase in recruitment was, in part, the result of a favorable larval advection from the deep offshore spawning to shallow inshore juvenile nursery grounds, conditions that appear commonly, particularly during strong El Niño events (Bailey and Picquelle 2002). Intervention analysis applied to many of the available GOA recruitment time series indicates that large flatfish recruitment (i.e. halibut and arrowtooth flounder) has significantly increased after the 1976 PDO shift, while pollock and cod recruitment has significantly increased during El Niño North years (Hollowed et al. 2001). As mentioned above, the frequency of El Niño North events in the GOA region can increase during positive PDO phases, thereby rendering more difficult the distinction between the effect of PDO or ENSO forcing in the North Pacific community dynamics.

The initial increase of macronekton biomass after the late 1970s may have triggered a series of food web interactions that directly and indirectly affected other species of the GOA community. For example, with regard to pollock, the post 1985 biomass decline was the result of mostly poor recruitment from the mid-1980s to the late 1990s. During this time frame, pollock had some occasional strong year-classes (e.g. 1984, 1988, and 1994), but never at the level of those observed before the 1980s. Bailey (2000) has shown that during the years of the decline, pollock recruitment shifted from being controlled at the larval stage (1970s and initial part of the 1980s) to being controlled at the juvenile stage (late 1980s and 1990s). This change in recruitment control was due to the gradual buildup of piscivorous macronekton and consequent increase of juvenile pollock predation mortality. Ciannelli et al. (2004) have shown that macronekton predators, besides having a direct effect on juvenile pollock survival (via predation), can also indirectly affect their dynamics by amplifying the density-dependent mortality.

Micronekton species of the GOA community, such as capelin and pandalid shrimps, might have suffered high predation mortality after the predators buildup, with consequent decline in abundance. To date, apart from the case of juvenile pollock described above, there is no direct evidence that predation was the primary cause of the decline of forage species in the GOA community. However, studies from other sub-Arctic ecosystems of the North Atlantic, point to the fact that macronekton species set off strong top-down control on their prey (Berenboim et al. 2000; Worm and Myers 2003; Hjermann et al., 2004; but see also Orensanz et al. 1998). Also, in the GOA community a top-down control of macronekton on micronekton is consistent with the changes of the groundfish diet observed in the last 20 years (Yang and Nelson 2000; Yang 2004). In addition to top-down forcing, during the years following the PDO regime shift (i.e. late 1970s and 1980s), capelin may also have been negatively influenced by strong competition with juvenile pollock. As indicated above, these two species have an almost complete diet and habitat overlap. In the GOA, capelin are at the southernmost range of their worldwide distribution, while, pollock are at the center of their range and will be more likely to out-compete capelin during warm climate phases.

Changes of the micronekton assemblage of the GOA food web had severe repercussions on apex predators, such as seabirds and pinnipeds. Springer (1993) hypothesized that food-related stresses have contributed to observed population declines of the seabirds. Among pinnipeds, the SSL decline in the western GOA is to these days one of the most interesting case studies in conservation biology (National Research Council 2003). Several hypotheses have been advanced to explain the decline and absence of recovery, including direct and indirect fishing effects (Alaska Sea Grant 1993), climate change (Benson and Trites 2002), nutritional stress (Trites and Donnelly 2003), parasites and disease agents (Burek et al. 2003), and, recently, top-down forcing (Springer et al. 2003). The majority of these hypotheses acknowledge the importance of direct and indirect food web interactions. For example, Springer et al. (2003) suggested that an increase of killer whale predation was responsible for the Steller's decline. According to this concept, killer whales turned to SSLs after the baleen whales of the GOA disappeared due to an ongoing legacy of the post Second World War whale hunt. In addition, one might speculate that baleen whale never fully recovered after the postwar decline due to a lack of sustainable and highly nutritious fish prey. Top-down foraging by killer whales has played a major role also on the decline of sea otter (*Enhydra lutris*) from western Alaska, with rather dramatic effects on the sea urchins (increase) and kelp (decrease) populations (Estes et al. 1998). In contrast to top-down forcing, the nutritional stress hypothesis (a.k.a., "junk-food" hypothesis) suggests that the SSL decline was due to a shift in their diet toward prey with lower energy and nutritional value (e.g. pollock, cod) as a consequence of the reduced availability of the high-energy forage species (Trites and Donnelly 2003).

Probably, to a greater extent than in the GOA food web, the pelagic community of the BS is dominated by a few very abundant species, resulting in strong interspecific interactions (Hamre 1994). Specifically, the relationship between cod, capelin, and young herring have been viewed as particularly important for the ecosystem functioning (Hamre 1994, 2000; Hjermann et al. in press). The recruitment of herring and cod is strongly associated with the temperature of the BS; specifically, in cold years, recruitment is always low, while in warm years, it may be low or high

(Ellertsen et al. 1989; Ottersen and Loeng 2000) (Figure 12.12). The increased westerly winds over the North Atlantic that are associated with a high (positive) NAO phase, has, at least for the most recent decades, affected BS water temperature by increasing (1) the volume flux of relative warm water from the southwest; (2) cloud cover; and (3) air temperature. Increased BS water temperature influences growth and survival of cod larvae both directly through increasing the development rate and indirectly through regulating *C. finmarchicus* production. Variation in availability of *C. finmarchicus* nauplii is an important factor for formation of strong cod year-classes. In fact, the match–mismatch hypothesis of Cushing (1982, 1990) states that the growth and survival of cod larvae depends on both the timing and magnitude of *C.finmarchicus* production. In addition, an increase of inflow from the zooplankton-rich Norwegian Sea further increases availability of food for the cod larvae (Ottersen and Stenseth 2001). High food availability for larval and juvenile fish results in higher growth rates and greater survival through the vulnerable stages when year-class strength is determined (Ottersen and Loeng 2000).

Cod and herring can potentially eat a large amount of capelin (herring eats larvae, cod eats larger stages). Therefore, capelin can be expected to experience high predation after a warm year with favorable conditions for cod and herring recruitment. This was confirmed by Hjermann et al. (2004), who found that capelin cohorts that are spawned two years after a warm year tend to be weak. It appears that the predation-mediated effect of climate has been the main mechanism of the two collapses of the capelin stock in the last two decades (1984–86 and 1992–94); a third collapse appears to be occurring at the time of writing. These collapses (stock reduction of >97%) had a huge impact on the BS community assemblage and food web structures. The most apparent impact was a drastic reduction in population of some seabird species (such as the common guillemot *U. aalge*; Krasnov and Barrett 1995; Anker-Nilssen et al. 1997) and mass migration of a huge number of harp seal toward the Norwegian coast; incidentally about 100,000 seals subsequently

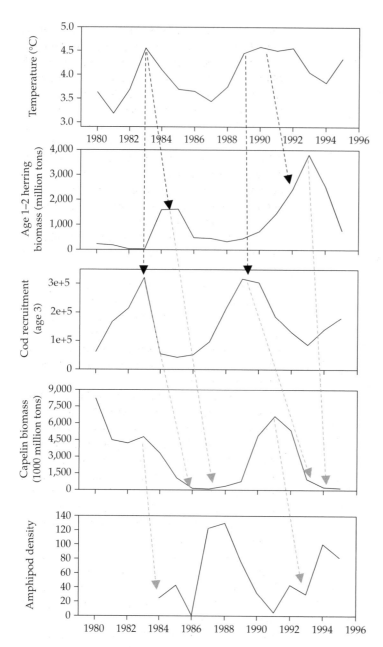

Figure 12.12 From top to bottom, time series of water temperature, herring (age 1–2) biomass, cod recruitment (age 3), capelin and amphipod (*P. libellula* and *P. abyssorum*) biomass in the BS. Arrows indicate climate (in black) and trophic (in gray) forcing on species dynamics. Water temperature in the BS (correlated with NAO index) has a positive effect on cod and herring recruitment. Herring (mostly age 1–2) and cod (mostly age 3–6) feed on capelin larvae, and adult, respectively, and consequently have a negative effect on capelin biomass. In turn, capelin feed on northern zooplankton, of which *P. libellula* and *P. abyssorum* are dominant components.

drowned in fish nets (Haug and Nilssen 1995). Also, observations of increased zooplankton biomass during capelin collapses (Figure 12.12) are indicative of the capelin impact at lower trophic levels. In the "collapse" years, herring replaced capelin as the main zooplankton feeder. By feeding also on capelin larvae, a low biomass of juvenile herring is able to block the rebuilding of the capelin stock, with the ultimate effect of replacing a large capelin biomass with a small herring biomass.

Based on these observations we conclude that the BS pelagic ecosystem appears, to some extent, to be a "wasp-waist" ecosystem, a term originally coined for upwelling regions such as the Benguela ecosystem (Cury et al. 2000). In such ecosystems,

1970–84

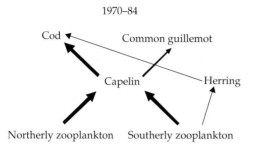

1986–88

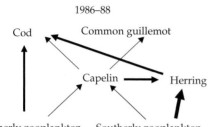

Figure 12.13 Conceptual model of changes in BS trophic links during years of high capelin abundance (1970–84) and capelin collapse (1986–88, "collapse" years). During "collapse" years cod diet switches from one based primarily on capelin, to another based on zooplankton and herring. Capelin decline, also has negative effects on seabird populations, such as the common guillemot. Herring replace capelin as a central forage species, and can exercise a large predation impact on capelin larvae (further delaying their recovery) and zooplankton, particularly from the Atlantic waters (southerly zooplankton). Because herring are not distributed as far north as capelin, most of the northerly zooplankton remains ungrazed during "collapse" years. Also, because herring live in the BS only for a limited period of their life cycle (see text), a large portion of the BS originated biomass is exported out of the system.

the crucial intermediate level of pelagic fish is dominated by a few species (capelin, juvenile cod, and herring in the BS), exerting top-down control on zooplankton and bottom-up control of predators (Figure 12.13).

Conclusions

The relation between structure and complexity of a food web and its stability has been a much debated issue within the field of ecology, starting with Mac Arthur (1955) and Elton (1958) who claimed that complexity begets stability—a view conveyed to many ecology students (see, for example, Odum

1963). May (1973) showed, through mathematical modeling, that the "complexity begets stability" idea might not necessarily be valid, although Maynard Smith (1974) cautioned about drawing too firm conclusions on the basis of pure theoretical analyses. It bears note that the set of empirical observations gathered in this review are not sufficient to address in full details of the issue of resilience in marine pelagic systems. In addition, the anthropogenic forcing (i.e. fishing), active in all three systems, may have synergic effect with climate, complicating the issue of community resilience even further. However, we feel that addressing the issue of community stability is particularly appropriate within the context of this book, and may offer the unique opportunity to test the traditional theoretical knowledge of complexity and stability in systems (i.e. marine pelagic) where such theories have, to date, been unexplored. By focusing on three ecological systems of different complexities and dynamics we can ask to what extent they exhibit different degrees of resilience.

A summary of selected metrics for the three inspected systems is presented in Table 12.1. Of the three, TP is the one where the ocean variability, associated with climate forcing, is most extreme. Community changes in the TP ecosystem occur likely with climate fluctuations, but high species diversity, high degree of omnivory, high connectivity, and weak species interactions contribute likely to its resilience and stability through time. In the GOA system, within the period in which community dynamics was monitored, we have witnessed only a single shift of the dominant climate forcing: a raise of the PDO index in the 1976, a brief return to pre-1976 after 1989, and what appears to be a more stable return of the PDO in recent years. Also, in recent years, it appears that the North Pacific climate is shifting toward a new mode of variability (Bond et al. 2004). The community has clearly changed after the 1976 climate change, but after a transitory shift in 1989, there was no sign of a return to the pre-1976 community state: piscivorous macronekton kept on rising (e.g. arrowtooth flounder), while forage species (e.g. shrimps and capelin), and apex predators (SSL and seabirds) kept on

Table 12.1 Descriptive metrics of the three pelagic ecosystems included in this review

Metrics	TP	Western GOA	BS
Extension (millions km^2)	80.0	0.38	1.4
Primary production (gC m^2 per year)	50–300	50–300	110
Dominant period of climate forcing (year)	4–10 (ENSO)	20–30 (PDO) 4–10 (Niño North)	Interannual with a weak period of about 8 years
Nodal micronekton	Epipelagic macronekton	Juvenile gadids	Capelin
Nodal macronekton	Small-size scombrids	Large flatfish	Cod
Nodal apex	Large-size scombrids, blue shark, toothed whales	Sea birds and pinnipeds	Sea birds and baleen whales
Nodal trophic links	Small-size scombrids—epipelagic micronekton Large-size scombrids—epi and mesopelagic micronekton	Large flatfish-micronekton	Cod-capelin-herring
Climate trace in food web	Bottom-up and middle-down	Bottom-up and top-down	Middle-out
Ranked complexity (including diversity, number of links and connectance)	High	Average	Low
Ranked resilience	High	Low	Average

decreasing. In recent years, immediately following the alleged 1999 change of PDO regime (warm to cold), pandalid shrimps, eulachon, and sandfish appear to be recovering, but there is no sign of change in other forage (e.g. capelin) or apex predator (e.g. arrowtooth flounder) dynamics. Thus, the evidence gathered so far indicates that the 1976 climate regime shift has led the GOA community toward a new equilibrium state. However, given the top-down control of apex predators on forage species and the relatively long life cycle of large macronekton, we may have to wait few more years to fully understand whether the GOA community can ever return to a pre-1976 state.

To a larger degree than the GOA, and certainly than other sub-Arctic systems (e.g. northwest Atlantic), the dynamics of dominant BS species has a tendency to recover after collapsing. This seems to be a system dominated by three species: cod, capelin, and herring. Changes of capelin and herring abundance (via top-down and bottom-up forcing), have led the entire community to profound phase-shifts in the food web structure. During the last 30 years we have witnessed two of these transitions, one in 1985–89 and another in 1994–99, with a possible third transition occurring at the time of writing (Hjermann et al. 2004). Thus the BS community appears to be con-

tinuously shifting among two states, albeit whose magnitude may vary, mainly depending on the abundance of the main forage species of the systems.

Our survey does suggest that the more complex system (TP) is more stable (or resilient)—thus supporting the "complexity begets stability" concept. However, much work is needed before we can reach firm and general conclusions for large marine ecosystems. Such work should cover both the empirical and the analytical facets of community dynamics studies. Within the empirical framework, to overcome the objective difficulty of manipulating marine pelagic ecosystems, climate forcing should be seen as a natural "large-scale perturbation experiment" (Carpenter 1990). Within the analytical framework, we stress the importance of statistical models with structure inferred from the observed patterns of population variability (e.g. Hjermann et al., 2004). We also stress the importance of model simulations in relation to different levels of internal structures (i.e. food web) and external forcing (e.g. Watters et al. 2003). Our hope is that what we have done here might serve as a spark leading to the development of more comprehensive analysis of food web and community dynamics in large marine pelagic ecosystems.

Acknowledgements

We are indebted to a number of people who provided data and information used in this chapter. They are: Paul Anderson, Mei-Sun Yang, Nate Mantua, Steven Hare, Ann Hollowed, and Matt Wilson. We are grateful to the editors of this book for inviting us to contribute with a chapter on marine pelagic food webs. Comments from Matt Wilson and Kevin Bailey improved an earlier version of this chapter. This research is contribution FOCI-0495 to NOAA's Fisheries-Oceanography Coordinated Investigations.

CHAPTER 13

Food-web theory provides guidelines for marine conservation

Enric Sala and George Sugihara

Introduction

Vast, unfathomable, and the major biome of our planet, the oceans had an assumed capacity that defied limitation. Unfortunately, this old presumption turned out to be terribly wrong as technological advancement reduced the world's oceans to mortal size by amplifying man's inputs (pollution) and magnifying his outputs (fishing). And the fact of this magnification combined with evidence of man's early impacts (Jackson and Sala 2001; Jackson et al. 2001; Pandolfi et al. 2003) has rapidly overtaken our historical myth of the infinite sea. The last couple of decades in particular have been hard on marine ecosystems and our myths about their robustness have all but faded. Witness the vast hypoxic zone (a 20,000 km^2 ecological desert in the Gulf of Mexico), the numerous other coastal fiascos (e.g. Dayton et al. 1998; Jackson et al. 2001), and the recently reported collapse of many major fisheries to 10% of preindustrialized fishing levels (Myers and Worm 2003).

Much to our dismay, ecological disasters in marine systems are in full bloom. These disasters are unintended and uncontrolled experiments, and can ironically, further our understanding of the functioning and structure of marine systems, and may eventually help to save them. Indeed examples of fisheries overexploitation can yield important ecological data that has been extremely useful to ecologists and marine fisheries scientists and may contribute to a general understanding of marine populations and food webs. These are valuable but costly data for marine science as some of these changes may not be easily reversible. Continuing on the negative side, this acceleration in environmental

misery careful the urgency for better understanding of marine ecosystems and for the need to connect this knowledge to conservation and better resource management. Given this background, it is the particular and immediate challenge of ecology and of food-web theory in general to contribute knowledge that can be used to assess, and ultimately manage, marine ecosystems toward desirable states and away from disaster. At the very least, one would like to have available some early quantitative indicators of when systems are being threatened.

Our objective here is to briefly review some of this very costly data with the aim of illustrating ecological generalities that should be useful as guidelines for detecting and understanding human impacts on marine ecosystems. That is, we shall attempt to summarize generalities concerning human impacts on marine food webs (broadly defined here to include properties of communities and ecosystems) as they relate to our theoretical understanding of these webs, and illustrate the generalities with specific examples. The time for scholasticism is past. It is time to step up, move beyond ivory tower theory, and state what we know (however humble that may be) in a way that can be useful for marine conservation.

Although the theme of this chapter concerns food webs, and the latest initiative in fisheries management is the ecosystems perspective, we wish to point out that it is important not to ignore the simple results from single-species management and conservation. For example, three simple patterns that occur regularly with overfishing and are emblematic of impacts in the fishery are: (1) a decrease in the average size of fish caught, (2) a decrease in the average age of fish caught, and (3) a decrease in the percentage abundance of super-spawners. We emphasize that guidelines based on facts such as these coming from single-species management should not be ignored, while plans are being made to manage from the ecosystem and food-webs perspective. That is to say, although the current momentum in fisheries management is toward the ecosystems perspective, we should not loose sight of what we have uncovered from single-species studies.

Dynamically propagated effects in food webs: interaction strength, and cascades

Simple cause and effect is comforting. Pull lever A and get result B. A world of simple direct effects is safe and knowable. Things get scary when the levers are connected with hidden wires. Marine ecosystems are known to contain such hidden wires, and the most well-studied configuration of wires is described by the so-called trophic cascade. This is a problem whose origins trace back to a classic paper by Hairston et al. (1960), which gave an explanation for why the earth is green, and not an overgrazed desert. Simply put, the idea is that overgrazing by herbivores is prevented by the action of predatory carnivores keeping herbivore numbers in check. This simple notion of the remote control of plants by top carnivores (via herbivores) was sharpened by Carpenter and Kitchell (1974) and became enshrined as the cascade effect. Typically, food chains with three or more nodes of odd length (length measured as the number of nodes starting with primary producers) are dominated by primary producers while chains of even length are not. Thus, in fresh waters, when secondary predatory fish are present in ponds (chains of even length (4)) the water tends to be clearer than when such top predators are removed leaving only primary predators, herbivores, and autotrophs (odd length (3)). Likewise, if predators-of-top-predators are introduced (e.g. fishing: producing chains of odd length with humans at the top), the water turns greener. These things seem to be true, both in models and in field experiments,

and in the marine cases to be discussed below. It is the remote-control aspect of this that is worrying in its ambiguity in complex networks where simple linear chains are the exception and not the rule. Nonetheless, when the very strongest interactions line up in a linear chain, food-web theory provides a reasonable cautionary guideline as to the expected propagation of indirect outcomes resulting from manipulation (harvesting) of some component. Thus, if a top predator is harvested, and there is a supporting linear chain of strong interactions below it, one should expect abundances in successive links in this chain to be alternatively augmented and reduced.

Topology and dynamics

The main caveat of the basic model result is that the chain is a simple linear one, rather than part of a more reticulated network. The latter describes a more realistic situation where the indirect effects are less clear. Studies with simple models that include omnivory links that bridge trophic levels and knit the simple chains into more complicated networks, confirm the intuitive hypothesis that these alternative pathways will tend to reduce the likelihood and magnitude of trophic cascades (e.g. McCann and Hastings 1997; McCann et al. 1998). These alternative pathways must be sufficiently weak so as to not be destabilizing in themselves, but strong enough to matter. This suggests that more highly connected systems with more omnivory links can be more resistant to trophic cascades. Note that this stabilizing effect of connectivity only superficially runs counter to classical dynamic arguments about stability–complexity (May 1973), where large numbers of species, higher connectivity, and strong interaction terms can be destabilizing. What harmonizes the result is the need for the secondary interactions to be sufficiently weak so as to not be destabilizing in themselves. That is, they cannot lead to Gershgorin disks with large radii. Recent analysis of the most detailed web to date for a Caribbean coral reef system (Bascompte et al. submitted) containing several hundred species and several thousand interaction coefficients shows that interaction strengths are often lognormally distributed so that relatively few interactions are strong and the vast majority are weak (supporting previous results obtained from smaller webs; for example, Paine 1992; Wootton 1997). One can assume that most real ecological systems are built this way. That is, they are the product of dynamic selection, in that the configurations we see are ones that can persist long enough to be seen. Likewise, system configurations that are dynamically fragile are rarely if ever seen. Although not well publicized, this idea was demonstrated in models fairly early on (Sugihara 1982), where it was shown that dynamic selection is capable of reproducing some difficult food-web regularities, including rigid circuits (triangulation), intervality, and tree-like guild patterns. That is to say, more often than not, randomly assembled simple model systems (large, random, and initially unstable) tend to decay dynamically into nonrandom configurations having the peculiar topologies found in natural food webs. That random model assemblages settle down by dynamic selection (via extinctions) to smaller stable assemblages having many topological attributes (triangulation, etc.) that are seen in nature is not surprising perhaps. It is an example of self-organization.

In similar vein one may speculate that in real systems, potential trophic cascades (linear chains of strongly interacting species) would be rare, and would be more commonly associated with omnivory than otherwise expected. This was found to be true in the highly resolved Caribbean coral reef system cited above (Bascompte et al. in preparation). The cooccurrence of strong interactions in tri-trophic food chains (three trophic levels connected hierarchically by two strong predatory links) in this Caribbean coral reef food web was less frequent than expected by chance. Moreover, a significant proportion of these strongly interacting chains had a strong omnivory link between the top predator and the basal resource. That is, tri-trophic chains containing two successive strong predatory links were far rarer than expected by chance; and when they did occur, they had a much higher likelihood of being associated with omnivory than expected by chance. This, combined with the observation that marine food webs are thought to have a slightly higher degree of connectivity than webs from other biomes (Link 2002), suggests that

there may be some structural buffering operating against cascades in marine systems. This buffering may promote community persistence and stability (Fagan 1997; McCann and Hastings 1997; McCann et al. 1998). Nonetheless, as Bascompte et al. point out, the strong predatory chains or potential cascades that exist are nonrandomly associated with commercial top predators (sharks, in their Caribbean web), and thus cascades remain a concern for ecosystem management. This is, perhaps a heavier concern in temperate marine food webs as these webs have been shown to be simpler and have lower connectivity than tropical marine webs. Therefore, it is not surprising that trophic cascades appear to be more ubiquitous in temperate marine food webs as compared to tropical webs (such as those described in Hughes 1994; McClanahan 1995; Dulvy et al. 2004).

Thus, the configuration of the strongest interactions in natural food webs is not random, and although current information suggests that in general it is biased toward more stable configurations, the danger remains that human activity in marine webs can be selective in a way that targets fragile nodes and links. Awareness of known principles relating dynamics to the configuration of strong linkages in webs is essential to informed conservation and management, and speaks to the acknowledged need for the ecosystems perspective in marine fisheries management.

Examples of dynamically propagated effects in marine food webs

We illustrate the relevance of these principles to marine conservation, with several cases of propagated effects involving strong interactors in marine food webs.

Coastal–intertidal

Starfishes (mostly *Pisaster*) in rocky intertidal food webs in the northeastern Pacific provide a paradigmatic example of cascades. They demonstrate the importance of strong interactors in the web and the consequences of their removal. In the coast of Washington, *Pisaster* is a top predator feeding mainly on the mussel *Mytilus californianus*, but also on barnacles and chitons. In the presence of

Pisaster, benthic dominance is shared among a number of species, including barnacles and algae. The classic experimental field work in this system conducted by Paine (1966, 1980) consisted of the removal of *Pisaster*, and showed that the removal of the top predator resulted in a successive replacement by more efficient occupiers of space. The endpoint of these successional changes was a community dominated overwhelmingly by *Mytilus*. Although the removal of the top predator was experimental, Paine's work was an early warning about the ecological risks and food-web impacts of species depletion.

Fishing provides many examples of propagated effects due to the removal of strong interactors from the web. A global-scale example is the often seen explosion in sea urchin populations that result from the commercial removal of a variety of large- and medium-sized sea urchin predators, such as sea otters and fishes. These cases show that following the removal of sea urchin predators, the abundant urchins can subsequently turn complex algal forests into marine barrens. Again, this is a global phenomenon which has happened, among other places, in the Mediterranean rocky sublittoral (Sala et al. 1998; Hereu 2004), and kelp forests in the north Atlantic (Witman and Sebens 1992; Vadas and Steneck 1995), Alaska (Estes and Duggins 1995; Estes et al. 1998), and New Zealand (Babcock et al. 1999; see also review by Pinnegar et al. 2000). In the relatively simple food webs in temperate seas, these trophic cascades generally result in a decrease of many other measures of ecological complexity (e.g. lower species richness and evenness).

Indeed, the hallmark example of trophic cascades in nature are the kelp forests in the Pacific northeast. In this system, sea otters are top predators, and these top predators feed upon benthic grazers, such as sea urchins and abalones. The consumption of sea urchins is especially important, since sea urchins are avid consumers of kelp. Kelps are the most important architectural species, providing a canopy that serves as refuge and provides microhabitats for an array of subordinate species. In the presence of sea otters, kelps will thrive and sea urchins generally shelter in crevices, feeding upon drift algae. The removal of the sea

otter by humans can result in increases of sea urchin abundance beyond a point where they leave shelter and eat the kelps themselves, eventually resulting in the destruction of the kelp canopy and the formation of a barren dominated by encrusting coralline algae (Estes and Duggins 1995).

Two major patterns that emerge for coastal marine food webs following the removal of strong interactors are as follows. Dramatic food-web-wide effects can ensue if: (1) the strong interactor is a predator on a food chain feeding upon a strong interactor which in turn feeds upon an architectural species (such as kelp), or (2) if the strong interactor is a predator feeding upon the dominant primary producer or architectural species. This should occur even though the strong interactor has low abundance.

Coastal–Pelagic

In some cases, dramatic changes in the abundance of the strongest interactors may not be caused by fisheries targeting them directly, but may result more indirectly from harvest of their prey. Here the effects may not be seen immediately and can involve significant time lags, especially with the tendency for longer response times (e.g. longer generation times) at higher trophic levels. An intriguing example is the delayed effect of the dramatic depletion (removal) of gray whales by oceanic fishing fleets in the 1960s and 1970s. These whales are a major prey item of transient orcas, both of which normally reside in the open ocean of the north Pacific. Decades after intensive hunting of gray whales, transient orcas moved to nearshore areas and shifted their feeding behavior to other marine mammals including sea otters (Estes et al. 1998; Springer et al. 2003). The reduction of sea otters are believed to have triggered the trophic cascade described above where kelp forests were eliminated and replaced by barrens (Estes et al. 1998).

Some generalities

When the food web contains an embedded three-species food chain (defined by two strong links) there is a higher likelihood of trophic cascades. For example, in a recent study of a Caribbean coral reef food web, where the top predators are sharks, groupers, and jacks; medium consumers are groupers; and basal species are herbivorous fishes, such as parrotfish and surgeonfish (Bascompte et al. in preparation), it was found that sharks and other top predators are overrepresented in strongly interacting food chains. Humans tend to selectively target these strong interactors first, because of their higher economic value (e.g. Pauly et al. 1998; Sala et al. 2004); and this has implications for food web structure and dynamics. The removal of sharks may have resulted in a decline of the parrotfish populations via trophic cascade (Bascompte et al. in preparation). The decline of the parrotfish has been implicated in the ecological shift of Caribbean reefs from coral to algal dominated (Hughes 1994). Thus the removal of reef sharks by fishing may have contributed to a trophic cascade that extends to lower trophic levels and involves competitive interactions between algae and corals. In this example, the selective depletion of the strongest interactors increased the likelihood of propagating trophic cascades (Bascompte et al. in preparation). Thus, although as a group marine food webs may have some structural buffering (higher connectance and omnivory), fishing remains a threat that promotes instability by virtue of its tendency toward targeted removal of species at higher trophic levels.

As previously mentioned, the effects of removing a strongly interacting species is more evident and immediate in simpler, less connected food webs (specifically webs having lower connectance). The intuitive explanation for the vulnerability of simple webs is their lack of ecological redundancy, with the caveat that the redundant secondary interactions cannot be too strong. This most likely accounts for the relative ubiquity of trophic cascades in temperate marine food webs as compared to tropical food webs.

Not all food web components are equally important with regard to web dynamics. Strong interactors can maintain food-web stability, and their removal should be averted if the goal is to preserve complex food webs. Experimental work in the eastern Pacific intertidal dealing with only four species interactions suggests that non-keystone species have only minor effects on the food web, although after keystone species removal

they can partly compensate for the reduced predation (Navarrete and Menge 1996). Other experimental work has indicated the potential importance of non-keystone predators in the absence of the keystone species (Dayton 1971; Paine 1971). Indeed, some weak interactors may have strong effects due to great field densities or to their link frequency, and their removal could also cause food-web instability (McCann et al. 1998; Kokkoris et al. 1999). How to distinguish these weak interactors with potentially destabilizing effects is another question. No matter what, pretending that the removal of small, seemingly insignificant species will have no significant effects on food webs is, at this point, an act of hubris and eventually risky. More studies are needed that explore the role of weak interactors in real, speciose food webs.

The effects of the removal of weak interactors are not as predictable as the effects of the removal of strong interactors (Paine 1980). On a per capita basis, the absence of a weak interactor in a species-rich food web should not have a significant effect on the web. From a real-world perspective and on a population basis, the removal of weak interactors can have unpredictable effects, these being basically a function of numbers. This uncertainty is enhanced by the fact that weak interactions generally show greater variance than strong interactions (Berlow 1999). Moreover, there is consensus that weak interactors may have important stabilizing roles in food webs (e.g. McCann et al. 1998). Although small species often have per capita interaction strengths similar or smaller than larger species, their tendency to have greater densities in the field increases their potential food-web impacts (Sala and Graham 2002). There is clear evidence of significant food-web-wide impacts due to striking increase in the population of weak interactors. Such increase can be caused by release from predation after the elimination of a strong predator, or to other factors such as environmental fluctuations. For example, amphipods are weak interactors in California kelp forests, and under usual densities their grazing has insignificant effects on the kelps (Sala and Graham 2002), but during El Niño events amphipod populations can exhibit explosive increases and eat out entire sections of the kelp forest (Graham 2000). Another example of the key role of species with relatively low biomass and weak per capita interaction strength is symbiont zooxanthellae in reef corals (Knowlton and Rower 2003). The loss of just a few species could enhance coral bleaching during prolonged warming events and shift the ecological state of coral reefs. Although these examples do not involve trophic cascades, they highlight the potential for trophic cascades when a species interacting with an architectural species is not a strong interactor but has a strong impact on the basis of extraordinary abundance.

Humans versus other marine top predators

In the northwest Mediterranean subtidal, it has been estimated that humans account for 11% of the trophic links for a food web with relatively low taxonomic resolution (Sala 2004). In the same web, top predators such as monk seals and sharks account for 24% of the links. Unlike other top predators, however, humans have the capability to reduce the complexity of food webs by (1) effectively eliminating other top predators, (2) reducing the abundance of many other species as a consequence of direct trophic effects, and (3) reducing the number of functional trophic levels (i.e. trophic levels containing species at abundances such that they still perform their ecological roles) (e.g. Dayton et al. 1998; Jackson et al. 2001; Pandolfi et al. 2003). Indeed, trophic levels are reduced to the point where even fish parasites are less diverse and abundant in fished areas than in marine reserves (C. F. Boudouresque, personal communication).

Food webs in the presence of marine top predators generally exhibit greater complexity than in their absence (e.g. Estes et al. 1998; Sala et al. 1998). Food webs that have been degraded by overfishing have lost the top-down regulation exerted by other predators and are increasingly influenced by bottom-up factors (Figure 13.1) (Jackson et al. 2001). Although humans exercise a formidable top-down control on marine food webs, unlike other top predators we increase the likelihood of large fluctuations and instability in food-web structure. For instance, the removal of herbivores in estuarine food webs subject to eutrophication can result in

planktonic algal blooms, and the removal of large herbivores in tropical coastal food webs can facilitate the appearance of sea grass disease (Jackson et al. 2001). One of the few examples of another top predator reducing the complexity of a food web is that of transient orcas in the north Pacific which, by reducing the abundance of sea otters, enhance an increase in sea urchin abundance and, by cascading effects, a decrease in kelp cover and a reduction of overall food-web complexity (Estes et al. 1998). Perhaps the reason in this case is that transient orcas, like humans, moved into coastal ecosystems relatively recently (Springer et al. 2003). Humans also have the advantage of agricultural subsidies, in turn subsidized by oil. This might be the single most important reason why humans are unlike marine top predators and their effects on food webs are remarkably different. Lotka-Volterra dynamics and population regulation can operate for low-mobility marine predators, but human populations so far do not appear to be regulated by changes in the abundance of lower trophic level marine prey. Ironically, humans are using fossil food webs to exploit present marine food webs.

Overfishing and inverting food-web structure

Ecological textbooks portray the classic pyramidal food-web structure, where the biomass of a trophic level is always lower than the biomass of the trophic level immediately underneath. The reasons are purely energetic. While this seems to be a general pattern in terrestrial food webs, in aquatic food webs the biomass of an upper trophic level can be higher than the biomass of a lower trophic level upon which it feeds, if the lower trophic level has sufficiently rapid turnover. This is the case of the coral reef food web of the northwest Hawaiian archipelago, a large marine protected area where the biomass of top predatory fishes is 54% of total fish biomass (Friedlander and DeMartini 2002). In contrast, in the main Hawaiian Islands, subject to intense fishing, the biomass of top predators is only 3%. Top predatory fishes are not exclusively piscivorous, hence the differences in biomass between top predators and all prey will be smaller than those for fishes alone. An interesting result

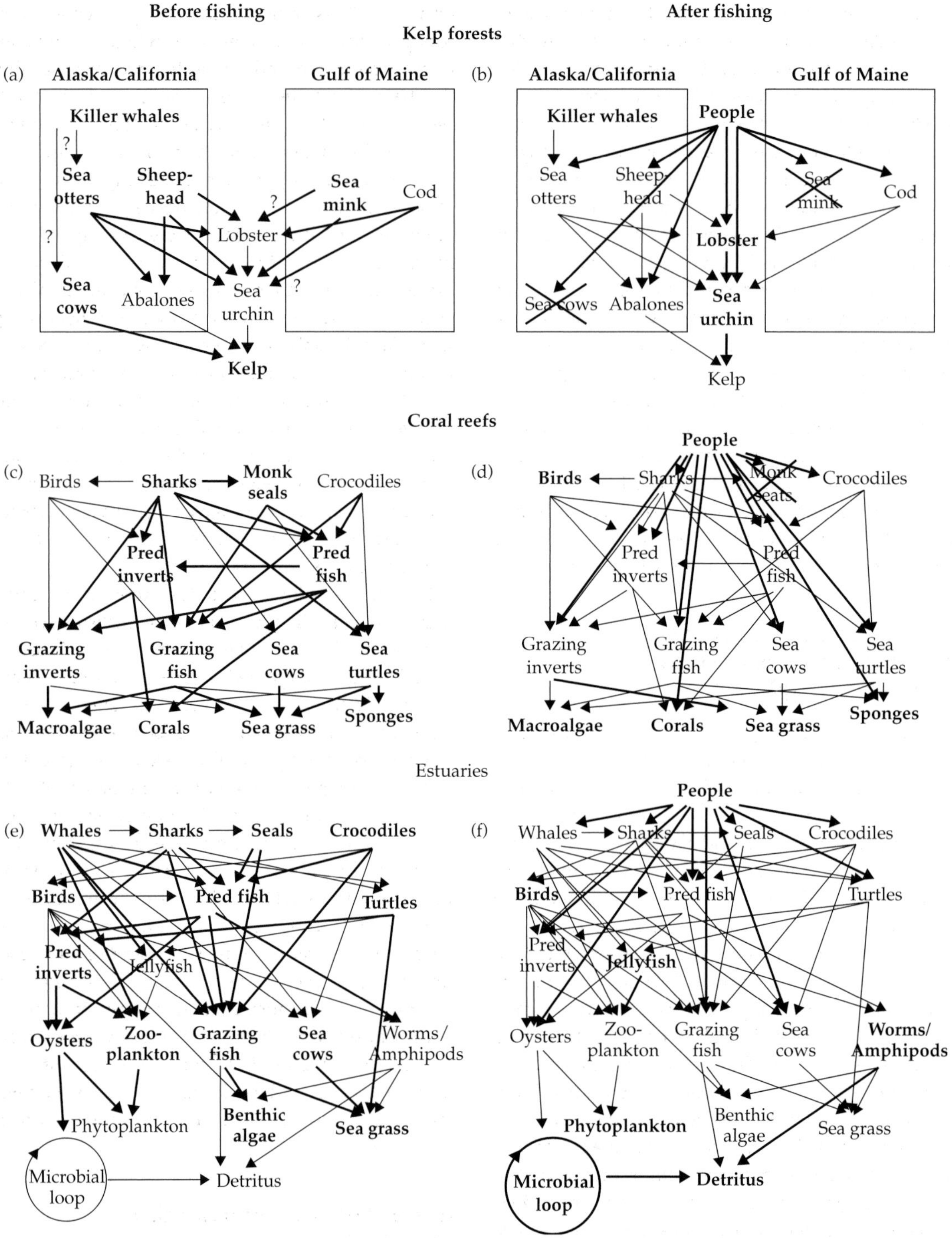

Figure 13.1 Schematical representation of selected marine food webs before and after intense fishing from Jackson et al. (2001).

from that study is that, although the biomass of top predators is overwhelmingly greater than that of lower trophic levels in the northwest Hawaiian archipelago, the biomass of its lower trophic levels is still greater than that in fished areas. This supports the idea that humans are often the strongest interactors in marine food webs. These extraordinary differences in food-web structure between areas with and without fishing suggest that our knowledge of marine food webs is probably biased toward simplified food webs, as our baselines for "pristine" systems shift. When top predators are eliminated, increases in abundance of weaker interacting prey may cause significant population impacts. The organization of the food web also shifts: the strength of top-down control is diminished and the importance of bottom-up control enhanced. We do not claim that this is necessarily a general pattern in marine food webs; nonetheless, it illustrates the danger of generalizing missing important functional components from webs.

Species relative abundance and evenness as an indicator of food-web complexity

The distribution of species relative abundance in food webs can also provide a simple means of measuring the health and vulnerability of ecosystems. Species abundance distributions reflect the complexity of the underlying food web and its associated dendrogram of niche similarities (Sugihara et al. 2003). Simpler food webs (e.g. two trophic levels based on single ecological resources) are expected to have very uneven abundance distributions and are dominated by a few species; whereas more complex systems will have more equitable relative abundance distributions (Sugihara et al. 2003). This applies to both between- and within-system comparisons. Thus, the overlying abundance pattern is a reflection of the functional organization of food webs. Equitable abundances imply symmetrical dendrograms of niche similarity. Such symmetrical or evenly branched dendrograms correspond to an underlying niche space that is complex and has many different structuring forces (e.g. a diverse resource base) that give rise to a partial ordering of niche similarities (apples and

oranges). This reflects the many more-or-less independent ways of making a living, which is an expression of the underlying heterogeneity in the "realized" or "functional" structure of the food web. Thus, the connectivity of the "effective" linkages has a more spread-out look as in scale-free networks, as opposed to uniformly dense as in random networks. The heterogeneity implied by scale-free structure in food webs corresponds to more evenly branched niche similarity dendrograms. Using the approach advocated earlier, along succession and for a particular food web belonging to a relatively stable environment, we would expect dendrograms to become more evenly branched to reflect the partial ordering of niches (apples and oranges aspect), abundance distributions to become more equitable, and food webs to become more complex (less homogeneously connected). However, this may not be true for high-energy systems where the endpoint of succession is characterized by the dominance of one or few species.

Evenness declines when the underlying functional portrait simplifies, and the niches become homogenized in that they revert from a "partial ordering" to a simplified "perfect ordering." Such homogenization occurs when a single overwhelmingly strong factor is imposed on the system. Thus, if a previously complex system is subjected to a strong homogenizing force (e.g. intense pollution), a formerly complex partially ordered system can become a perfect ordering. Species can now be appropriately ranked in relation to how they respond to the single dominant structuring force.

The predicted decrease in equitability in abundances that comes from simplification of the underlying niche space via homogenization of niches is a phenomenon that has enormous empirical support. The reduction of evenness that accompanies human disturbance (such as nutrient enrichments or other forms of homogenizing stress) is one of the most robust generalizations in ecology.

Acceleration and homogenization of marine food webs

Odum (1969) and Margalef (1997) suggested the general pattern that average growth rates of organisms will decrease with successional maturity, and that mature communities will have smaller production:biomass (P:B) ratios. In this scheme, early successional stages, characterized by fast growing opportunistic species, are replaced by slower growing, tough competitors. There are exceptions of course, but this is a good generality. Fishing acts the opposite way, selectively removing the top predators first, and gradually moving down to target lower levels in the food web (Pauly et al. 1998; Sala et al. 2004). The removal of predators accelerates the turnover of food webs by (1) reducing the biomass in species with lower turnover (P:B ratio), and (2) eventually triggering an increase in the biomass of prey with higher turnover. The end result is an increase in the turnover of the entire food web.

Accelerated growth rates often lead to destabilization of systems. A very general and robust result from ecological theory concerns the destabilizing effect of accelerating growth rates. This idea was discussed early on by Rosenzweig (1971) in his paradox of enrichment, and has emerged as one of the more robust generalities of ecological theory (May 1974). Resource enrichment (e.g. nutrient loading) provoke higher growth rates that destabilize the system and lead to species loss and reduced equitability. In the process of selectively culling species with lower growth rates, fishing increases the average growth rates in food webs, which are also more likely to be regulated by bottom-up processes, and exhibit greater fluctuation and instability. Chronic fishing pressure does not allow marine food webs to exhibit lengthy successional dynamics, and it can lock food-web structure into one dominated by high turnover species. The endpoint of the degradation dynamics may be (as it has been already observed in estuarine systems) a food web overwhelmingly dominated by planktonic organisms, such as jellyfish and protists (Jackson et al. 2001). The functional structure of the web is lost after the ecological extinction of entire trophic levels. While on land higher turnover means greater agricultural production, in the sea it is not clear whether higher turnover always means higher production available to humans. We believe that understanding the relationship between food web turnover and food availability is going to be a crucial step in fisheries management and marine conservation in general.

The decline of top predators, the rise of high turnover species, and the microbial loop represent an acceleration and functional homogenization of the food web. An increasingly important factor in this homogenization of nature is the spread of invasive species. Biological invasions have the potential to homogenize entire food webs, and shift species abundance distributions to more skewed ones. Although connectivity and number of trophic links in an invaded food web might be maintained, the community-wide ecological effects of affected species can be dramatically altered. Invasive species including crabs (Grosholz et al. 2000), snails (Steneck and Carlton 2001), and algae (Meinesz 2002; Boudouresque and Verlaque 2002) can affect all trophic levels. An extreme example of the homogenization and acceleration of a marine food web after a biological invasion is the small planktonic ctenophore *Mnemiopsis* in the Black Sea. The introduction of this exotic species caused dramatic reductions in fish biomass and explosive increases of gelatinous zooplankton (Shiganova and Bulgakova 2000) resulting in overall reduced equitability in abundances in the web.

In conclusion, humans universally accelerate marine food webs by (1) eliminating food-web components with slow dynamics (low turnover) and enhancing dominance of high-turnover species, and (2) accelerating the dispersal of locally weak interactors (but potentially strong somewhere else), which probably would have failed to disperse and colonize on their own during ecological or evolutionary timescales. We are also accelerating evolution by increasing by several orders of magnitude the number of natural experiments on species interactions in ecological timescales (Palumbi 2002). In other words, enhancing the links between food webs entails the acceleration of the dynamics within the webs. Inevitably, the accretion of structure and information in human webs (mostly urban) has to cause an acceleration and homogenization of the biosphere, including marine food webs, besides the acceleration of the oxidation of the necrosphere (Margalef 1991). Such are the consequences of ecological globalization.

Human impacts reverse successional trends

There are regularities commonly found in eco-logical successions. Some general changes occur-ring from early to mature successional stages are increases in species richness, number of trophic levels, biomass of higher trophic levels, total bio-mass, and three-dimensional biogenic structure (Margalef 1997). Although these regularities have been well documented in phytoplankton, we know very little for most other marine systems (but see Grigg and Maragos 1974; Grigg 1983). The old diversity concept, understood as the distributions of abundances into species in food webs (also called "ecodiversity," to distinguish it from "biodiversity"), often increases along terrestrial succession as well (Margalef 1997). In marine food webs, ecodiversity may show a unimodal rela-tionship. In an underwater lava flow in Hawaii, ecodiversity of colonizing coral communities increased with time but decreased before reaching the successional endpoint, due to extreme compe-tition for space, which led to monopolization by a few dominant species (Grigg and Maragos 1974). In any case, the increase in species richness in a food web over successional time will inevitably result in the multiplication of the number of spe-cies dependent or subordinate to those already present, and therefore in an increase of food-web complexity.

Although we know very little of changes in marine food-web properties in successions, there is evidence of short-term (annual) successions in planktonic and benthic communities. In a Medi-terranean rocky sublittoral food web, algal assemblages undergo striking seasonal changes in structure (Ballesteros 1991; Sala and Boudouresque 1997). Although algal species composition does not change significantly, total biomass shifts from an annual low where biomass is partitioned quite evenly among a large number of species (high ecodiversity), to a peak where biomass is mono-polized by a few large species (low ecodiversity) (Ballesteros 1991). These seasonal changes in algal biomass are immediately followed by similar changes in epifaunal invertebrates (Sala 1997). The biomass of sea urchins and fishes, in contrast, does not exhibit significant seasonal changes, although their prey do (García-Rubies 1996; Sala and Zabala 1996). Although the species composition and topology of this food web does not change, the biomass of many of its components changes dra-matically throughout the year. We would expect that changes in prey items cause subsequent changes in interaction strengths. Therefore the structure of the food web exhibits seasonal chan-ges in structure and complexity. While these changes occur at a scale of months, longer-term changes can occur due to invasions of exotic spe-cies and interannual variations in recruitment of important species, such as sea urchins, among other factors (e.g. Sala 2004).

Similar changes in food-web structure occur at pluriannual scales in Californian giant kelp forests. The most mature successional stage of these kelp forests is virtually a monoculture of the giant kelp, *Macrocystis pyrifera* (Dayton et al. 1984, 1992). Every few years, strong storms or El Niño South-ern Oscillation episodes virtually destroy the giant kelp canopy, resetting the food web to early successional stages dominated by undercanopy or turf algae (e.g. Dayton et al. 1992). In both the Mediterranean and Californian food webs, successional trajectories involve an increase in three-dimensional biogenic structure, a recreation of ecological niches that allows the recruitment of new species in the food web, and an increase in total biomass and total production. Regardless of what triggers these asymmetrical changes of food-web complexity (e.g. disturbances or seasonal environmental cycles), complexity clearly increases along succession.

Human impacts (as well as other kinds of catastrophic disturbance) nearly always reverse successional trends. In the temperate food webs described above, fishing can cause sea urchin population explosions and the virtual elimination of complex algal forests (Estes et al. 1998; Sala et al. 1998; Steneck 1998). In the Mediterranean food web, increased sea urchin grazing reduces the biomass and diversity of the benthic community, and reduces the differences between the endpoints of the annual succession: less total algal biomass, less structural complexity, and much smaller epi-faunal biomasses (Verlaque 1987; Hereu 2004).

Most of the benthic dynamics in sea urchin barrens lay on a turf of microscopic primary producers, such as benthic diatoms, which is continuously grazed by the urchins. Fishing marine food webs thus accelerates the system by increasing the global P:B ratio. The number of species, the number of trophic links, and complexity in general also decrease with increased fishing. In addition, three-dimensional biogenic structure can drop dramatically because of trophic cascades. These cascades have the strongest food-web-side impacts when the removal of a predator results in the elimination of architectural species. The removal of the architectural species in turn has effects on the recruitment and feeding behavior of many other species. In the case of coral reefs, a decline in coral cover has been shown to cause striking declines in reef fish species richness and abundance (Jones et al. 2004).

Invasive species homogenize food webs by truncating the frequency distribution of species abundance, eventually turning it into an extremely skewed distribution dominated by the invader. Despite the introduction of one more species, changes arising from invasion generally end up in local extinctions and decline in species richness. A paradigmatic example is the transformation of Mediterranean sublittoral habitats after the introduction of the tropical green alga *Caulerpa taxifolia*. In the absence of the invader, Mediterranean sublittoral food webs are composed of an extremely diverse (up to >100 algal species in only 400 cm^2) and dynamic benthos, exhibiting annual and pluriannual successional dynamics (see above). After the invasion, the same habitats become a green carpet, and diversity plummets to very low values (Meinesz 2002). The dominance of space inhibits significant successional changes on the local communities. Finally, because *C. taxifolia* is a chemically defended species, the number and the diversity of trophic links also decreases.

The difference between nonanthropogenic and anthropogenic disturbance is that the former are generally pulse disturbances and seldom cause local extinctions, allowing the food web to restart ecological succession post-disturbance. In contrast, most human disturbances are chronic and cause local and even global extinctions. In many cases, this chronic pressure does not allow the ecological

communities to move along succession and locks them in early successional states as long as pressure is applied. But the removal of the disturbance may not be sufficient to allow the community to exhibit significant successional changes. This is a key point that has tremendous consequences for the conservation of marine food webs, as we will discuss below.

Environmental gradients and human impacts

Environmental gradients can also produce changes in the structure and dynamics of marine food webs and hence determine the strength and scale of human impacts. We would expect the complexity of food webs to be greater in oligotrophic systems (e.g. coral reefs) than in systems subject to high-energy/nutrient inputs (e.g. upwelling areas). The number of species and trophic levels, and thus the number of trophic links and functional subwebs are expected to be greater in low-energy systems, while the likelihood of monocultures or dominance of a few architectural species is greater in high-energy systems. Food-web structural diversity and evenness will hence be higher in low-energy systems. For example, food web complexity is lower in a high-nutrient Californian kelp forest overwhelmingly dominated by the biomass of one species, the giant kelp, than in an oligotrophic Mediterranean algal community with greater species richness and equitability (Graham 2004; Sala 2004). There also exist differences within systems. For instance, the lower trophic levels of Caribbean coral reefs were dominated by single species of *Acropora* (a coral with relatively high growth rate) in shallow habitats subject to strong wave energy, whereas in deeper, calmer habitats coral abundance was shared more evenly among tens of species (e.g. Goreau 1959). Anthropogenic disturbances have turned the shallow reefs previously dominated by *Acropora* into algal beds also dominated by a few species with yet higher turnover (Hughes 1994; Knowlton 2001). In the Mediterranean sublittoral, benthic communities exhibit an amazing gradient in structure and dynamics over a mere 30-m distance along a vertical wall. Shallow food webs are dominated by a few species of algae with strong seasonal dynamics, while deeper webs are dominated by a diverse community of suspension feeders (including sponges, ascidians, and cnidarians) with slow, pluriannual dynamics (Garrabou et al. 2002). This strong biological gradient is explained by a strong gradient in physico-chemical conditions: shallow communities are exposed to higher-energy inputs (more light and wave motion) than deeper ones.

Fluctuations in food-web complexity are expected to be higher in high-energy systems. The above differences in food-web structure, specially for benthic webs, partly occur because high-energy systems allow for dominance of species with high turnover, whereas in low-energy systems surface-dependent strategies are based mainly on the slow accretion of biomass and nonfunctional structures, such as the biogenic matrix of a coral reef (Zabala and Ballesteros 1989). Food webs in higher-energy systems have a faster turnover (P : B ratio) than these in low-energy systems. This links structure and dynamics of food webs within a particular set of environmental conditions: complex food webs are "decelerated" and exhibit slow dynamics, while degraded food webs have simpler structure and are "accelerated" (higher turnover and instability).

What systems are thus more susceptible to greater rates of change and acceleration due to human activities? We would expect that human activities of similar intensity cause greater damage in systems with less complexity because they have reduced functional redundancy. Therefore, tropical oligotrophic food webs such as coral reefs would be more resistant to disturbance and less prone to fluctuation due to the reasons explained above. For instance, tropical systems would be less likely to exhibit trophic cascades. However, based on the environmental constraints, low-energy systems will suffer a greater rate of change than high-energy systems under a similar disturbance. In other words, the loss of complexity and information will be relatively greater in a "decelerated" system. Therefore, it seems that the best question to ask is not what food webs can we exploit, but "how are food webs going to respond to human activities on the basis of their intrinsic and environmental constraints?" The answer to this question will help determine

potential impacts of human activities and help prevent them.

Conservation of marine food webs

It should be said that conservation (or management) is a value-laden concept. Nonexploitative uses of marine ecosystems are optimal with complex food webs (e.g. a diver will spend a vacation in an unfished coral reef rather than in an overfished reef). In contrast, social groups with sole interest in industrial exploitative activities, such as fishing, may aim at less complex food webs in order to target single species with large P : B ratios. The simplest approach is to eliminate marine top predators to reduce competition for prey (Yodzis 2001), following the disastrous example of previous wildlife management in the continental US. A downside is that simpler food webs are also more prone to instability and fluctuations. Moreover, even though simplified webs may not be a goal in itself, under current fishing practices marine food webs will inevitably be simplified and degraded. It is still not clear which food-web complexity levels produce optimal catches. Will fishing be more productive by exploiting accelerated food webs or decelerated ones? Accelerated food webs are dominated by high P : B ratios, but total production available to humans may be lower, than in complex webs composed of more trophic levels. This is an unexplored area that deserves serious attention.

Do we want to preserve homogeneous ecosystems with complex food webs, or mosaics of patches at different successional stages? Conservation could be interpreted as the preservation of something stable, but food webs may be stable only at short temporal scales. Thus the goal of marine conservation should not be the preservation of a particular food-web structure, because this structure will inevitably change over time, and the cost : benefit expectations of the public may be lowered. For instance, a coral reef can be restored to former levels of complexity by an expensive and time consuming management process, with the involvement of local communities whose goal is to enjoy both the intrinsic and instrumental values of the reef. However, the reef can be torn to pieces by a hurricane, and unless the local species pool and the environmental conditions are right, it will not recover without further intervention. The public may be disappointed, blame managers and scientists, and be skeptical of further restoration efforts. Regardless of the goals of conservation efforts, the public and the decisionmakers need to understand the dynamic nature of ecological communities.

A better, albeit complex, approach for identifying marine conservation goals is to integrate ecological succession and food-web dynamics, as we have attempted here. Anthropogenic degradation of food webs is similar to catastrophic disturbances, although anthropogenic disturbances are chronic. In any event, anthropogenic disturbances disassemble marine food webs, and conservation efforts that reduce or eliminate these disturbances eventually result in the reassembly of the webs. The process can be viewed as a typical ecological succession where a food web acquires complexity over long-time periods, asymptotically decreasing the rate of change, to lose it more or less catastrophically, following anthropogenic disturbance in shorter timescales. When the disturbance is past, the food web slowly regains complexity, the timescale depending on the post-disturbance starting point. A good example is the changes in fish biomass in the presence and the absence of fishing. The recovery of predatory fish involves much longer timescales than their removal by fishing (Russ and Alcala 1996). And, although small species with fast turnover can recover to former unexploited biomass in coastal reserves within a few years, the complete recovery of higher trophic level predators can take more than 25 years (Micheli et al. in press). Marine conservation science, as successional dynamics, is asymmetrical: we know a great deal about the loss of complexity, but relatively little about the slow recovery of that complexity.

The goal of marine conservation should therefore be to preserve the global initial conditions so that food webs can self-organize and respond naturally to environmental change. Global initial conditions include availability of subsidies, favorable environments, and low level of chronic disturbance, among others. Food webs isolated from all possible subsidies are less resistant to disturbance and indeed more likely to be simplified. These subsidies include trophic subsidies as well as the availability of

species for colonization at the right successional time, including strong interactors and keystone species. For instance, species with relatively low dispersal cannot recover locally (e.g. in a marine reserve) after being eliminated if the regional pool is exhausted. However, this simple fact has been ignored in many conservation works, and used as an argument by marine reserve opponents to conclude that reserves do not work and should be eliminated. The metapopulation and metacommunity aspects of marine food webs are essential for understanding successional dynamics in degraded webs. In some specific cases, the recovery of top predators can be enhanced through human intervention (e.g. sea otter reintroductions in central California), but in many others these interventions will not prove cost-effective. Future studies should explore the rates of recovery of marine food webs, not just single species, embedded in a spatial mosaic of webs at varying levels of complexity. Meanwhile, we should aim at ensuring that strong interactors are available, and to preserve functional trophic levels or food-web modules (*sensu* Paine 1980). Diagnosing health and measuring success of management can be carried out using simple measures of food-web complexity. We believe developing a restoration ecology based on successional dynamics, and with practical implications at relevant scales, is an important challenge for food-web students and conservation biologists.

Future directions: the linkage between marine food webs and human networks

Science alone will not solve the conservation problems of marine ecosystems. Food-web theory and practice can provide decisionmakers with useful information and recommendations vis-à-vis the impacts of anthropogenic activities on marine ecosystems. However, these data and recommendations will be weighted against socioeconomic considerations before decisions are made. A major problem, which in many cases has resulted in counterproductive actions, is that food webs and human societal networks have seldom been analyzed in an integrated way. In most food-web studies which include humans, the socioeconomic intricacies that regulate the interaction strength among humans tend to be ignored. In other words, although economies and marine food webs are linked, food-web models and economic models are not linked. We urgently need to develop an integrated network science that links human dynamic actions and sociopolitical networks to ecological networks and other earth science systems.

Acknowledgments

We are grateful to J. Bascompte, L. Bersier, C. Boudouresque, R. Bradbury, F. Courchamp, P. Dayton, M. Graham, A. Hobday, J. Jackson, N. Knowlton, C. Melián, R. Paine, and R. Southwood for interesting discussions on many aspects of food webs.

CHAPTER 14

Biodiversity and aquatic food webs

Helmut Hillebrand and Jonathan B. Shurin

Introduction

Many fundamental ideas about food webs and the control of species diversity have roots in aquatic ecology. These include the trophic control of biomass (Forbes 1887; Lindeman 1942), the importance of top-down versus bottom-up regulating mechanisms (Lubchenco 1978; McQueen et al. 1989; Leibold et al. 1997), the mechanics of competition (Tilman 1982; Sommer 1985; Keddy 1989), and the topology of food webs (Martinez 1991; Havens 1992). Studies of food webs and species diversity diverge in that food-web ecologists often aggregate groups of species that share resources and consumers into guilds. Even very detailed food-web studies with information on many species contain a number of unresolved nodes (e.g. "bacteria," Martinez 1991). By contrast, species diversity is often defined monotrophically as the number of species within a guild such as primary producers or herbivores (Duffy 2002). While this division has historical and conceptual roots, it is becoming apparent that studies of food webs and species diversity have much relevance to one another. For instance, diversity and species heterogeneity within guilds can affect biomass partitioning and the control of trophic structure (Leibold et al. 1997). Similarly, trophic architecture may have major effects on diversity within guilds (e.g. Worm et al. 2002). Here we discuss avenues for integrating the two approaches, and the relationships among concepts pertaining to food webs and species diversity.

We begin by analyzing both the causes and consequences of diversity in simple consumer—prey systems (Figure 14.1(a,b)). First, we investigate how consumer presence affects prey diversity (prediction 1) and how prey density affects consumer diversity (prediction 2) (Figure 14.1(a)). Second, we reverse the question and review the evidence on how consumer and prey diversity affects the "strength" of the trophic interaction (i.e. the biomass effect of predators on resources, and vice versa, predictions 3, 5) and diversity at adjacent trophic levels (predictions 4, 6) (Figure 14.1(b)). We then continue by adopting a more holistic view, asking how effects from single trophic interactions propagate through food webs (prediction 7) and how diversity at one level affects the diversity at other trophic levels (prediction 8) (Figure 14.1(c)). Finally, we extend our scope and look at the regional setting of the food web (Figure 14.1(d)), addressing the importance of species additions by dispersal and invasion on trophic interactions (prediction 9).

The effect of single trophic interactions on diversity

The effect of consumer presence on prey diversity (prediction 1)

The effect of consumers on the coexistence of competing prey species has been a central focus in community ecology (Paine 1966; Lubchenco 1978; Gurevitch et al. 2000; Chase et al. 2002). Two counteractive mechanisms mediate the effects of consumption: first, consumers reduce prey biomass by increasing prey mortality and/or suppressing prey reproduction. In so doing, consumers potentially drive prey species locally extinct, thus reducing prey diversity. Second, consumers also prevent competitive exclusion of prey species and thus maintain local diversity.

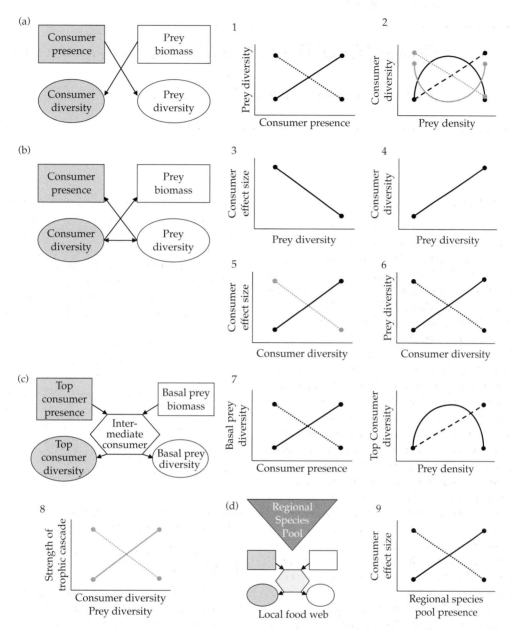

Figure 14.1 Predictions on the relation between diversity and trophic interactions. Effects are visualized as positive, negative, unimodal, or U-shaped trends. Gray symbols indicate less important or less well-studied relations. The 9 predictions are arranged in 4 clusters as outlined in the chapter. (a) **Predictions on the effect of single trophic interactions on consumer or prey diversity**. (1) Consumer presence affects prey diversity, (2) prey density affects consumer diversity. (b) **Predictions on the effect of consumer or prey diversity on the biomass or diversity of the adjacent trophic level in single trophic interactions**. (3) Consumer diversity affects consumer effect size, (4) consumer diversity affects prey diversity, (5) prey diversity affects consumer effect size, (6) prey diversity affects consumer diversity. (c) **Predictions on the propagation of effects on (or of) diversity in multiple trophic interactions**. (7) (Top) consumer effects on diversity cascade downwards and prey density effects on diversity propagate upwards, (8) diversity in consumer or prey levels affect the strength of the trophic cascade. (d) **Predictions on the importance of regional diversity in diversity–food-web relations**. (9) Regional species richness affects local consumer effect size.

These contrasting forces result in widely divergent consumer effects on prey diversity, ranging from positive to negative effects (Chase et al. 2002; Figure 14.1(a)). Positive effects of consumer presence were found in plant—herbivore interactions in both pelagic (Proulx et al. 1996) and benthic habitats in lakes (Hillebrand 2003), for large carnivores such as fish (Diehl 1992; Shurin 2001) as well as for small ones such as predatory nematodes (Michiels et al., 2003). But even within the same study systems, consumer presence decreased prey diversity under different circumstances (Proulx et al. 1996; Shurin 2001; Hillebrand 2003). It is thus of major importance to describe the circumstances under which consumer-mediated coexistence occurs (Chase et al. 2002). These factors pertain to the consumer–prey interaction itself (intrinsic factors) or to abiotic or biotic features of the habitat (extrinsic factors).

Intrinsic factors
Consumers are expected to promote prey diversity when their impact is greater on the dominant competitor. Actively selecting consumers can switch toward the most abundant prey and act as a stabilizing factor maintaining prey diversity (Chesson 2000). The same outcome—maintenance of prey diversity—also prevails when consumers are specialists and consumers of abundant prey are more common, whereas rare prey species suffer low per capita-loss to consumers (Pacala and Crawley 1992).

Consumers can also prevent prey exclusion by feeding mainly upon the dominant competitor species and limiting its growth (Chase et al. 2002), an effect known as keystone predation (Paine 1966; Lubchenco 1978; Leibold 1996). Keystone predation does not require actively selecting consumers, but may occur when dominance correlates to higher susceptibility to consumption due to growth form or chemical constitution (Hillebrand et al. 2000; Chase et al. 2002). Prominent examples of keystone predation are from marine intertidal areas, where primary space holders are often dominated by a single species or growth form in the absence of consumers (Paine 1966; Lubchenco 1978; Worm et al. 1999; Hillebrand et al. 2000). Other systems with more symmetric

competitive hierarchies may be less prone to keystone predation effects (Chase et al. 2002). The probability of consumer-mediated coexistence is moreover affected by intrinsic factors such as consumer-to-prey size ratio, which determines the consumer effect on the spatial structure (and therewith diversity) of the prey assemblage (Steinman 1996; Hillebrand 2003).

Additionally, consumer presence can weaken the competitive interactions between prey species by indirect effects on the maintenance of resource diversity (Abrams 2001), the regeneration of nutrients (Attayde and Hansson 1999), or on prey behavior (Eklöv and VanKooten 2001; Werner and Peacor 2003). These indirect effects promote prey diversity if they reduce the impact of the competitively dominant prey species.

Prey diversity also increases with consumer presence if the effect of consumers is distributed heterogeneously in space and time (Levins 1979; Pacala and Crawley 1992). Prey can persist by repeated dispersal to open spaces in a metacommunity (see prediction 9 for details), depending on their dispersal rates and effects on prey extinction (Shurin and Allen 2001). Consumers also remove the requirement for a trade-off between colonization and competition for coexistence between competing prey species: a prey that is an inferior colonizer *and* competitor can persist if the consumer is sufficiently abundant.

Extrinsic factors
Consumer density and prey productivity are two biotic factors that mediate consumer effects on prey diversity. Consumer density and feeding rate regulates how strongly consumers affect prey mortality. In analogy to the intermediate disturbance hypothesis (Connell 1978; Flöder and Sommer 1999), the highest prey diversity has been observed at intermediate consumer density (Lubchenco 1978; Sommer 1999). Rare or inefficient consumers do not prevent competitive exclusion of their prey, whereas abundant or efficient consumers can increase mortality to levels that inhibit the local existence of prey. Second, productivity affects the rate of competitive exclusion, which determines whether consumers promote prey coexistence (Huston 1994). Positive consumer

effects on prey diversity are most likely when competitive exclusion is fast (i.e. at high productivity). Productivity increases the population growth rates of competing species and intensifies competitive dominance (see below for more on productivity–diversity relationships). Consumers and disturbances play similar roles in maintaining diversity in that the response depends on whether their effects fall on dominant or subordinate species, and the rate at which processes which limit diversity (e.g. competitive exclusion) occur. The roles of productivity and disturbance have been incorporated in both nonequilibrium (Huston 1979, 1994) and equilibrium (Kondoh 2001) models. Both predict interactive unimodal relationships between diversity and both disturbance and productivity. These models offer concise predictions for consumer–prey dynamics: at high productivity, consumers enhance prey diversity, which suffers from fast exclusion rates, by preventing exclusion. At low productivity, consumers reduce prey diversity by adding mortality in a situation where few prey species are able to exist.

These predictions have been tested in two quantitative reviews, both of which corroborated the predicted patterns. In studies from freshwater, marine, and terrestrial systems, a preponderance of positive effects of consumers on prey diversity was found in highly productive systems, whereas negative effects of consumers dominated at low productivity (Proulx and Mazumder 1998). Worm et al. (2002) conducted a quantitative meta-analysis on freshwater and marine experiments that manipulated both consumer presence and resource availability. Corroborating the model predictions, consumers promoted prey diversity at high nutrient availability and decreased prey diversity at low nutrients (Worm et al. 2002). The interaction term between consumption and nutrient availability was stronger than the main effect of either factor, suggesting that the feedback between consumers and resources is of major importance for regulating prey diversity.

In addition to productivity, other physical characteristics of the habitat that influence the intensity of species interactions moderate the effects of consumers on prey diversity. Spatial heterogeneity in the physical habitat strongly affects consumer–prey dynamics in experiments and models. For instance, the presence of vegetation in ponds alters the effect of predatory fish on invertebrate diversity (Diehl 1992). Moreover, consumer presence often depends on the presence of vegetation structure within the habitat (e.g. macroalgae for marine invertebrates, Worm et al. 1999) or outside the habitat (e.g. trees as landmarks for swarming insects, whose larvae remain close to the egg deposition site, Harrison and Hildrew 2001). Temporal heterogeneity (i.e. periodic disturbance) also mediates the effects of consumers on prey diversity (Weithoff et al. 2000). Adverse physical conditions such as frequent disturbances or stress were proposed to reduce the importance of biotic interactions (Menge and Sutherland 1976; but see Chesson and Huntly 1997 for a critique).

The effect of prey density on consumer diversity (prediction 2)

Here, we ask—in analogy to the effects of consumers on prey diversity—how resource density affects consumer diversity (Figure 14.1(a)). This question is central to the debate about the relationships between productivity and diversity. Increasing productivity has been proposed to generate a unimodal or monotonic increasing response of diversity (Rosenzweig and Abramsky 1993; Abrams 1995). The increasing part of the productivity–diversity relationship has a strong theoretical basis. Increasing productivity results in larger population sizes and thus lower extinction risk (Rosenzweig and Abramsky 1993; Abrams 1995). Additionally, diversity increases with productivity if higher productivity promotes higher density of rare prey, allowing more specialized consumers to exist, or if productivity results in higher intraspecific density-dependence (Abrams 1995). Theoretical predictions for the negative part of the productivity–diversity curve are less straightforward. Rosenzweig and Abramsky (1993) reviewed nine hypotheses for the decrease of diversity at highest productivity and proposed that exclusion of consumer species increased at highest productivity because of reduced temporal and spatial heterogeneity in prey supply at higher prey density. By contrast, Abrams (1995) predicted that monotonic increasing functions should prevail and argued that empirical evidence for reduced prey heterogeneity at high supply rates was poor. Recently, Wilson et al. (2003) showed that enrichment may decrease evenness and species richness if interspecific variance in carrying capacity increases with productivity (i.e. mean carrying capacity). In that case, few species come to dominate the community at high productivity (Wilson et al. 2003). Thus, the relationship between diversity and productivity depends on how the effects of abiotic factors (e.g., nutrients) are distributed among species.

The diverse theoretical predictions are reflected by a lack of any general empirical pattern relating diversity and productivity. Surveys of the literature have found that unimodal, positive relations, and nonsignificant patterns are common, but negative and U-shaped relationships are also observed (Waide et al. 1999; Dodson et al. 2000; Mittelbach et al. 2001). The measures of productivity used in these studies varied broadly and comprise direct measures of productivity (such as actual or potential evapotranspiration, primary productivity, or resource availability) and indirect measures (such as biomass, rainfall), which may already affect the shape of the relationship (Groner and Novoplansky 2003).

Aquatic systems harbor mostly unimodal functions of diversity and productivity. Waide et al. (1999) found preponderance of unimodal functions for aquatic vertebrates and invertebrates. Plants showed fewer significant relationships in general, but the patterns in aquatic systems were more often unimodal than on land. In a survey of 33 lakes, Dodson et al. (2000) found unimodal relationships between species richness of different taxonomic groups and lake productivity. Mittelbach et al. (2001) used original data (rather than the published results as in Waide et al. 1999) for a meta-analysis of the productivity–diversity relation and found that the average linear regression term was positive and the average quadratic regression coefficient was negative. This analysis indicated unimodal nonlinearity, and unimodality was the pattern most often observed, followed by monotonic increases. Among different groups of aquatic bacteria, diversity was related unimodally to productivity, although

significant U-shaped functions were observed too (Horner-Devine et al. 2003).

This lack of generality suggests that diversity–productivity patterns depend on system-specific environmental conditions. The effect of productivity on diversity at one trophic level depends on the presence of a consumer (Worm et al. 2002). Coexistence at one level is also enhanced if productivity fluctuates in time (Levins 1979; Sommer 1985) or is heterogeneously distributed in space (Amarasekare and Nisbet 2001; Chase et al. 2001). The history of community assembly can also influence the shape of diversity–productivity patterns, with different invasion histories showing distinct relationships (Fukami and Morin 2003).

Chase and Leibold (2002) found that different relations operate on different scales in ponds, with a unimodal relationship between system productivity and local richness, and a monotonic positive relationship with regional richness. This pattern indicates that productivity effects on species coexistence are scale dependent; increasing nutrient supply enhances local diversity in oligotrophic systems and decreases it under eutrophic conditions, while greater productivity always promotes regional diversity. Thus, more productive "regions" have higher beta diversity or more turnover in species composition between local sites. Chase and Leibold (2002) propose a number of plausible explanations for this observation, including greater variability in abiotic conditions among sites in productive regions or local interactions giving rise to alternative stable states only under high productivity.

The effect of diversity on trophic interactions (predictions 3–6)

Diversity is under strong direct control by trophic interactions, but evidently also from other local and historical factors. It is therefore warranted to reverse the question and to ask whether variation in diversity has consequences for the control of trophic structure and partitioning of biomass in communities. This question is highly relevant in the face of globally increasing rates of biodiversity loss. These two aspects are not independent, but tightly linked by feedback loops, that is, the relation between species diversity and trophic interactions is bidirectional (Worm and Duffy 2003). Recent experimental and theoretical advances revealed that productivity and community stability may change with diversity within trophic groups (Loreau 2000; Tilman et al. 2001). Notwithstanding the much debated difficulties in the design and proper analysis of these experiments, these studies raise the question of how the diversity of prey or consumers affects (1) the strength of trophic interactions and (2) diversity at the corresponding higher or lower trophic level (Figure 14.1(b)).

The effect of prey diversity

Effect on consumer control of prey biomass (prediction 3)
The impact of prey diversity on consumer control over prey biomass has received surprisingly little theoretical attention, and experimental manipulations are scarce (Duffy 2002). Theory predicts that high prey diversity corresponds to higher variance in edibility and thus higher probability of including inedible species (Duffy 2002). Alternatively, consumers can derive nutritional benefits from a diverse prey community (the balanced diet hypothesis, DeMott 1998), and therefore be more abundant in areas with more varied resources. Thus, two opposite predictions emerge. Either increasing prey diversity enhances consumption resistance (and reduce the consumer efficiency, Figure 14.1(b)), or increasing prey diversity increases consumer abundance (and enhance consumer effects).

Steiner (2001) found weaker consumer control over algal biomass by zooplankton at higher algal diversity, which he attributed to the presence of inedible phytoplankton at high diversity that were able to compensate for the reduction in vulnerable taxa. In a quantitative meta-analysis of grazer–periphyton experiments, grazer effects decreased significantly with increasing algal diversity (Hillebrand and Cardinale 2004). This relation of consumer effects to prey diversity remained consistent even after carefully accounting for several confounding variables such as habitat type, experiment design, or consumer and prey biomass.

By contrast, a meta-analysis of trophic cascade experiments across ecosystems found that producer diversity did not explain variability in the strength of top-down control (Borer et al., in press). Thus, high prey diversity may dampen top-down control over community biomass within systems, however these effects may be obscured by other sources of variability among systems.

Effect on consumer diversity (prediction 4)
Prey diversity is generally predicted to increase consumer diversity through greater opportunities for niche specialization and differentiation. This prediction applies to both nonsubstitutable resources (nutrients, light) and substitutable resources such as biological prey for heterotrophic consumers. However, the mechanisms may differ for these two prey types.

Non-substitutable resources. Classical competition theory predicts that the number of coexisting species at equilibrium cannot exceed the number of limiting resources in homogenous environments (Tilman 1982), a prediction that has been corroborated in controlled experiments (Tilman 1982; Sommer 1985). Consequently, Interlandi and Kilham (2001) found that phytoplankton diversity in lakes was greatest when more resources (light and nutrients) were below limiting levels. Equilibrium-dynamics thus suggest a linear link between resource diversity and consumer diversity. Analyses of nonequilibrium models show that nonlinear dynamics allow many species to coexist on few resources (Armstrong and McGehee 1980). Using models with three limiting resources and a large number of competing species, nonequilibrium dynamics with oscillations and chaotic fluctuations reduced competitive exclusion and maintained high diversity (Huisman and Weissing 1999).

Substitutable resources. By ingesting organic particles, heterotrophic consumers acquire resource packages, which are (partly) substitutable but differ in quality. While producer species vary tremendously in nutrient content, invertebrate and vertebrate consumers are less variable in their biochemical body composition. Therefore, the growth and survivorship of consumers is determined by the match or mismatch between consumer nutrient demand and prey nutrient composition (Sterner and Elser 2002). Loladze et al. (2004) showed theoretically that these stoichiometric constraints can result in coexistence of two consumer species on one prey. In addition, increased prey diversity may increase consumer survivorship, growth, and reproduction due to a more balanced diet (DeMott 1998; Anderson and Pond 2000). Finally, higher diversity of substitutable prey allows more consumers to coexist, given that consumers show specialization for one prey type and the proportion of such specialists remains constant or increases with prey diversity.

More detailed theoretical or empirical studies on the effect of prey diversity on consumer diversity are scarce. Petchey (2000) proposed three mechanisms promoting consumer persistence at higher prey diversity, "prey composition," which is identical to a more balanced diet, "prey biomass" and "prey reliability." In addition to providing a more balanced diet, higher prey diversity may produce more abundant prey through sampling effects or over-yielding at higher diversity (more prey biomass) or reduced temporal variability of prey biomass (more reliable prey). Petchey (2000) tested these predictions in aquatic microcosms and found that the persistence of consumers was enhanced by prey diversity via the prey reliability mechanism. A number of empirical tests in terrestrial systems also point to positive effects of resource diversity on the number of coexisting consumers (Siemann et al. 1998; Haddad et al. 2001), although other results are more equivocal (Koricheva et al. 2000; Hawkins and Porter 2003). Also the diversity of marine phytoplankton and zooplankton were uncorrelated at the global scale (Irigoien et al. 2004). Whereas high prey diversity promotes consumer coexistence by providing more opportunities for niche differentiation, the importance of this mechanism relative to others in regulating heterotrophic diversity remains unknown in most systems.

The effect of consumer diversity

Effect on consumer control of prey biomass (prediction 5)

Most theory and experimental work indicates that more diverse consumer assemblages should be more effective at reducing prey biomass (Holt and Loreau 2001; Figure 14.1(b)). The proposed mechanisms parallel biodiversity effects on process rates of primary producers (Loreau 2000; Tilman et al. 2001). Consumer diversity is positively related to consumer biomass and production (Moorthi 2000; Duffy et al. 2003) due to selection or complementarity effects (Loreau 2000). Selection effects increase the chance that a certain trait (here consumer efficiency) is present at higher diversity (Loreau 2000). Complementarity mechanisms (Loreau 2000) may prevail when different consumer species have different food requirements or when positive interactions among consumer species increase consumer efficiency.

The available empirical evidence for evaluating these predictions has yielded generally supportive, but also some mixed results. In aquatic microcosms, phytoplankton biomass decreased with increasing consumer diversity (Naeem and Li 1998). Downing and Leibold (2002) manipulated species richness across three functional groups in pond food webs (invertebrate grazers, invertebrate predators, macrophytes) and analyzed the response of phyto-, zooplankton, and periphyton biomass. Phytoplankton biomass increased, and zooplankton and periphyton biomass decreased with increasing overall richness, consistent with higher consumption of periphyton and zooplankton by either invertebrate grazers or predators. Suspension feeding invertebrates in streams showed significantly increased filtration rates at higher species diversity, which was also higher than expected from monoculture performance (Cardinale et al. 2002). Sommer et al. (2003) found increased consumption of phytoplankton in mesocosms stocked with copepods after invasion of cladocerans into the mesocosms. Convincing evidence for positive effects of consumer species richness on process rates comes from a series of well-connected experimental and field studies on shredder species in streams. In a field survey of streams in Sweden, shredder species richness was positively associated with consumption (leaf litter breakdown) rates (Jonsson et al. 2001). Laboratory experiments showed that high species richness increased leaf processing rates (Jonsson and Malmqvist 2000). Replacing individuals of one species with individuals from a different species also increased process rates (Jonsson and Malmqvist 2003*b*).

However, not all studies have found diverse consumers to be more efficient at reducing prey biomass. Equivocal effects of diversity on process rates were found for filter-feeding collectors in streams, whereas process rates even decreased with increasing species richness for both grazers and predators (Jonsson and Malmqvist 2003*a*). In experiments with multispecies algal assemblages, manipulating the number of *Daphnia* species did not affect phytoplankton biomass (Norberg 2000). However, consumer richness was low (maximum four species) and the species were closely related. The importance of the range of diversity manipulated becomes evident from the only marine study on consumer diversity to date: grazer diversity and identity in eelgrass communities affected neither epiphyte grazing nor eelgrass biomass when 1–3 consumer species were manipulated (Duffy et al. 2001). Subsequent experiments with up to six species resulted in a clear decline in algal biomass at high grazer diversity, driven mainly by the highest diversity treatment (Duffy et al. 2003).

Examples of species interactions that can result in strong top-down effects at high consumer diversity are physical habitat modification and behavioral flexibility. Cardinale et al. (2002) found that increasing diversity of suspension-feeding caddisfly larvae led to reduced current shading by increasing biophysical complexity and thus increased the fraction of suspended material available to filter feeders. In cases of prey species able to seek refuges, the presence of two consumers may facilitate consumption (Eklöv and VanKooten 2001): if only one predatory fish is present in either pelagic or littoral habitats, prey move to the less risky habitat, whereas the presence of both pelagic and littoral predators reduces the possibility for predation avoidance and increases predation success of both species. Similar predator facilitation has also been shown between trout and stonefly predators in streams that modify prey (mayfly) behavior (Soluk and Richardson 1997). Such positive effects or indirect mutualism may be highly asymmetric. Both snails and tadpoles consume periphyton and prefer microalgae

over filamentous algae. In experimental containers, snail presence had positive effects on tadpole development, growth, and weight (Brönmark et al. 1991). Since tadpoles were competitively dominant over snails, snails were forced to feed on low-quality algae. By doing so, they increased nutrient turnover rates, which enhanced microalgal productivity and food supply to tadpoles.

However, species interactions provide mechanisms not only for positive, but also for negative effects of consumer diversity on the strength of consumer effects. Increasing the number of consumer species can decrease consumption if direct behavioral interactions, territoriality, or aggression play important roles. Amarasekare (2003) proposed that decreasing consumption rates can be expected if consumer species compete not only by exploitation of prey but also engage in interference competition. When inferior exploiters are superior interference competitors, consumer efficiency will be reduced at higher diversity—a prediction corroborated by experimental results on a terrestrial parasitoid system (Amarasekare 2003). Finke and Denno (2004) provide another example where intraguild predation and aggression among different spider species reduced their cascading effects on herbivorous insects and plants. Thus, diverse predator assemblages exerted weaker top-down control over lower trophic levels.

Effect on prey diversity (prediction 6)

Whereas the effect of prey diversity on consumer diversity is relatively well known (see above), there is only scarce information whether a single consumer species has qualitatively different effects than more diverse consumer assemblages. We showed above (prediction 1) that consumers enhance prey diversity if competitively dominant prey suffer higher mortality or if consumers maintain resource diversity for the prey. Both mechanisms are potentially affected by consumer diversity. Higher consumer diversity can result in more complete utilization of the prey spectrum. In that case, added consumer species reduce prey diversity by increasing the consumption of subdominant prey species, which reduces the probability of consumer-mediated coexistence. Alternatively, higher consumer diversity can increase the diversity of the prey's resources through nutrient regeneration, space opening, or increased spatial heterogeneity. These mechanisms are not mutually exclusive, but may interactively shape the response of prey diversity to increased consumer diversity.

Assembling the food web: cascades and propagating effects (predictions 7–8)

The scenarios described above for the effects of consumer and resource control over diversity become more complex once we imagine a world with more than two trophic levels, and the potential for omnivorous feeding links. Extending the previous sections, we ask below the questions how top-consumer presence cascades down to basal prey diversity or conversely, how basal resource enrichment propagates up to higher trophic levels (prediction 7) and how diversity changes on one or several trophic levels affect propagating trophic interactions (prediction 8) (Figure 14.1(c)).

Propagating trophic effects on the diversity of basal prey or top consumers (prediction 7)

Predictions are still straightforward for chain-like interactions with predator, intermediate consumer, and basal prey. Such chains represent a cornerstone of community ecology, implemented by concepts such as the *Trophic Cascade Model* (Carpenter et al. 1987), the *Exploitation Ecosystem Model* (Oksanen et al. 1981), or the *Bottom-up Top-down Model* (McQueen et al. 1989). The main focus of these concepts has been to explain the distribution and control of biomass on different trophic levels (Leibold 1989; Leibold et al. 1997), whereas the control of diversity received less attention. From the previous sections we derive that predator presence will alter the biomass and diversity of the intermediate consumers, which in turn affects the diversity of the basal prey. Similarly, basal prey density will also affect predator diversity via changes in the intermediate consumer level. The sign of these indirect effects depends on the mechanisms and environmental factors outlined above (predictions 1–6).

The cascading effects of predator presence on basal prey diversity were shown in simple aquatic food webs found in the pitcher plant *Sarracenia purpurea*, where the addition of a top predator reduced the abundance of the intermediate consumer, cascading into increased bacterial abundance and diversity (Kneitel and Miller 2002). The bottom-up propagation of increasing basal prey density on predator diversity has mainly been shown in correlational studies suggesting non-linear relationships. Leibold (1999) analyzed 21 fishless ponds and found hump-shaped curves of primary producer (phytoplankton) diversity and consumer (zooplankton) diversity with resource level (either nitrogen or phosphorus). Jeppesen et al. (2000) surveyed Danish lakes in five productivity (total P) classes and found that species richness of zooplankton and submerged macrophytes declined with increasing productivity, whereas unimodal relationships were observed for phytoplankton, fish, and floating macrophytes. In a third study, lake productivity was unimodally related to the diversity of both primary producers and consumers (Dodson et al. 2000).

A critical factor relating propagating resource effects and predator diversity in chain-like interactions is the increasing preponderance of non-edible prey at high resource supply (Leibold 1989; Grover 1995). These predictions correspond well to experimental results on pelagic communities indicating that the degree of resource and consumer control depends on the proportion of inedible algae (Bell 2002; Steiner 2003). Thus, enrichment may inhibit rather than benefit top consumers due to interactions with inedible prey. However, different enrichment–edibility trade-offs may be important in other aquatic communities. In benthic communities, species profiting most from enrichment are also most susceptible to herbivores (Hillebrand et al. 2000).

The possible effects of top consumers on basal prey diversity become much more complex once more reticulate feeding relations such as omnivory are considered. Omnivory is common in freshwater food webs (Diehl 1992, 1995), both as lifelong or life-history omnivory (Polis and Strong 1996; Persson 1999). Omnivory may uncouple consumer and prey abundance (Polis and

Strong 1996), or lead to complex population dynamics (Persson 1999). Intraguild predation may lead to nonlinear consumer–prey dynamics and nonlinear effects of productivity on the food web (Diehl and Feissel 2001). During ontogeny of large consumers, size varies and so does prey size, resulting in complex physiology–population relations, mutual predation, or intraspecific predation (cannibalism) between size classes (Woodward and Hildrew 2002; Persson et al. 2003). Moreover, indirect consumer effects such as nutrient regeneration also propagate through food webs (Persson 1999). Thus, while predictions from chain-like trophic interactions allow addressing the possible effects of top-consumers on basal prey diversity, more realistic and reticulate food webs prevent simple assessments of cascading effects on basal prey diversity.

Cascading effects of diversity and diversity loss (prediction 8)

Propagating effects of diversity in top or basal trophic groups on the biomass and diversity on the other trophic level are not well studied in multitrophic assemblages (Raffaelli et al. 2002). Theoretical predictions for diversity effects on biomass in multitrophic situations are highly complex (Holt and Loreau 2001; Thebault and Loreau 2003). This complexity arises already in simple food-web models with 2–3 trophic levels without omnivory (Abrams 1993): increasing diversity within a trophic level simply from 1 to 2 produced virtually any outcome for propagating effects depending on food-web structure.

Experimental tests of these complex predictions are scarce and focus on the role of the intermediate consumer. Duffy (2002) suggested that increasing diversity at intermediate consumer levels would reduce the strength of trophic cascades since probability of resistant species being present increases with diversity. A terrestrial study including plants, herbivores, and their parasitoids corroborated this suggestion (Montoya et al. 2003), whereas Shurin (2001) found that high grazer diversity did not weaken the strength of the trophic cascade.

Increasing diversity of basal prey can also enhance consumption resistance (Hillebrand and Cardinale 2004) and therewith reduce the propagation of enrichment effects through the food webs. To our knowledge, this prediction has not been tested. Moreover, the propagation of effects is complicated by indirect effects such as apparent competition (Holt et al. 1994) and preference of associated prey species (Wahl et al. 1997; Karez et al. 2000). Both patterns may result in increasing consumer density at higher prey diversity.

Much of the diversity–ecosystem functioning debate is motivated by the current high rate of anthropogenic species loss and it is important to ask whether the diversity within the food web affects the probability of secondary losses following the deletion of one or several species. Borrvall et al. (2000) found that more species per functional group reduced the risk of cascading extinctions in model food webs. The probability of secondary losses was also affected by the trophic position of the species lost and the distribution of interaction strengths. However, a subsequent study with more complex food webs addressing connectance as well as diversity revealed that the probability of secondary extinctions decreased with connectance, but was independent of species richness (Dunne et al. 2002a).

Emergent properties of food webs

The obvious complexity of predictions on diversity effects in reticulate food webs (predictions 7–8) resulted in early attempts to describe general properties of entire food webs in order to derive general conclusions on food-web function and assembly (Cohen 1978; Martinez 1991; Havens 1992). Several of these emerging properties of food webs are supposed to change with diversity, such as the overall number of links (Havens 1992; Martinez 1993), the number of links per species (Martinez 1991, 1993; Havens 1992; Schmid-Araya et al. 2002b) and the mean and maximum length of food chains (Martinez 1991; Schmid-Araya et al. 2002b). Other properties have been more controversial. It remains unclear whether connectance (the proportion of possible links realized) is insensitive to web size (Martinez 1991, 1993) or if it decreases with diversity (Havens 1992;

Schmid-Araya et al. 2002*b*). The general question whether more diverse food webs are more stable (persistent, resistant against perturbations and extinctions) has a long history rich in reversals of paradigms (see McCann 2000 for a recent review).

More recent food-web analyses predicted the topology of food webs from species number and connectance (Williams and Martinez 2000), analyzed the relationship between diversity and the distance of nodes over which trophic effects propagate (Williams et al. 2002) and the division of food webs into compartments (Krause et al. 2003). Brose et al. (2004) unified the scaling of diversity to spatial and food-web properties, which allows investigating the impact of spatial patterns on food-web structure. New analysis methods come from network theory, which has become an important tool to analyze food webs (Dunne et al. 2002*b*, Garlaschelli et al. 2003). Whereas the analysis of emergent properties may not be suitable to elude mechanisms of the relation between food-web structure and biodiversity, it results in the description of important patterns which give rise to testable hypotheses on the response and effects of diversity in trophic interactions. Moreover, there is clear potential to increase the strength of such analyses (Borer et al. 2002).

Extending the scope

The regional perspective (prediction 9)

Much of our thinking about aquatic food webs is focused on processes that take place within communities. This perspective dates back to Forbes' (1887) notion of lakes as "microcosms," or isolated communities, separate and apart from the surrounding landscape. Indeed, the important roles of many local biotic and abiotic constraints on food webs have been unambiguously demonstrated. However, considerable evidence has also accumulated that all ecological communities are influenced by forces that operate over broader scales than just within habitats. These regional forces include dispersal among habitats, speciation, and subsequent geographic spread, and large-scale climatic effects. Early thinking about the roles of local and regional processes treated regional

processes and local interactions as mutually exclusive alternative explanations for community structure (e.g. Cornell and Lawton 1992). Much effort has been directed at estimating the relative importance of local interactions and regional-scale processes in shaping assemblages (Hillebrand and Blenckner 2002; Shurin and Srivastava, in press). Recent results indicate several ways in which local and regional forces jointly influence communities and interact in generating patterns in food webs (Leibold et al. 2004). Integrating the two schools of thought has led to important novel insights into dynamics and patterns in communities. Here we discuss the implications of regional dynamics for the structure and function of aquatic food webs.

One way to view the dichotomy between local interactions and regional-scale processes is that dispersal and speciation provide the supply of species to colonize habitat patches (Figure 14.1(d)), while local interactions serve as "filters" selecting which species coexist. By this view, diversity within a habitat can be limited either by the supply of propagules or local exclusion by biotic or abiotic interactions. A variety of evidence suggests that diversity within a guild or trophic level can have major implications for energy flow and trophic structure. If dispersal constrains local diversity, then it may in turn affect food-web processes. For instance, phytoplankton species richness was found to interact with microbial richness in determining community production in laboratory microcosms (Naeem et al. 2000). Here, the mechanism of diversity effects was a mutualism between producers and decomposers driven by the abilities of species to utilize and produce different organic and inorganic compounds. Species-specific mutualism via nutrient cycles is one possible route to diversity effects. Thus, diversity within guilds of organisms at the same trophic position can have major implications for food-web structure.

Another potential role for dispersal in shaping food webs is in limiting the length of food chains. Several lines of evidence suggest that organisms at higher trophic positions in lakes are more constrained in their distributions by dispersal than more basal species. Producers, decomposers, and basal consumers like rotifers and protozoans are likely to be relatively effective dispersers because of their abilities to reproduce asexually and utilize dormant life stages that may disperse through space or time (e.g. diapause, Bohonak and Jenkins 2003; Havel and Shurin 2004). By contrast, many top predators such as fishes, amphibians, and insect predators are obligately sexual (and therefore subject to Allee effects), have lower local population sizes (which should produce fewer propagules), and lack resting stages to withstand transport through the terrestrial matrix. If organisms at higher trophic positions are generally poor dispersers, then the number of trophic levels in a given habitat may often be limited by colonization of top predators. For instance, Magnuson et al. (1998) found that lake isolation and the presence of stream corridors had major effects on fish communities, indicating that the inability to colonize excludes some fishes. Another indication that dispersal ability declines with increasing trophic position is that many phytoplankton and zooplankton assemblages show rapid species turnover and high resilience in response to perturbation such as acidification (Vinebrooke et al. 2003) or predator introduction (Knapp et al. 2001). New species invade rapidly when environmental conditions change, indicating a large supply of potential colonists from the region or dormant pool. By contrast, fishes and crustacean zooplankton at higher trophic positions showed less species turnover in response to acidification, indicating lower ability to disperse (Vinebrooke et al. 2003). Cottenie et al. (in preparation) derived at a similar conclusion from the stronger distance-dependent decay of similarity in larger organisms on higher trophic positions compared to smaller organisms representing the base of the food web. Thus, the identity of species at high trophic positions may often be sensitive to dispersal at the regional scale.

Strong dispersal limitation among predators also has the potential to influence local and regional diversity of organisms at lower positions in the food webs that support them. For instance, species of odonates often segregate between lakes with and without fish (McPeek 1998). Since many fishless lakes are invasible by fish when dispersal opportunities present themselves (e.g. Knapp et al. 2001), this suggests that metapopulation behavior of fish promotes regional diversity among insects, and possibly other prey as well. Shurin and Allen (2001) used metacommunity models to show that predators can promote regional coexistence among competing prey, and either increase or decrease local diversity depending on the traits of the species. Homogenizing predator distributions can have the effect of regionally excluding species that rely on spatial refugia for persistence. For instance, trout stocking reduced the number of fishless lakes in the Sierra Nevada mountains, with major implications for other species, particularly amphibians such as Red Legged Frogs (Knapp et al. 2001). Habitat connectivity can also influence top-down control of prey diversity and biomass by predators. Shurin (2001) found that introduction of fish and insect predators reduced zooplankton diversity in isolated ponds, but that these effects were either reduced or reversed when colonization from the regional pool was allowed. Similarly, responses of marine macroalgae to grazers and nutrients depend on the availability of the seed bank. The presence of propagules allowed for earlier recruitment which resulted in a seasonal escape from grazing (Lotze et al. 2000). Also competition between algal species was affected by propagule supply, a process altered by herbivore presence due to a trade-off between profiting from the seed bank and being susceptible to grazing (Worm et al. 2001). These observations indicate that local processes such as keystone predation interact with dispersal among habitats in shaping local diversity and composition.

Another potential indication of the role for regional dispersal in food-web structure is the relationship between food-chain length (as measured by the trophic position of top predators) and lake surface area (Vander Zanden et al. 1999; Post et al. 2000). These stable isotope studies found that piscivorous fishes feed at higher trophic positions in lakes with large surface areas. One possible explanation for this pattern is that larger lakes contain more diverse prey assemblages of zooplankton, phytoplankton, and benthic prey (Dodson et al. 2000). A greater number of prey species provides more potential trophic pathways from producers to predators than in less speciose assemblages. If large lakes contain more species due to higher colonization rates (bigger surface areas provide larger targets to propagules), then this suggests that biogeographic processes influence energy flow throughout a food web. This explanation for the relationship between lake surface area and trophic position of fishes remains to be tested.

Conclusions

An ongoing challenge for both studies of biodiversity within trophic levels and of food webs has been an apparent lack of generalities that apply across ecosystems or organisms. On the scale of two interacting trophic levels (consumer and prey), the response of diversity to changes in the biomass of the adjacent trophic level can be predicted reasonably well according to a few basic principles. Empirical studies have identified a number of conditions, particularly consumer preference, productivity, and consumer density, that influence the types of effects on biodiversity that are observed. At the same time, model and empirical studies predict pathways for diversity on one trophic level to affect the interaction between consumers and prey. By contrast, few generalities apply across multiple systems. The relation of biodiversity and food-web structure increases dramatically in complexity as soon as the food web comprises more than two trophic levels and reticulate interactions. Here, the topological analyses of emergent properties of food webs may help to reveal relevant and testable hypotheses. As studies are performed in more ecosystems, and as theory develops in parallel, more regularities will likely emerge. However, we argue that recognizing the considerable heterogeneity among members of a guild, and the inherently multitrophic nature of all communities, are essential steps if our understanding of biodiversity and food webs are to progress.

Outlook

Our review on biodiversity and freshwater food webs spanned several organizational levels and showed that biodiversity and food-web ecology are tightly coupled. Nevertheless, several areas demand increased research efforts. Here, we outline four major issues extending the approach of this chapter.

Diversity of the small

The results presented here are based on a limited subset of communities from aquatic habitats. Most research has been done in "classical food webs" comprising phytoplankton, zooplankton, planktivorous, and piscivorous fish in the pelagial or benthic algae, macrozoobenthos, and fish in the benthos. However, much diversity in aquatic systems occurs in the pelagic microbial food web (Berninger et al. 1993; Finlay et al. 1997) and in the meio- and microbenthos (Schmid-Araya et al. 2002a; Traunspurger 2002). This lack of information is astonishing in the light of clear trophic links between macroscopic and microscopic food webs (Wickham 1995; Adrian et al. 2001). For instance, cascading effects of predatory fish on invertebrate detritivores and fungal biomass and detritivore diversity have been observed in lakes (Mancinelli et al. 2002). Invertebrates grazing on biofilm consume a complete microbial food web and first experimental results indicate that invertebrates have different effects on the diversity of the different autotrophic and heterotrophic components (Hillebrand 2003; Wickham et al. 2004). Including the microbial and detritus-based food webs in studies of biodiversity will therefore change the perception of patterns between diversity and trophic interactions.

Evolution and food webs

In order to understand the relation between diversity and food webs, we have to extend the temporal framework to include evolutionary processes. The importance of speciation and microevolution is highlighted by two recent aquatic examples. Eklöv and Svanbäck (submitted) found that predators in different habitats separated single-species prey populations (perch) into subpopulations, which developed different morphology and exhibited contrasting fitness parameters in different habitats. This study suggests that not only competition but also trophic interactions may drive speciation (see also Rundle et al. 2003). The second example is on simple consumer–prey microcosms, where consumers showed clear microevolutionary trends (Fussmann et al. 2003) and diversification of algal prey into several clones changed dramatically the consumer–prey dynamics (Yoshida et al. 2003). Evolutionary dynamics clearly drive patterns of diversity by providing new species, and may occur on short enough timescales to influence the outcome of ecological interactions.

Macroecology and food webs

The description of large-scale diversity patterns and their underlying causes has received wide attention, emerging into the field of macroecology (Gaston and Blackburn 1999). There is an astonishing lack of consideration from trophic ecology in the discussion of macroecological patterns. For example, the importance of ecosystem size on food-chain length corresponds well to theoretical analysis of species area relationships (SARs). Holt et al. (1999) modeled closed communities of specialized consumers and found that the SAR will be steeper in slope with increasing trophic rank. Although empirical evidence is scarce and model outcomes may be altered by system openness and generalist consumption, this would in turn mean that richness of top predators increases faster with ecosystem size than for basal species, resulting in an increase in mean chain length (cf. previous results on food-chain length). A recent meta-analysis of nearly 600 latitudinal gradients of diversity revealed that the decline of diversity from the equator to the poles varied with trophic position (Hillebrand 2004). The gradient was weakest for autotrophs and strongest for carnivores, suggesting that trophic structure alters the importance of possible causes for the latitudinal gradient (such as area, productivity, and net

speciation). Brose et al. (2004) presented a first unification of diversity scaling to macroecological and food-web patterns.

Anthropogenic effects on food webs

Most ecosystems are dominated by humans and the anthropogenic effect on species within these ecosystems is rarely evenly distributed over trophic levels. Cultural eutrophication of lake ecosystems may shift the balance from benthic toward pelagic food webs and result in the loss of benthic pathways of organic matter processing (Vadeboncoeur et al. 2003). Human activities that decrease biological diversity are most strongly felt at top predator levels. Fisheries are prominent examples where humans reduce the number of large fish species with unforeseen consequences for diversity (Myers and Worm 2003). Even more complex anthropogenic effects such as climate change will probably affect especially top and intermediate consumers (Petchey et al. 1999), because the generally lower overall diversity and smaller population sizes at higher trophic positions skew any proposed biodiversity loss toward stronger effects on top species (Duffy 2002). Therefore, food-web ecology has to provide answers how the anticipated further loss in important consumer species affects food-web structure and ecosystem functioning.

Acknowledgments

This review was read, corrected, and commented upon by U.-G. Berninger, A. Ehlers, M. Feiling, A. Longmuir, H. K. Lotze, B. Matthiessen, S. Moorthi, R. Ptacnik, U. Sommer, B. Worm. All of them tremendously improved this chapter. HH acknowledges financial support from the Erken Laboratory (Uppsala University, Sweden) and Institute for Marine Research (Kiel University, Germany).

PART IV

Concluding remarks

CHAPTER 15

Ecological network analysis: an escape from the machine

Robert E. Ulanowicz

Introduction

The scientific community is abuzz over the use of networks to describe complex systems (Watts 1999; Barabási 2002). Recently, the leading journals have reported the rediscovery of the fact that collections of processes and relationships in complex systems often deviate from conventional statistics (Jeong et al. 2000; Montoya and Sole 2002). Many networks of natural systems are said to be "scale-free" in that their elements are distributed according to non-normal power laws (Ulanowicz and Wolff 1991; Barabási and Albert 1999). Others appear to be nested in hierarchical fashion, while still others are dominated by chains of what Almaas and Barabási (2004) have called "hot links" (see also Ulanowicz and Wolff 1991). Because scientists have been conditioned for over 300 years to regard nature as a grand clockwork, the race is now on to elucidate the "mechanisms" behind why the elements of natural systems are arranged in these ways.

Before getting too caught up in the search for mechanisms, it might be wise for investigators to consider if they are pointing their flashlights in the right direction (Popper 1977), for as the late media pundit, Mashall McLuhan (1964) once wrote,

The hybrid or the meeting of two media is a moment of truth and revelation from which new form is born ... The moment of the meeting ... is a moment of freedom and release from the ordinary trance and numbness imposed by them on our senses.

By this he meant that when confronted by new ways of seeing things, observers too often are numbed into interpreting what they see according to old, habitual ways. McLuhan's favorite example was that of International Business Machine, Inc, which began its existence building machinery for businesses. The corporation floundered in unexceptional fashion, until it dawned upon someone that the nature of their enterprise was really more like the transfer of information. Thereupon, the company fortune exploded with such vigor that its consequent meteoric rise is studied till today by business students and stockbrokers alike. (Eventually, the association with machinery was hidden within their acronym, IBM, which one now associates more with the technology of computation than with machines per se).

If some readers remain unimpressed by McLuhan's admonition, perhaps they might heed the castigating words of the mathematician, John Casti (2004). Casti builds upon a children's story, "Little Bear," by Else Minarik (1957) wherein the principal character tries to get to the moon by climbing a tree. Casti contends that using the conventional methods of physics to improve one's understanding of complex systems is akin merely to climbing a taller tree. He cites how, when complex systems are approached using conventional tools, they almost invariably give qualitatively *contradictory* prognoses (and not ones that are merely quantitatively inaccurate).

If some readers should find Casti's admonition a little too pessimistic, they would do well to recall how Eugene and Howard Odum, Robert Rosen, Stanley Salthe, this author and others have labored to point out how the dynamics of living systems is qualitatively distinct from that of purely physical systems. In fact, if one examines closely the fundamental postulates upon which science has operated for the past 300 years, one discovers that each axiom in its turn is violated by one or another ecosystem behavior (Ulanowicz 1999.) Given such disparity, it comes as no surprise that conventional approaches to ecosystems behavior have not yielded any more progress than might come from "climbing a taller tree." Partly out of frustration, a number of investigators have embarked upon a phenomenological search for new ways to quantify ecosystem dynamics and have keyed on quantifying networks as a possibly fruitful approach (Ulanowicz 1986; Wulff et al. 1989; Higashi and Burns 1991). In light of these considerations, it should prove helpful to study in more detail exactly how ecosystem dynamics transcend the usual scientific metaphysic and to explore more fully how quantifying ecosystem networks provides a completely new perspective on the natural world.

Normal science

The problem with writing about the "conventional" approach to science is that no single image exists. Rather, as Kuhn (1962) has suggested, each individual scientist weights differently the various criteria that he or she uses to delimit legitimate science. To deal with such diversity it is helpful to focus on a set of fundamental postulates that once formed a broad consensus about nature around the turn of the nineteenth century (ca. 1800). This "strawman" is not intended to describe the beliefs of scientists today—no one still adheres to the truth of all the classical postulates. On the other hand, virtually every contemporary approach to natural problems still depends upon one or more of these assumptions. The argument made here is that *none* of the postulates remains inviolate within the domain of ecosystem dynamics, and it is the magnitude of such discrepancies that has forced the current phenomenological turn toward describing ecosystems in terms of networks.

While descriptions of the scientific method are legion, one rarely encounters attempts to enumerate the fundamental assumptions upon which the method is based. One exception is that of Depew and Weber (1994), who articulated four

fundamental postulates about nature according to which Newtonian investigations were pursued:

Newtonian systems are causally *closed*. Only mechanical or material causes are legitimate.
Newtonian systems are *deterministic*. Given precise initial conditions, the future (and past) states of a system can be specified with arbitrary precision.
Newtonian systems are *reversible*. Laws governing behavior work the same in both temporal directions.
Newtonian systems are *atomistic*. They are strongly decomposable into stable least units, which can be built up and taken apart again.

In addition, Prigogine and Stengers (1984, see also Ulanowicz 1999) alluded to a fifth article of faith, namely that

Physical laws are *universal*. They apply everywhere, at all times and scales.

Ecosystem dynamics

Although it might at first seem somewhat removed from the subject of networks, determinism is the most convenient assumption with which to begin the discussion of ecosystem dynamics. Every ecologist is aware of the significant role that the aleatoric plays in ecology. Chance events occur everywhere in ecosystems. Stochasticity is hardly unique to ecology, however, and the entire discipline of probability theory has evolved to cope with contingencies. Unfortunately, few stop to consider the tacit assumptions made when invoking probability theory—namely that chance events are always simple, generic, and recurrent. If an event is not simple, or if it occurs only once for all time (is truly singular), then the mathematics of probabilities would not apply.

It may surprise some to learn that ecosystems appear to be rife with singular events. To see why, it helps to recall an argument formulated by physicist Walter Elsasser (1969). Elsasser sought to delimit what he called an "enormous" number. By this he was referring to numbers so large that they should be excluded from physical consideration, because they greatly exceed the number of physical events that possibly could have occurred since the Big Bang. To estimate a threshold for enormous numbers Elsasser reckoned the number of simple protons in the known universe to be about 10^{85}. He then noted as how the number of nanoseconds that have transpired since the beginning of the universe have been about 10^{25}. Hence, a rough estimate of the upper limit on the number of conceivable events that could have occurred in the physical world is about 10^{110}. Any number of possibilities much larger than this value simply loses any meaning with respect to physical reality.

Anyone familiar with combinatorics immediately will realize that it does not take very many distinguishable elements or processes before the number of their possible configurations becomes enormous. One does not need Avagadro's number of particles (10^{23}) to produce combinations in excess of 10^{110}—a system with merely 80 or so distinct components will suffice. In probabilistic terms, any event randomly comprising more than 80 separate elements is virtually certain to never have occurred earlier in the history of the universe. Such a constellation is unique over all time past. It follows, then, that in ecosystems with hundreds or thousands of distinguishable organisms, one must reckon not just with occasional unique events, but rather with a legion of them! Unique, singular events are occurring all the time, everywhere! In the face of this reality, all talk of determinism as a universal characteristic is futile, and the argument for reversibility collapses as well.

Despite the challenge that rampant singularities pose for the Baconian pursuit of science, it still can be said that a degree of regularity appears to characterize such ecological phenomena as succession. The question then arises as to the origins and maintenance of such order? An agency that both creates and maintains regularities is embedded in the patterns of processes that are represented by trophic *networks*. In particular, the key to how living systems act differently from purely physical systems appears to reside in the adjunction of autocatalytic loops (or cycles of mutualism, that can be found in ecosystem networks) with frequent aleatoric events (Ulanowicz 1997*a*). Here autocatalysis will be defined as any manifestation of a positive feedback loop whereby the direct effect of every link on its downstream neighbor

is positive. Without loss of generality, one may focus attention on a serial, circular conjunction of three processes A, B, and C (Figure 15.1). Any increase in A is likely to induce a corresponding increase in B, which in turn elicits an increase in C, and whence back to A.

A didactic example of autocatalysis in ecology is the community that builds around the aquatic macrophyte, *Utricularia* (Ulanowicz 1995). All members of the genus *Utricularia* are carnivorous plants. Scattered along its feather-like stems and leaves are small bladders, called utricles (Figure 15.2(a)) Each utricle has a few hair-like triggers at its terminal end, which, when touched by a feeding zooplankter, opens the end of the bladder, and the animal is sucked into the utricle by a negative osmotic pressure that the plant had maintained inside the bladder. In nature the surface of *Utricularia* plants is always host to a film of algal growth known as periphyton. This periphyton in turn serves as food for any number of species of small zooplankton. The autocatalytic cycle is closed when the *Utricularia* captures and absorbs many of the zooplankton (Figure 15.2(b)).

In chemistry, where reactants are simple and fixed, autocatalysis behaves just like any other

Figure 15.1 A simple example of autocatalysis.

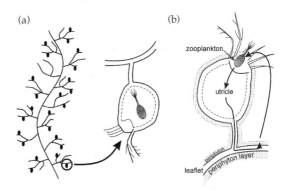

Figure 15.2 (a) *Utricularia*, a carnivorous plant. (b) The cycle of rewards in the *Utricularia* system.

mechanism. As soon as one must contend with organic macromolecules and their ability to undergo small, incremental alterations, however, the game changes. Especially when the effect of any catalyst on the downstream element is fraught with contingencies (rather than being obligatorily mechanical), a number of decidedly nonmechanical behaviors can arise (Ulanowicz 1997a). These emergent attributes of complex systems render the remaining Newtonian postulates inappropriate for ecosystem dynamics (Ulanowicz 2004).

Perhaps most importantly, autocatalysis is capable of exerting *selection* pressure upon its ever-changing, malleable constituents. To see this, one considers a small spontaneous change in process B. If that change either makes B more sensitive to A or a more effective catalyst of C, then the transition will receive enhanced stimulus from A. Conversely, if the change in B either makes it less sensitive to the effects of A or a weaker catalyst of C, then that perturbation will likely receive diminished support from A. That is to say that there is a preferred *direction* inherent in autocatalysis— that of increasing autocatalytic participation. This preferred direction can be interpreted as a breaking of symmetry, and such asymmetry, like the singular events just discussed, also transcends the assumption of reversibility. Furthermore, with such increasing autocatalytic engagement, or mutual adaptation, elements lose their capability of acting on their own; or, should they remain capable of persisting in isolation, it would be with behavior radically different from that exhibited as part of the autocatalytic scheme. That is, the full cycle manifests an *organic* nature that is refractory to the assumption of atomism.

To see how another very important attribute of living systems can arise, one notes in particular that any change in B is likely to involve a change in the amounts of material and energy that are required to sustain process B. As a corollary to selection pressure one immediately recognizes the tendency to reward and support any changes that serve to bring ever more resources into B. Because this circumstance pertains to any and all members of the feedback loop, any autocatalytic cycle becomes the epi-center of a *centripetal* flow of resources toward which as many resources as possible will converge (Figure 15.3). That is, an autocatalytic loop embedded in a network *defines itself* as the focus of centripetal flows.

It is important to note as how autocatalytic selection pressure is exerted in top–down fashion— that is, action by an integrated cluster of processes upon its constituent elements. Centripetality, in its turn, is best described as an agency that acts at the focal level. Both selection and centripetality violate the restriction of causal *closure,* which permits only mechanical actions at smaller levels to ramify up the hierarchy of scales. In autocatalytic selection, causal action resembles the final causality of Aristotle, which was explicitly excluded from Newtonian discourse, while centripetality bears all the trappings of Aristotelian formal cause (by virtue of the agency being exerted by a *configuration* of processes), which concept likewise atrophied in the wake of Newton.

Centripetality also guarantees that whenever two or more autocatalytic loops exist in the same network and draw from the same pool of finite resources, *competition* among the loci will necessarily ensue. In particular, whenever two loops share pathway segments in common, the result of this competition is likely to be the exclusion or radical diminution of one of the nonoverlapping sections. For example, should a new element D happen to appear and to connect with A and C in parallel to their connections with B, then if D is more sensitive to A and/or a better catalyst of C, the ensuing dynamics should favor D over B to the extent that B will either fade into the background or disappear altogether (Figure 15.4). That is, the

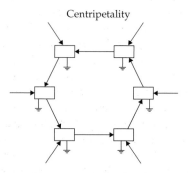

Figure 15.3 Centripetal action as engendered by autocatalysis.

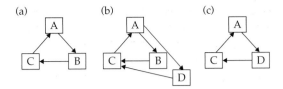

Figure 15.4 The selection of new element D to replace B.

selection pressure and centripetality generated by complex autocatalysis (as embedded in the macroscopic network of processes) is capable of influencing the replacement of elements.

The reader will note the above emphasis upon a causality that arises out of a *configuration of processes*, and which is able to influence significantly which objects will remain and which will pass from the scene. Unlike the conventional attitude that all events are the consequences of actions by objects, the reverse now becomes possible: objects themselves can become the products of constellations of processes. In other words, networks themselves can become legitimate agents of change. Hence, to describe network dynamics, it is no longer mandatory that one search for constituent mechanisms that are facilitated by objects residing in the nodes. One is now free to develop what might be called *process ecology* (Ulanowicz 2004).

Finally, it is worthwhile to note how autocatalytic selection can act to stabilize and regularize behaviors across the hierarchy of scales. Unlike with Newtonian universality, a singular event anywhere rarely will ramify up and down the hierarchy without attenuation. The effects of noise at one level are usually subject to autocatalytic selection at higher levels and to energetic culling at lower levels. Nature as a whole exhibits regularities, but in place of the universal effectiveness of all natural laws, one discerns instead a *granularity* about the real world. That is, models of events at any one scale can explain matters at another scale only in inverse proportion to the remoteness between them. Obversely, the domain within which irregularities and perturbations can damage a system is usually circumscribed. Chance does not necessarily unravel a system, which is held together by the (flexible) lattice of network interactions itself.

A new metaphysic

It begins to appear as though Casti was not exaggerating after all. As Popper (1990) suggested, a wholly new perspective on how things happen in nature may be required in order to achieve an adequate understanding of development and evolution. The topsy-turvy realm of ecological dynamics must seem strange indeed to those educated as biologists always to look to smaller scales for the causes behind phenomena, but in hindsight the appeal of reductionism now seems but a chimera.

In order for reductionism to work, the simplest and most enduring elements must all be at the very bottom of the spatio-temporal scales, so that the more complicated and less durable objects can be built up from them. Such is the case with the nested hierarchy of quarks, mesons, electrons, atoms, and chemical compounds. When one reaches the ecosystem, however, one encounters a significant inversion of these assumptions. The ecosystem per se is not as well-organized as the individual organisms that comprise it (Ulanowicz 2001). Furthermore, the network constituents (and mechanisms) come and go, while the configuration of ecosystems processes endures (and some would even say, preceded the current forms of its constituents (Odum 1971)).

Such inversion notwithstanding, no one should regard the causal agency inherent in networks as a triumph for Holism as it was once depicted. Certainly, no one is contending that configurations of processes *fully determine* the fate and nature of each constituent. One must always bear in mind that singular events loom significantly in the dynamical picture. In the overwhelming number of cases, however, singular events occur and disappear, leaving no trace upon the overall system makeup. Occasionally, they will exert detrimental effects upon autocatalytic action, and the system will respond by reconfiguring itself so as to ameliorate such disturbance. In a very small number of instances, a singular event can serve to enhance the autocatalytic functioning of the system and will become incorporated as an enduring (historical) change to the larger network structure (which thenceforth will exert somewhat different selection pressure upon subsequent singular events).

The realm of ecosystem behavior is certainly different from that of classical mechanical dynamics. Instead of a world that is closed, atomistic, reversible, deterministic, and universal; one now perceives a domain that is (respectively) open, organic, historical, contingent, and granular.

A network dynamic

If the reader studies closely the scenario described above, he or she will discern the interplay of two antagonistic tendencies. In one direction there is what might be called a *probabilistic drift* that ratchets the system in a direction of ever-greater autocatalytic activity. Opposing this drift is the entropic tendency resulting from the unpredictable occurrences of singular events, which, on one hand, act to disrupt system organization, but on the other could also provide a source for diversity and novel behaviors. Fortunately, these two tendencies can both be tracked as changes in quantitative network properties.

The probabilistic drift toward greater organization has long been characterized as "increasing network ascendency" (Ulanowicz 1980, 1986, 1997a). The ascendency of a network is defined as the product of its total activity (as measured by the sum of all the arc weights) times the average mutual information inherent in the linkage structure (Rutledge et al. 1976; Hirata and Ulanowicz 1984). This mutual information of the flow structure measures, on the average, how definitively transfers act in the system. That is, if a transfer is but one of a number of similar, parallel processes, it contributes little to the mutual information; but if a process plays a unique role in sustaining another node or subgraph, then the contribution of that key link to the mutual information becomes significant. Zorach and Ulanowicz (2003) showed how this latter attribute is captured by the network's mutual information, which turns out to be the logarithm of the effective number of *distinct roles* embedded in the network.

To quantify the ascendency, one must know the magnitude, T_{ij}, of each flow from arbitrary node i to any other node j. The total activity then becomes the sum of all the T_{ij}, or $T_{..}$, where a dot in place of a subscript indicates summation over that index.

Hirata and Ulanowicz (1984) showed how the ascendency can then be expressed as

$$A = \sum_{i,j} T_{ij} \log\left[\frac{T_{ij}T_{..}}{T_{.j}T_{i.}}\right] \geq 0,$$

and Zorach and Ulanowicz (2003) showed how the geometric mean number of roles in the network can be estimated as b^A, where b is the base used to calculate the logarithms in the formula for A.

In opposition to this drift toward increasing ascendency is the spontaneous tendency to increase what has been called the network "overhead," Φ. Overhead is the encapsulation of all ambiguity, incoherence, redundancy, inefficiency, and indeterminacy inherent in the network (Ulanowicz and Norden 1990). It can be quantified by an information-theoretic property called the "conditional entropy," which is complementary to the mutual information that was used to quantify the ascendency. In terms of the T_{ij}, Φ can be written as

$$\Phi = -\sum_{i,j} T_{ij} \log\left[\frac{T_{ij}^2}{T_{.j}T_{i.}}\right].$$

As with the ascendency, Zorach and Ulanowicz (2003) have demonstrated how the logarithm of the geometric mean of the network link-density, L_D, is equal to one-half of the overhead. That is, $L_D = b^{\phi/2}$. (Link-density is the effective number of arcs entering or leaving a typical node. It is one measure of the connectivity of the network.)

Experience shows that the effective numbers of roles and the connectivities of real ecosystems are not arbitrary. It has long been known, for example, that the number of trophic roles (levels) in ecosystems is generally fewer than 5 (Pimm and Lawton 1977). Similarly, the effective link-density of ecosystems (and a host of other natural systems) almost never exceeds 3 (Pimm 1982; Wagensberg et al. 1990). Regarding this last stricture, Ulanowicz (2002) suggested how the May–Wigner stability criterion (May 1972) could be reinterpreted in information-theoretic terms to identify a threshold of stability at $e^{e/3}$, or ca. 3.015 links per node.

Both limits appear quite visibly when one plots the number of roles versus the effective connectivity of

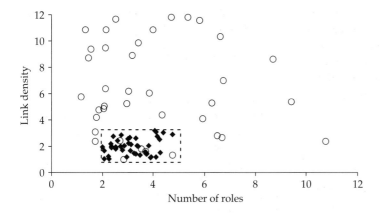

Figure 15.5 Combinations of link densities and numbers of roles pertaining to random networks (open circles) and actual ecosystems networks (solid squares). The "window of vitality" is indicated by the dotted lines.

a collection of 44 estimated ecosystem flow networks (Figure 15.5). Whereas the pairs of numbers generated by randomly constructed networks are scattered broadly over the positive quadrant, those associated with actual ecosystem networks are confined to a small rectangle near the origin. Ulanowicz (1997*b*) has labeled this rectangle the "window of vitality," because it appears that the entire drama of ecosystem dynamics plays out within this small theatre: as mentioned above, the *endogenous* tendency of ecosystems is to drift toward the right within this window (i.e. toward ever-increasing ascendency, or higher system performance). At any time, however, singular events can appear as exogenous perturbations that shift the network abruptly to the left. (Whether the link-density rises or falls during this transition depends upon the nature and severity of the disturbance). In particular, whenever the system approaches one of the outer edges of the window, the probability increases that it will fall back toward the interior. Near the top, horizontal barrier (LD = ~3.015 links per node) the system lacks sufficient cohesiveness and disintegrates spontaneously. As the system approaches the right-hand frame (# roles = 4.5 to 5.0), it presumably undergoes something like a "self-organizing catastrophe" (Bak 1996) as described by Holling (1986, see also Ulanowicz 1997*a*).

As one follows the historical dance of an ecosystem within the window of vitality, it is important to hold firmly in mind that any description of a trajectory solely in terms of mechanisms and the actions of individual organisms will perforce remain inadequate. Rather, the prevailing agencies at work are the tendency of configurations of processes (subgraphs) to increase in ascendency acting in opposition to the entropic tendency generated by complex, singular events. It is only by focusing on these larger actors that one can discover, as Karl Popper (1990) once put it,

we are not things, but flames. Or a little more prosaically, we are, like all cells, *processes of metabolism*; nets of chemical processes . . . (Italics by Popper.)

That is, by embarking upon a serious examination of the nature of ecological networks, ecologists are not simply climbing trees; they are attempting to go *beyond* rocket science!

Acknowledgments

The author was supported in part during the writing of this essay by the National Science Foundation's Program on Biocomplexity (Contract No. DEB-9981328.) Mr Michael J. Zickel helped with the drawing of Figure 15.5.

A prospectus for future aquatic food web studies

Mathew A. Leibold

Aquatic ecologists have been food web ecologists from an early date. In limnology, Forbes viewed lakes as microcosms in which food web interactions determined how entire ecosystems were structured. Thieneman, Ivlev, Hutchinson and Lindeman continued this perspective in limnology as did others who made more comparative studies of lakes such as Birge and Juday. Other areas of aquatic ecology have similarly strong traditions related to food web interactions. Many of the early ideas have been modified, but food web interactions still remain as one of the strongest threads in the study of lakes, streams, estuaries, intertidals, and oceans as well as in the smaller phytotelmata and in the lab studies with microorganisms.

For most researchers who have studied food web interactions in aquatic systems, it is hard not to be conscious of just how important food web interactions can be in shaping communities and ecosystems. Not only are the magnitude of some food web interactions in all of these systems large, but there are numerous cases where any given trophic interaction has strong and obvious ramifications that alter entire suites (or cascades) of ecological processes and the populations of organisms that are often quite distant (in causal terms) from the modified interaction. While there may be reasons to think that indirect effects may have limits on their propagation, it is clear from many studies, especially those conducted in aquatic systems, that they are an important part of the overall dynamics of ecosystems. Regardless of the form that it takes, the study of food webs is a reflection on the causes and consequences of these indirect effects in networks of interacting species.

Many aquatic ecologists (though maybe not all of them) would be more puzzled by the lack of indirect effects in aquatic food webs, than they would be by their presence.

The prevalence and significance of indirect effects, and especially of feedback mediated via indirect effects, means that the study of food webs is almost necessarily the study of complex systems. Food web scientists are in a continuous struggle to identify ways to resolve questions about this complexity with answers that almost invariably simplify potentially crucial elements of it. Thus food web diagrams conventionally focus on tropho-species (groups of organisms potentially involving multiple species, that have unique sets of links to other such tropho-species). This simplifies many theoretical and empirical aspects of food web research but theory tells us that quantitative differences among the species that are members of the same tropho-species could be just as important in affecting food web phenomena as are the qualitative differences implied by the concept of tropho-species. Similarly, trophic relations are not the only important ones in real communities that affect food web phenomena. Nevertheless food web studies are among those that most directly and realistically confront the complexity that exists in most natural ecosystems. Work to date on aquatic food webs shows that it is possible to gain insights about important ecological phenomena without so drastically simplifying this complexity that our theories seem like pale caricatures of natural systems.

As an example, food web studies on the indirect interactions and feedback between 'bottom-up' and 'top-down' food web interactions seem to be critical to our current explanations of trophic structure in aquatic systems. Surveys of plant and herbivore biomass in lakes supported a 'bottom-up' perspective because they showed that both increased in lakes with higher levels of limiting nutrients and consequent productivity. This strong pattern however was in conflict with numerous experiments and other studies that showed strong top-down regulation of both plant and herbivore biomass when fish stocks were altered. Theory indicated that these two observations were largely incompatible in simple food chains with only one species per trophic level. More recent work indicates that it is the more complex set of interactions and feedbacks that occur in food webs that can resolve this apparent contradiction where there are multiple species involved in trophic transfers at any single trophic level. As in many other cases where paradoxical patterns have been found in aquatic ecosystems, it was the food web perspective that allowed for the synthesis and resolution of the problem.

Additionally, many other aspects of food web structure and dynamics are perhaps better documented in aquatic systems than in terrestrial ones. While terrestrial ecologists may (arguably) still be able to make significant progress taking strict 'bottom up' perspectives that ignore how 'top-down' effects work, the simultaneous action and interaction of top-down and bottom-up interactions in aquatic systems is hard to deny. Aquatic systems provide some of the best examples for many of the ideas that have emerged from consideration of these effects. The contributions in this book continue in this vein and address some of the most important current questions about food webs:

– How do structural properties of food webs affect the dynamic properties of the ecosystems they are a part of?
– How can we explain the amazing complexity of food webs?
– How are food webs assembled by processes of colonization and extinction?
– How does this assembly affect other dependent properties of the ecosystem?

– How do food webs and consequently their effects on ecosystems respond to environmental changes?

As illustrated by the contributions to this book, the answers to these questions in aquatic systems are starting to resolve themselves with novel ideas about these issues, novel ways to evaluate these ideas, and novel insights about how these ideas relate to other phenomena in ecology, biology, and environmental sciences as a whole.

Nevertheless food web ecology seems to be an area that continuously generates as many questions as it answers. An important aspect of this book is that the contributions suggest many new directions that could or should be taken in future work. There seem to be at least three important issues that emerge:

First, how much can we extrapolate from aquatic systems to other systems? Some authors have argued that such extrapolation is unwarranted for example in the study of trophic cascades. While they recognized how significant the finding of trophic cascades in aquatic systems were, they argued that trophic cascades were less likely to be important in terrestrial systems because terrestrial food webs were claimed to be more complex, with individual populations that were more highly structured, and with more complications arising from factors such as omnivory etc. Clearly, such claims would come as a surprise to many aquatic ecologists to begin with, but additionally such claims beg the question of why such differences might exist and why they would negate the importance of trophic cascades. Since then, examples of trophic cascades in terrestrial systems have been found. And while it is clear that the effects can differ in magnitude from some of the findings in lakes, it is also clear that there is tremendous variation in trophic cascades in aquatic systems as well. Somehow, there is a need to better investigate how and why some patterns extrapolate well from aquatic systems to terrestrial ones and others do not. Part of resolving this has to be in identifying the ways complex processes interact and this can really only be resolved with a strong theoretical framework. This book goes a long way in this direction by having a strong basis in ecological theory but this is an area that still requires further development.

Second, at what scale do different interpretations of food web dynamics apply? Clearly, food web models that are built on locally interacting populations in closed systems can make predictions that differ strongly from food web models that allow for spatial structure or for other meta-community level processes to occur. There is, for example, a much smaller problem in explaining the diversity and complexity of food webs at larger spatio-temporal scales where such metacommunity dynamics exist than at the smaller scale typified by models of locally interacting populations. A number of the contribution to this book address some elements of this issue, but we are still a long way from understanding the dynamics of food webs at larger spatio-temporal scales even though this is maybe of greater ultimate relevance. Furthermore, there still seem to be significant differences in how food web scientists interpret the links between food web attributes and principles that apply to ecological processes that operate over different scales.

Finally, one element that is only beginning to be addressed is the role of evolutionary or adaptive dynamics in food webs. A number of the contributions here deal either directly or indirectly with some form of adaptive dynamics (in the broadest sense of the word) and it seems clear that there are important implications for food web theory and for future food webs studies. There are at least three types of adaptive dynamics that might be important. The first involves adaptive plasticity, the second involves adaptive genetic changes within populations of a single species, and the third involves changes in species composition that result in adaptation in the sense that they yield an improved correspondence between the traits of the organisms found in a food web and the environment in which the food web occurs. These three types of adaptive dynamics are all likely to be important, at least under some scenarios, and they are likely to interact with each other. While the importance of non-random organization of food webs has been recognized since the work of May and Pimm, it seems likely that more careful thinking about how these three processes affect food webs will improve our understanding of food web dynamics.

Aquatic ecologists as a group are likely to make among the strongest contributions to food web thinking. In part this is because some aquatic systems are 'model systems' for the study of food web dynamics (e.g., streams, intertidal, lakes and ponds), in part this is because it is possible to conduct experiments that involve major aspects of realistic food webs (e.g., work with protists in microcosms, mesocosms in lakes and streams, cage experiments in intertidals), and in part because aquatic systems span the gamut of complexity ranging from relatively simple phytotelmata to entire oceans. This book illustrates that answers to complex questions that arise from food web ecology are most satisfying when they appear useful in the interpretation of phenomena across such diverse systems and using such diverse methods.

References

Introduction

Azam, F., and A. Z. Worden. 2004. Microbes, molecules, and marine ecosystems. *Science* **303**: 1622–1624.

Azam, F. 1998. Microbial control of oceanic carbon flux: the plot thickens. *Science* **280**: 694–696.

Cohen, J. E. 1998. *Food Webs and Niche Space*. Princeton University Press, Princeton, NJ.

Christensen, V. and D. Pauly. (eds). 1993. *Trophic Models of Aquatic Ecosystems*. ICLARM, Manila, p. 390.

Dunne, J. A., R. J. Williams, and N. D. Martinez. 2002a. Network structure and biodiversity loss in food webs: robustness increases with connectance. *Ecology Letters* **5**: 558–567.

Dunne, J. A., R. J. Williams, and N. D. Martinez. 2002b. Food-web structure and network theory: the role of connectance and size. *Proceedings of the National Academy of Sciences USA* **99**: 12917–12922.

Garlaschelli, D., G. Caldarelli, and L. Pietronero. 2003. Universal scaling relations in food webs. *Nature* **423**: 165–168.

MacArthur, R. H. 1955. Fluctuations of animal populations, and a measure of community stability. *Ecology* **36**: 533–536.

May, R. M. 1973. *Stability and Complexity in Model Ecosystems*, 2nd edn. Princeton University Press, Princeton, NJ.

Paine, R. T. 1966. Food web complexity and species diversity. *American Naturalist* **100**: 65–75.

Pimm, S. L. 1982. *Food Webs*. Chapman and Hall, London.

Polis, G. A., and K. O. Winemiller. 1996. Food Webs: *Integration of Patterns and Dynamics*. Chapman and Hall.

Ulanowicz, R. E. 1997. *Ecology, The Ascendent Perspective*. Columbia University Press, New York, 224 pp.

Ulanowicz, R. E., and L. G. Abarca-Arenas. 1997. An informational synthesis of ecosystem structure and function. *Ecology Modelling* **95**: 1–10.

Ulanowicz, R. E. 1996. The propensities of evolving systems. In: E. L. Khalil and K. E. Boulding (eds), *Evolution, Order and Complexity*. Routledge Publishers, London, pp. 217–233.

Williams, R. J., and N. D. Martinez. 2000. Simple rules yields complex food webs. *Nature* **404**: 180–183.

Williams, R. J., E. L. Berlow, J. A. Dunne, A.-L. Barabási, and N. D. Martinez. 2002. Two degrees of separation in food webs. *Proceedings of the National Academy of Sciences USA* **99**: 12913–12917.

Wulff, F., J. G. Field, and K. H. Mann. (eds). 1989. *Network Analysis in Marine Ecology*. Springer-Verlag, New York, p. 284.

Chapter 1

Andersen, T. 1997. *Pelagic Nutrient Cycles: Herbivores as Sources and Sinks*. Springer-Verlag, Berlin, Heidelberg, New York.

Andersen, T., and D. O. Hessen. 1991. Carbon, nitrogen, and phosphorus content of freshwater zooplankton. *Limnology and Oceanography* **36**: 807–814.

Baudouin-Cornu, P., Y. Surdin-Kerjan, P. Marliere, and D. Thomas. 2001. Molecular evolution of protein atomic composition. *Science* **293**: 297–300.

Carrillo, P., M. Villar-Argaiz, and J. M. Medina-Sanchez. 2001. Relationship between N:P ratio and growth rate during the life cycle of copepods: an *in situ* measurement. *Journal of Plankton Research* **23**: 537–547.

Cebrian, J. 1999. Patterns in the fate of production in plant communities. *American Naturalist.* **154**: 449–468.

Cebrian, J., and C. M. Duarte. 1995. Plant growth-rate dependence of detrital carbon storage in ecosystems. *Science* **268**: 1606–1608.

Cebrian, J., M. Williams, J. McClelland, and I. Valiela. 1998. The dependence of heterotrophic consumption and C accumulation on autotrophic nutrient content in ecosystems. *Ecology Letters* **1**: 165–170.

Darchambeau, F., P. J. Faerøvig, and D. O. Hessen. 2003. How *Daphnia* copes with excess carbon in its food. *Oecologia* **136**: 336–346.

DeMott, W. R. 2003. Implications of element deficits for zooplankton growth. *Hydrobiologia* **491**: 177–184.

DeMott, W. R., and R. D. Gulati. 1999. Phosphorus limitation in *Daphnia*: evidence from a long-term study of three hypereutrophic Dutch lakes. *Limnology Oceanography* **44**: 1557–1564.

DeMott, W. R., R. D. Gulati, and K. Siewertsen. 1998. Effects of phosphorus-deficient diets on the carbon and phosphorus balance of *Daphnia magna*. *Limnology and Oceanography* **43**: 1147–1161.

Diehl, S. 2003. The evolution and maintenance of omnivory: dynamic constraints and the role of food quality. *Ecology* **84**: 2557–2567.

Dykhuizen, D. E., A. M. Dean, and D. L. Hartl. 1987. Metabolic flux and fitness. *Genetics* **115**: 25–31.

Eickbush, T. H., W. D. Burke, D. G. Eickbush, and W. C. Lathe. 1997. Evolution of R1 and R2 in the rDNA units of the genus *Drosophila*. *Genetica* **100**: 49–61.

Elser, J. J., and J. Urabe. 1999. The stoichiometry of consumer-driven nutrient cycling: theory, observations, and consequences. *Ecology* **80**: 735–751.

Elser, J. J., K. Acharya, M. Kyle, J. Cotner, W. Makino, T. Markow, T. Watts, S. Hobbie, W. Fagan, J. Schade, J. Hood, and R. W. Sterner. 2003. Growth rate–stoichiometry couplings in diverse biota. *Ecology Letters* **6**: 936–943.

Elser, J. J., D. Dobberfuhl, N. A. MacKay, and J. H. Schampel. 1996. Organism size, life history, and N : P stoichiometry: towards a unified view of cellular and ecosystem processes. *BioScience* **46**: 674–684.

Elser, J. J., W. F. Fagan, R. F. Denno, D. R. Dobberfuhl, A. Folarin, A. Huberty, S. Interlandi, S. S. Kilham, E. McCauley, K. L. Schulz, E. H. Siemann, and R. W. Sterner. 2000a. Nutritional constraints in terrestrial and freshwater food webs. *Nature* **408**: 578–580.

Elser, J. J., H. Hayakawa, and J. Urabe. 2001. Nutrient limitation reduces food quality for zooplankton: *Daphnia* response to seston phosphorus enrichment. *Ecology* **82**: 898–903.

Elser, J. J., R. W. Sterner, E. Gorokhova, W. F. Fagan, T. A. Markow, J. B. Cotner, J. F. Harrison, S. E. Hobbie, G. M. Odell, and L. J. Weider. 2000b. Biological stoichiometry from genes to ecosystems. *Ecology Letters* **3**: 540–550.

Faerovig, P. J., T. Andersen, and D. O. Hessen. 2002. Image analysis of *Daphnia* populations: non-destructive determination of demography and biomass in cultures. *Freshwater Biology* **47**: 1956–1962.

Fagan, W. F., E. H. Siemann, R. F. Denno, C. Mitter, A. Huberty, H. A. Woods, and J. J. Elser. 2002. Nitrogen in insects: implications for trophic complexity and species diversification. *American Naturalist* **160**: 84–802.

Fussmann, G. F., S. P. Ellner, and J. Hairston, N. G. 2003. Evolution as a critical component of plankton dynamics. *Proceedings of the Royal Society of London Series B* **270**: 1015–1022.

Gorokhova, E., T. A. Dowling, L. J. Weider, T. Crease, and J. J. Elser. 2002. Functional and ecological significance of rDNA IGS variation in a clonal organism under divergent selection for production rate. *Proceedings of the Royal Society of London B* **269**: 2373–2379.

Hairston, J., N. G., and N. G. Hairston, Sr. 1993. Cause–effect relationships in energy flow, trophic structure, and interspecific interactions. *American Naturalist* **142**: 379–411.

Hassett, R. P., B. Cardinale, L. B. Stabler, and J. J. Elser. 1997. Ecological stoichiometry of N and P in pelagic ecosystems: comparison of lakes and oceans with emphasis on the zooplankton—phytoplankton interaction. *Limnology and Oceanography* **41**: 648–662.

Hastings, A., and M. Conrad. 1979. Length and evolutionary stability of food chains. *Nature* **282**: 838–839.

Hessen, D. O. 1992. Nutrient element limitation of zooplankton production. *American Naturalist* **140**: 799–814.

Hessen, D. O., and B. Bjerkeng. 1997. A model approach to planktonic stoichiometry and consumer–resource stability. *Freshwater Biology* **38**: 447–471.

Hessen, D. O., and A. Lyche. 1991. Inter- and intraspecific variations in zooplankton element composition. *Archives of Hydrobiology* **121**: 343–353.

Hessen, D. O., and J. Skurdal. 1987. Food consumption, turnover rates and assimilation in the noble crayfish (*Astacus astacus*). *Freshwater Crayfish* **7**: 309–317.

Hessen, D. O., B. A. Faafeng, and P. Brettum. 2003. Autotroph: herbivore biomass ratios; carbon deficits judged from plankton data. *Hydrobiologia* **491**: 167–175.

Hessen, D. O., P. J. Faerovig, and T. Andersen. 2002. Light, nutrients, and P : C ratios in algae: Grazer performance related to food quality and quantity. *Ecology* **83**: 1886–1898.

Hobbie, S., Q. Hong, D. A. Beard, and S. Liang. 2003. Stoichiometric network theory for nonequilibrium biochemical systems. *European Journal of Biochemistry* **270**: 415–421.

Holt, R. D. 1995. Linking species and ecosystems: where's Darwin? In: C. G. Jones and J. H. Lawton (eds), *Linking Species and Ecosystems*. Chapman and Hall, New York, pp. 273–279.

Jaenike, J., and T. A. Markow. 2002. Comparative elemental stoichiometry of ecologically diverse *Drosophila*. *Functional Ecology* **17**: 115–120.

Lavigne, D. M. 1996. Ecological interactions between marine mammals, commercial fisheries, and their prey: unravelling the tangled web. In: W. A. Montevecchi (ed.) *Studies of High-Latitude Seabirds. 4. Trophic Relationships and Energetics of Endotherms in Cold Ocean Systems.* Occasional Paper Number 91. Canadian Wildlife Service, St John's Newfoundland.

Lima, S. L., and L. M. Dill. 1990. Behavioral decisions made under the risk of predation—a review and prospectus. *Canadian Journal of Zoology—Reviews in Canadian Zoology* **68**: 619–640.

Loladze, I., Y. Kuang, and J. J. Elser. 2000. Stoichiometry in producer–grazer systems: linking energy flow and element cycling. *Bulletin of Mathematical Biology* **62**: 1137–1162.

Loladze, I., Y. Kuang, J. J. Elser, and W. F. Fagan. 2004. Competition and stoichiometry: coexistence of two predators on one prey. *Journal of Theoretical Biology* **65**: 1–15.

Lotka, A. J. 1925. *Elements of Physical Biology.* Williams and Wilkins, Baltimore, MD.

Loxdale, H. D., and G. Lushai. 2003. Rapid changes in clonal lines: the death of a "sacred cow." *Biological Journal of the Linnean Society* **79**: 3–16.

Lushai, G., and H. D. Loxdale. 2002. The biological improbability of a clone. *Genetics Research* **79**: 1–9.

Lushai, G., H. D. Loxdale, and J. A. Allen. 2003. The dynamic clonal genome and its adaptive potential. *Biological Journal of the Linnean Society* **79**: 193–208.

Main, T., D. R. Dobberfuhl, and J. J. Elser. 1997. N:P stoichiometry and ontogeny in crustacean zooplankton: a test of the growth rate hypothesis. *Limnology and Oceanography* **42**: 1474–1478.

Makino, W., and J. B. Cotner. 2003. C:N:P stoichiometry of a heterotrophic bacterial community in a freshwater lake: implications for substrate- and growth-dependent variations. *Aquatic Microbial Ecology* **34**: 33–41.

Makino, W., J. B. Cotner, R. W. Sterner, and J. J. Elser. 2003. Are bacteria more like plants or animals? Growth rate and substrate dependence of bacterial C:N:P stoichiometry. *Functional Ecology* **17**: 121–130.

Markow, T. A., B. Raphael, D. Dobberfuhl, C. M. Breitmeyer, J. J. Elser, and E. Pfeiler. 1999. Elemental stoichiometry of *Drosophila* and their hosts. *Functional Ecology* **13**: 78–84.

Muller, E., R. M. Nisbet, S. A. L. M. Kooijman, J. J. Elser, and E. McCauley. 2001. Stoichiometric food quality and herbivore dynamics. *Ecology Letters* **4**: 519–529.

Munch, S. B., and D. O. Conover. 2003. Rapid growth results in increased susceptibility to predation in *Menidia menidia*. *Evolution* **57**: 2119–2127.

Murdoch, W. W., C. J. Briggs, and R. M. Nisbet. 2003. *Consumer–Resource Dynamics.* Princeton University Press, Princeton, NJ.

Perkins, M. C., H. A. Woods, J. F. Harrison, and J. J. Elser. 2003. Dietary phosphorus affects the growth of larval *Manduca sexta*. *Archives of Insect Biochemistry and Physiology* **55**: 153–168.

Plath, K., and M. Boersma. 2001. Mineral limitation of zooplankton: stoichiometric constraints and optimal foraging. *Ecology* **82**: 1260–1269.

Schade, J., M. M. Kyle, S. S. Hobbie, W. W. Fagan, and J. J. Elser. 2003. Stoichiometric tracking of soil nutrients by a desert insect herbivore. *Ecology Letters* **6**: 96–101.

Schulz, K. L., and R. W. Sterner. 1999. Phytoplankton phosphorus limitation and food quality for *Bosmina*. *Limnology Oceanography* **44**: 1549–1556.

Sommer, U. 1992. Phosphorus-limited Daphnia: intraspecific facilitation instead of competition. *Limnology Oceanography* **37**: 966–973.

Sterner, R. W. 1986. Herbivores' direct and indirect effects on algal populations. *Science* **231**: 605–607.

Sterner, R. W. 1995. Elemental stoichiometry of species in ecosystems. In: C. Jones and J. Lawton (eds), *Linking Species and Ecosystems.* Chapman and Hall, New York, pp. 240–252.

Sterner, R. W., and J. J. Elser. 2002. *Ecological Stoichiometry: The Biology of Elements from Molecules to the Biosphere.* Princeton University Press, Princeton, NJ.

Sterner, R. W., and D. O. Hessen. 1994. Algal nutrient limitation and the nutrition of aquatic herbivores. *Annual Review of Ecological Systems* **25**: 1–29.

Sterner, R. W., J. J. Elser, E. J. Fee, S. J. Guildford, and T. H. Chrzanowski. 1997. The light:nutrient ratio in lakes: the balance of energy and materials affects ecosystem structure and process. *American Naturalist* **150**: 663–684.

Sullender, B. W., and T. J. Crease. 2001. The behavior of a *Daphnia pulex* transposable element in cyclically and obligately parthenogenetic populations. *Journal of Molecular Evolution* **53**: 63–69.

Sutcliffe, W. H. 1970. Relationship between growth rate and ribonucleic acid concentration in some invertebrates. *Journal of the Fisherics Research Board of Canada* **27**: 606–609.

Tilman, D. 1982. *Resource Competition and Community Structure.* Princeton University Press, Princeton, NJ.

Turchin, P. 2003. *Complex Population Dynamics. A Theoretical/Empirical Synthesis.* Princeton University Press, Princeton, NJ.

Urabe, J., and R. W. Sterner. 1996. Regulation of herbivore growth by the balance of light and nutrients. *Proceedings of the National Academy of Sciences, USA* **93**: 8465–8469.

Urabe, J., J. J. Elser, M. Kyle, T. Sekino, and Z. Kawabata. 2002*a*. Herbivorous animals can mitigate unfavorable ratios of energy and material supplies by enhancing nutrient recycling. *Ecology Letters* **5**: 177–185.

Urabe, J., M. Kyle, W. Makino, T. Yoshida, T. Andersen, and J. J. Elser. 2002*b*. Reduced light increases herbivore production due to stoichiometric effects of light: nutrient balance. *Ecology* **83**: 619–627.

Vadstein, O., B. O. Harkjerr, A. Jensen, Y. Olsen, and H. Reinertsen. 1989. Cycling of organic carbon in the photic zone of a eutrophic lake with special reference to the heterotrophic bacteria. *Limnology and Oceanography* **34**: 840–855.

Varma, A., and B. O. Palsson. 1994. Metabolic flux balancing: basic concepts, scientific and practical use. *Bio/Technology* **12**: 994–998.

Vrede, T., J. Persson, and G. Aronsen. 2002. The influence of food quality (P:C ratio) on RNA:DNA ratio and somatic growth rate of *Daphnia*. *Limnology and Oceanography* **47**: 487–494.

Weider, L. J., K. L. Glenn, M. Kyle, and J. J. Elser. 2004. Associations among ribosomal (r)DNA intergenic spacer length, growth rate, and C:N:P stoichiometry in the genus *Daphnia*. *Limnology and Oceanography* **49**: 1417–1423.

Wildermuth, M. C. 2000. Metabolic control analysis: biological applications and insights. Genome Biology **1**: Reviews 1031.1–Reviews 1031.5.

Yoshida, T., L. E. Jones, S. P. Ellner, G. F. Fussmann, and J. Hairston, N. G. 2003. Rapid evolution drives ecological dynamics in a predator–prey system. *Nature* **424**: 303–306.

Chapter 2

Bascompte, J., and R. V. Solé. (eds). 1998. *Modeling Spatiotemporal Dynamics in Ecology*. Springer, Berlin.

Bascompte, J., C. J. Melián, and E. Sala. 2004. Interaction strength motifs and the overfishing of marine food webs. (submitted).

Caldarelli, G., P. G. Higgs, and A. J. McKane. 1998. Modelling coevolution in multispecies communities. *Journal of Theoretical Biology* **193**: 345–358.

Carr, M. H., T. W. Anderson, and M. A. Hixon. 2002. Biodiversity, population regulation, and the stability of coral-reef fish communities. *Proceedings of the National Academy of Sciences, USA*, **99**: 11241–11245.

Caswell, H., and J. E. Cohen. 1993. Local and regional regulation of species–area relations: a patch-occupancy model. In: R. E. Ricklefs and D. Schluter (eds), *Species Diversity in Ecological Communities*. The University of Chicago Press. Chicago, IL.

Cottenie, K., E. Michelis, N. Nuytten, and L. De Meester. 2003. Zooplankton metacommunity structure: regional vs. local processes in highly interconnected ponds. *Ecology* **84**: 991–1000.

Cottenie, K., and L. De Meester. 2004. Metacommunity structure: synergy of biotic ineractions as selective agents and dispersal as fuel. *Ecology* **85**: 114–119.

Dunne, J. A., R. J. Williams, and N. D. Martinez. 2002. Food web structure and network theory: the role of connectance and size. *Proceedings of the National Academy of Sciences, USA* **99**: 12917–12922.

Fisher, W.(ed). 1978. *FAO Species Identification Sheets for Fishery Purposes. Western Central Atlantic (Fishing area 31)*. Vol I–VII. Food and Agriculture Organization of the United Nations, Rome.

Froese, R., and D. Pauly. 2003. FishBase World Wide Web electronic publication. http://fishbase.org, version 24 Septemper 2003.

Gotelli, N. J. 1991. Metapopulation models: the rescue effect, the propagule rain, and the core-satellite hypothesis. *American Naturalist* **138**: 768–776.

Hanski, I. 1983. Coexistence of competitors in a patchy environment. *Ecology* **64**: 493–500.

Hanski, I., and M. E. Gilpin. (eds). 1997. *Metapopulation Biology: Genetics, and Evolution*. Academic Press, San Diego.

Hastings, A. 1980. Disturbance, coexistence, history and the competition for space. *Theoretical Population Biology* **18**: 363–373.

Holt, R. D. 1996. Food webs in space: an island biogeographic perspective. In: G. A. Polis, and K. O. Winemiller (eds), *Food Webs: Integration of Patterns and Dynamics*. Chapman and Hall, London.

Holt, R. D. 1997. From metapolulation dynamics to community structure: some consequences of spatial heterogeneity. In: I. Hanski, and M. Gilpin (eds), *Metapopulation Biology*. Academic Press, San Diego, pp. 149–164.

Hori, M., and T. Noda. 2001. Spatio-temporal variation of avian foraging in the rocky intertidal food web. *Journal of Animal Ecology* **70**: 122–137.

Hubbell, S. P. 2001. *The Unified Neutral Theory of Biodiversity and Biogeography*. Princeton University Press, Princeton, NJ.

Jennings, S., and S. Mackinson. 2003. Abundance–body mass relationships in size-structured food webs. *Ecology Letters* **6**: 971–974.

Kareiva, P. 1987. Habitat fragmentation and the stability of predator–prey interactions. *Nature* **326**: 388–390.

Karlson, R. H., and H. V. Cornell. 2002. Species richness of coral assemblages: detecting regional influences at local spatial scales. *Ecology* **83**: 452–463.

Kneitel, J. M., and T. E. Miller. 2003. Dispersal rates affect species composition in metacommunities of *Sarracenia purpurea* inquilines. *American Naturalist* **162**: 165–171.

Levins, R. 1969. Some demographic and genetic consequences of environmental heterogeneity for biological control. *Bulletin of the Entomological Society of America* **15**: 237–240.

Loreau, M., Mouquet, N., and R. D. Holt. 2003. Meta-ecosystems: a theoretical framework for a spatial ecosystem ecology. *Ecology Letters* **6**: 673–679.

Margalef, R. 1963. On certain unifying principles in ecology. *American Naturalist* **97**: 357–374.

Melián, C. J., and J. Bascompte. 2002. Food web structure and habitat loss. *Ecology Letters* **5**: 37–46.

Melián, C. J., and J. Bascompte, 2004. Food web cohesion. *Ecology* **85**: 352–358.

Miklós, I., and J. Podani. 2004. Randomization of presence–absence matrices: comments and new algorithms. *Ecology* **85**: 86–92.

Mouquet, N., and M. Loreau. 2002. Coexistence in metacommunities: the regional similarity hypothesis. *American Naturalist* **159**: 420–426.

Mouquet, N., and M. Loreau. 2003. Community patterns in source–sink metacommunities. *American Naturalist* **162**: 544–557.

Mumby, P. J., A. J. Edwards, J. E. Arias-González, K. C. Lindeman, P. G. Blackwell, et al. 2004. Mangroves enhance the biomass of coral reef fish communities in the Caribbean. *Nature* **427**: 533–536.

Opitz, S. 1996. Trophic Interactions In Caribbean Coral Reefs. Technical Report 43. International Centre for Living Aquatic Resource Management (now World-Fish Center), Penang, Malaysia.

Polis, G. A., W. B. Anderson, and R. D. Holt. 1997. Toward an integration of landscape and food web ecology: the dynamics of spatially subsidized food webs. *Annual Review of Ecology and Systematics* **28**: 289–316.

Randall, J. E. 1967. Food habits of reef fishes of the West Indies. *Studies in Tropical Oceanography* **5**: 665–847.

Ricklefs, R. E. 1987. Community diversity: relative roles of local and regional processes. *Science* **235**: 167–171.

Shurin, J. B. 2001. Interactive effects of predation and dispersal on zooplankton communities. *Ecology* **82**: 3404–3416.

Thompson, J. N. 1999. Specific hypotheses on the geographic mosaic of coevolution. *American Naturalist* **153**: S1–S14.

Tilman, D. 1994. Competition and biodiversity in spatially structured habitats. *Ecology* **75**: 2–16.

Tilman, D., and P. Kareiva. (eds) 1997. *Spatial Ecology. The Role of Space in Population Dynamics and interspecific Interactions.* Princeton University Press, Princeton, NJ.

Wilson, D. S. 1992. Complex interactions in meta-communities, with implications for biodiversity and higher levels of selection. *Ecology* **73**: 1984–2000.

Chapter 3

Almunia, J., G. Basterretxea, J. Arístegui, and R. E. Ulanowicz. 1999. Benthic–pelagic switching in a coastal subtropical lagoon. *Estuarine, Coastal and Shelf Science* **49**: 363–384.

Baird, D. 1998. Orientors and ecosystem properties in coastal zones. In: F. Muller and M. Leupelt (eds), *Ecotargets, Goal Functions, and Orientors.* Springer-Verlag, Heidelberg. pp. 232–242.

Baird, D. and J. J. Heymans. 1996. Assessment of the ecosystem changes in response to freshwater inflow of the Kromme River Estuary, St Francis Bay, South Africa: a network analysis approach. *Water SA* **22**: 307–318.

Baird, D., J. M. Mcglade, and R. E. Ulanowicz. 1991. The comparative ecology of six marine ecosystems. *Philosphical Transactions of the Royal Society London B* **333**: 15–29.

Baird, D., and R. E. Ulanowicz. 1989. Seasonal dynamics of the Chesapeake Bay ecosystem. *Ecological Monographs* **59**: 329–364.

Baird, D. and R. E. Ulanowicz. 1993. Comparative study on the trophic structure, cycling and ecosystem properties of four tidal estuaries. *Marine Ecology Progress Series* **991**: 221–237.

Baird, D., R. E. Ulanowicz, and W. R. Boynton. 1995. Seasonal nitrogen dynamics in the Chesapeake Bay: a Network Approach. *Estuarine, Coastal and Shelf Science* **41**: 137–162.

Baird, D., J. Luczkovich, and R. R. Christian. 1998. Assessment of spatial and temporal variability in ecosystem attributes of the St Marks National Wildlife Refuge, Apalachee Bay, Florida. *Estuarine, Coastal and Shelf Science* **47**: 329–349.

Baird, D., R. R. Christian, C. H. Peterson, and G. Johnson. 2004. Consequences of hypoxia on estuarine ecosystem function: energy diversion from consumers to microbes. *Ecological Applications* **14**: 805–822.

Batagelj, V., and A. Mrvar. 2002. Pajek 0.85 Network Software. Available at: http://vlado.fmf.uni-lj.si/pub/networks/pajek/

Borgatti, S. P., M. G. Everett, and L. C. Freeman. 2002. *Ucinet for Windows: Software for Social Network Analysis.* Analytic Technologies, Harvard.

Brando, V. E., R. Ceccarelli, S. Libralato, and G. Ravagnan. 2004. Assessment of environmental management effects in a shallow water basin using mass-balance models. *Ecological Modelling* 172: 213–232.

Christensen, V. 1995. Ecosystem maturity—towards quanitification. *Ecological Modelling* 77: 3–32.

Christensen, V., and D. Pauly. 1992. ECOPATH II—a software for balancing steady-state ecosystem models and calculating network characteristics. *Ecological Modelling* 61: 169–185.

Christensen, V., and D. Pauly. (eds). 1993. *Trophic Models of Aquatic Ecosystems.* ICLARM Conference Proceedings 26, Manila, Phillipines.

Christensen, V., and D. Pauly. 1995. Fish production, catches and the carrying capacity of the world oceans. *Naga, The ICLARM Quarterly* 18: 34–40.

Christian, R. R., and J. J. Luczkovich. 1999. Organizing and understanding a winter's seagrass food web network through effective trophic levels. *Ecological Modelling* 117: 99–124.

Christian, R. R., and R. E. Ulanowicz. 2001. Network ecology. In: A. El-Shaarawi and W. W. Pierogorsch (eds), *Encyclopedia of Environmetrics*, Vol. 3. John Wiley and Sons Ltd, Chichester, UK, pp. 1393–1399.

Christian, R. R., and C. R. Thomas. 2003. Network analysis of nitrogen inputs and cycling in the Neuse River estuary, North Carolina, USA. *Estuaries* 26: 815–828.

Christian, R. R., E. Forès, F. Comín, P. Viaroli, M. Naldi, and I. Ferrari. 1996. Nitrogen cycling networks of coastal ecosystems: influence of trophic status and primary producer form. *Ecological Modelling* 87: 111–129.

Christian, R. R., J. K. Dame, G. Johnson, C. H. Peterson, and D. Baird. 2004. Monitoring and Modeling of the Neuse River Estuary, Phase 2: Functional Assessment of Environmental Phenomena through Network Analysis. Final Report, UNC Water Resources Research Institute, Raleigh, North Carolina, USA. Report No. 343-E.

Christensen, V., C. J. Walters, and D. Pauly. 2000. *ECOPATH with ECOSIM: A Users Guide*, October 2000 edn. Fisheries Centre, University of British Columbia, Vancouver, Canada, and International Center for Living Aquatic Resources Management, Penang, Malaysia, 130 pp.

Elmgren, R. 1989. Man's impact on the ecosystem of the Baltic Sea: energy flows today and at the turn of the century. *Ambio* 18: 326–332.

Fath, B. D., and B. C. Patten. 1999. Review of the foundations of network analysis. *Ecosystems* 2: 167–179.

Finn, J. T. 1976. Measures of ecosystem structure and function derived from analysis of flows. *Journal of Theoretical Biology* 56: 363–380.

Finn, J. T., and T. Leschine. 1980. Does salt marsh fertilization enhance shellfish production? An application of flow analysis. *Environmental Management* 4: 193–203.

Gordon, D. C., P. R. Boudreau, K. H. Mann, J.-E. Ong, W. L. Silvert, S. V. Smith, G. Wattayakom, F. Wulff, and T. Yanagi. 1996. LOICZ Biogeochemical Modelling Guidelines. LOICZ Reports and Studies No. 5, 96 p.

Heymans, J. J., and D. Baird. 2000. Network analysis of the northern Benguela ecosystem by means of NETWRK and ECOPATH. *Ecological Modelling* 131: 97–119.

Holligan, P. M., and H. de Boois. (eds). 1993. Land-Ocean Interactions in the Coastal Zone [LOICZ] Science Plan. IGBP Report No. 25, 50 pp.

Johnson, J. C., S. P. Borgatti, J. J. Luczkovich, and M. G. Everett. 2001. Network role analysis in the study of food webs: an application of regular role coloration. *Journal of Social Structure* 2(3).

Johnson, J. C., J. Boster, and L. Palinkas. 2003. Social roles and the evolution of networks in isolated and extreme environments. *The Journal of Mathematical Sociology* 27: 89–122

Kay, J. J., L. A. Graham, and Ulanowicz, R. E. 1989. A detailed guide to network analysis. In: F. Wulff, J. G. Field, and K. H. Mann (eds), *Network Analysis in Marine Ecology*. Springer Verlag, Berlin, Heidelberg, pp. 15–61.

Kremer, J. N. 1989. Network information indices with an estuarine model. In: F. Wulff, J. G. Field, and K. H. Mann (eds), *Network Analysis in Marine Ecology: Methods and Applications.* Coastal and Estuarine Studies 32, Springer-Verlag, Heidelberg, pp. 119–131.

Lin, H.-J., J.-J. Hung, K-.T. Shao, and F. Kuo. 2001. Trophic functioning and nutrient flux in a highly productive tropical lagoon. *Oegologia* 129: 395–406.

Luczkovich, J. J., S. P. Borgatti, J. C. Johnson, and M. G. Everett. 2003. Defining and measuring trophic role similarity in food webs using regular equivalence. *The Journal of Theoretical Biology* 220: 303–321.

Monaco, M. E., and R. E. Ulanowicz. 1997. Comparative ecosystem trophic structure of three U. S. mid-Atlantic estuaries. *Marine Ecology Progress Series* 161: 239–254.

Odum, E. P. 1969. The strategy of ecosystem development. *Science* 164: 262–270.

Pauly, D., and V. Christensen. 1995. Primary production required to sustain global fisheries. *Nature* 374: 255–257.

Pernetta, J. C., and J. D. Milliman. (eds). 1995. Land–Ocean Interactions in the Coastal Zone [LOICZ] Implementation Plan. IGBP Report No. 33, 215 pp.

Polovina, J. J. 1984. Model of a coral reef ecosystem. Part I. The ECOPATH model and its applications to French Frigate Shoals. *Coral Reefs* 3: 1–11.

Richardson, D. C., and J. S. Richardson. 1992. The kinemage: a tool for scientific communication. *Protein Science* **1**, 3–9. Available at: http://kinemage.biochem. duke.edu/

Scharler, U. M., and D. Baird. 2003. The influence of catchment management on salinity, nutrient stochiometry, and phytoplankton biomass of Eastern Cape estuaries, South Africa. *Estuarine, Coastal and Shelf Science* **56**: 735–748.

Scharler, U. M., and D. Baird. A comparison of selected ecosystem attributes of three South African estuaries with different freshwater inflow regimes, using network analysis. *Journal of Marine Systems* (in press).

Smith, S., D. Swaney, L. Talaue-McManus, S. Bartley, P. Sandhei, C. McLacghlin, V. Dupra, C. Crossland, R. Buddemeier, B. Maxwell, and F. Wulff. 2003. Humans, hydrology, and the distribution of inorganic nutrient loading to the ocean. *Bioscience* **93**: 239–245.

Szyrmer, J., and R. E. Ulanowicz. 1987. Total flows in ecosystems. *Ecological Modelling* **35**: 123–136.

Ulanowicz, R. E. 1984. Community measures of marine food networks and their possible applications. In: M. J. R. Fasham (ed.), *Flows of Energy and Materials in Marine Ecosystems*. Plenum, London, pp. 23–47.

Ulanowicz, R. E. 1986. *Growth and Development: Ecosystem Phenomenology*. Springer-Verlag, New York.

Ulanowicz, R. E. 1995*a*. Network growth and development: ascendency. In: B. C. Patten and S. E. Jørgensen (ed), *Complex Ecology: The Part–Whole Relation in Ecosystems*. Prentice Hall, PTR, New Jersey, pp. 643–655.

Ulanowicz, R. E. 1995*b*. Ecosystem trophic foundations: Lindeman Exonerata. In: B. C. Patten and S. E. Jorgensen (eds), *Complex Ecology: The Part–Whole Relation in Ecosystems*. Prentice Hall, Englewood Cliffs, NJ, pp. 549–560.

Ulanowicz, R. E., and D. Baird. 1999. Nutrient controls on ecosystem dynamics: the Chesapeake Bay mesohaline community. *Journal of Marine Science* **19**: 159–172.

Ulanowicz, R. E., and J. H. Tuttle. 1992. The trophic consequences of oyster stock rehabilitation in Chesapeake Bay. *Estuaries* **15**: 298–306.

Ulanowicz, R. E., C. Bondavalli, and M. S. Egnotovich. 1999. Network Analysis of Trophic Dynamics in South Florida Ecosystem, FY 97: The Florida Bay Ecosystem. Annual Report to the United States Geological Service Biological Resources Division University of Miami Coral Gables, FL 33124.

Walters, C., V. Christensen, and D. Pauly. 1997. Structuring dynamic models of exploited ecosystems from mass-balance assessments. *Reviews in Fish Biology and Fisheries.* **7**: 139–172.

Wulff F, J. G. Field, and K. H. Mann. (eds). 1989. *Network Analysis in Marine Ecology: Methods and Applications*. Coastal and Estuarine Studies 32, Springer-Verlag, Heidelberg.

Wulff, F., and R. E. Ulanowicz. 1989. A comparative anatomy of the Baltic Sea and Chesapeake Bay ecosystems. In: F. Wulff, J. G. Field, and K. H. Mann (eds), *Network Analysis in Marine Ecology*. Coastal and Estuarine Studies Series. Springer-Verlag, Berlin, pp. 232–256.

Chapter 4

Adrian, R., N. Walz, T. Hintze, S. Hoeg, and R. Rusche. 1999. Effects of ice duration on the plankton succession during spring in a shallow polymictic lake. *Freshwater Biology* **41**: 621–623.

Anderson, W. L., D. M. Robertson, and J. J. Magnuson. 1996. Evidence of recent warming and El-Niño-related variations in ice break-up of Wisconsin lakes. *Limnology Oceanography* **41**: 815–821.

Anneville, O., S. Soussi, S. Gammeter, and D. Straile. 2004. Seasonal and inter-annual scales of variability in phytoplankton assemblages: comparison of phytoplankton dynamics in three peri-alpine lakes over a period of 28 years. *Freshwater Biology* **49**: 98–115.

Anneville, O., S. Soussi, F. Ibañez, V. Ginot, J.-C. Druart, and N. Angeli. 2002. Temporal mapping of phytoplankton assemblages in Lake Geneva: annual and interannual changes in their patterns of succession. *Limnology Oceanography* **47**: 1355–1366.

Bäuerle, E., and U. Gaedke. (eds). 1998. Lake constance—characterization of an ecosystem in transition. *Archives of Hydrobiology Special Issues Advances Limnology* **53**: 1–610.

Beardall, J., and J. A. Raven. 2004. The potential effects of global climate change on microalgal photosynthesis, growth and ecology. *Phycologia* **43**: 26–40.

Benndorf, J., H. Kneschke, K. Kossatz, and E. Penz. 1984. Manipulation of the pelagic food web by stocking with predacious fishes. *Internationale Revue der Gesamten Hydrobiologie* **69**: 407–428.

Brett, M. T., and C. R. Goldman. 1996. A meta-analysis of the freshwater trophic cascade. *Proceedings of the National Academy of Sciences USA* **93**: 7723–7726.

Briand, J. F., C. Leboulanger, J. F. Humbert, C. Bernard, and P. Dufour. 2004. *Cylindrospermopsis raciborskii* (Cyanobacteria) invasion at mid-latitudes: selection, wide physiological tolerance, or global warming? *Journal of Phycology* **40**: 231–238.

Cáceres, C. 1998. Seasonal dynamics and interspecific competition in Oneida Lake *Daphnia*. *Oecologia* **115**: 233–244.

Carpenter, S. R., and J. F. Kitchell. 1993. *The Trophic Cascade in Lakes*. Cambridge University Press.

Carpenter, S. R., S. G. Fisher, N. B. Grimm, and J. F. Kitchell. 1992. Global change and freshwater ecosystems. *Annual Review of Ecological Systems* 23: 119–139.

Carpenter, S. R., J. F. Kitchell, and J. R. Hodgson. 1985. Cascading trophic interactions and lake productivity. *Bioscience* 35: 634–639.

Chen, C. Y., and C. L. Folt. 1996. Consequences of fall warming for zooplankton overwintering success. *Limnology and Oceanography* 41: 1077–1086.

Cohen, J. E., T. Jonsson, and S. R. Carpenter. 2003. Ecological community description using the food web, species abundance, and body size. *Proceedings of the National Academy of Sciences USA* 100: 1781–1786.

Conover, D. O. 1992. Seasonality and the scheduling of life-history at different latitudes. *Journal of Fish Biology* 41B: 161–178.

Cushing, D. H. 1990. Plankton production and year-class strength in fish populations: an update of the match/mismatch hypotheses. *Advances Marine Biology* 26: 249–293.

Eckmann, R., U. Gaedke, and H. J. Wetzlar. 1988. Effects of climatic and density-dependent factors on year-class strength of *Coregonus lavaretus* in Lake Constance. *Canadian Journal of Fisheries and Aquatic Sciences* 45: 1088–1093.

Gaedke, U. 1992. The size distribution of plankton biomass in a large lake and its seasonal variability. *Limnology and Oceanography* 37: 1202–1220.

Gaedke, U. 1993. Ecosystem analysis based on biomass size distributions: a case study of a plankton community in a large lake. *Limnology and Oceanography* 38: 112–127.

Gaedke, U. 1995. A comparison of whole-community and ecosystem approaches (biomass size distributions, food web analysis, network analysis, simulation models) to study the structure, function and regulation of pelagic food webs. *Journal of Plankton Research* 17: 1273–1305.

Gaedke, U. and D. Straile. 1994a. Seasonal changes of the quantitative importance of protozoans in a large lake. An ecosystem approach using mass-balanced carbon flow diagrams. *Marine Microbial Food Webs* 8: 163–188.

Gaedke, U. and D. Straile. 1994b. Seasonal changes of trophic transfer efficiencies in a plankton food web derived from biomass size distributions and network analysis. *Ecological Modelling* 75/76: 435–445.

Gaedke, U., S. Hochstädter, and D. Straile. 2002. Interplay between energy limitation and nutritional deficiency: empirical data and food web models. *Ecological Monographs* 72: 251–270.

Gaedke, U., D. Ollinger, E. Bäuerle, and D. Straile. 1998. The impact of the interannual variability in hydro-dynamic conditions on the plankton development in Lake Constance in spring and summer. *Archives in Hydrobiology Special Issues in Advances Limnology* 53: 565–585.

Gaedke, U., D. Straile, and C. Pahl-Wostl. 1996. Trophic structure and carbon flow dynamics in the pelagic community of a large lake. In: G. A. Polis and K. O. Winemiller (eds), *Food Webs: Integration of Pattern and Dynamics*. Chapmann and Hall, pp. 60–71.

George, D. G. and G. P. Harris. 1985. The effect of climate on long-term changes in the crustacean zooplankton biomass of Lake Windermere, UK. *Nature* 316: 536–539.

George, D. G. and D. P. Hewitt. 1999. The influence of year-to-year variations in winter weather on the dynamics of *Daphnia* and *Eudiaptomus* in Estwaite Water, Cumbria. *Functional Ecology* 13(Suppl.1): 45–54.

George, D. G., and A. H. Taylor. 1995. UK lake plankton and the Gulf stream. *Nature* 378: 139.

Gerten, D., and R. Adrian. 2000. Climate-driven changes in spring plankton dynamics and the sensitivity of shallow polymictic lakes to the North Atlantic Oscillation. *Limnology and Oceanography* 45: 1058–1066.

Gerten, D. and R. Adrian. 2002a. Effects of climate warming, North Atlantic Oscillation, and El Niño-Southern Oscillation on thermal conditions and plankton dynamics in northern hemispheric lakes. *The Scientific World Journal* 2: 586–606.

Gerten, D., and R. Adrian. 2002b. Species-specific changes in the phenology and peak abundance of freshwater copepods in response to warm summers. *Freshwater Biology* 47: 2163–2173.

Gillooly, J. F. 2000. Effect of body size and temperature on generation time in zooplankton. *J Plankt Research* 22: 241–251.

Goldman, C. R., A. Jassby, and T. Powell. 1989. Inter-annual fluctuations in primary production: meteorological forcing at two subalpine lakes. *Limnology and Oceanography* 34: 310–323.

Güss, S., D. Albrecht, H.-J. Krambeck, D. C. Müller-Navarra, and H. Mumm. 2000. Impact of weather on a lake ecosystem, assessed by cyclo-stationary MCCA of long-term observations. *Ecology* 81: 1720–1735.

Hairston, N. G. Jr. 1998. Time travelers: what's timely in diapause reserach? *Archives in Hydrobiology Special Issues in Advances in Limnology* 52: 1–15.

Hairston, N. G. Jr., and C. Cáceres. 1996. Distribution of crustacean diapause: micro- and macroevolutionary pattern and process. *Hydrobiology* 320: 27–44.

Hairston, N. G. Jr., and N. G. Sr. Hairston. 1993. Cause–effect relationships in energy flow, trophic strcture, and interspecific interactions. *American Naturalist* 142: 379–411.

Hairston, N. G. Jr., A. M. Hansen, and W. R. Schaffner. 2000. The effect of diapause emergence on the seasonal dynamics of a zooplankton assemblage. *Freshwater Biology* 45: 133–146.

Hairston, N. G., F. E. Smith, and L. B. Slobodkin. 1960. Community structure, population control, and competition. *American Naturalist* 94: 421–425.

Hansson, L. 1996. Algal recruitment from lake sediments in relation to grazing, sinking, and dominance patterns in the phytoplankton community. *Limnology and Oceanography* 41: 1312–1323.

Houghton, J. T., Y. Ding, D. J. Griggs, M. Noguer, P. J. van der Linden, X. Da, K. Maskell, and C. A. Johnson. 2001. Climate Change 2001—The Scientific Basis. Contribution of Working Group I to the Third Assessment Report of the Intergovernmental Panel on Climate Change. Cambridge University Press.

Hrbáček, J. 1958. Typologie und Produktivität der teichartigen Gewässer. *Verhein Internationale Verein Limnologie* 13: 394–399.

Hurrell, J. W. 1995. Decadal trends in the North Atlantic Oscillation: regional temperatures and precipitation. *Science* 269: 676–679.

Jankowski, T. and D. Straile. 2004. Allochronic differentiation among *Daphnia* species, hybrids and backcrosses: the importance of sexual reproduction for population dynamics and genetic architecture. *Journal of Evolutionary Biology* 17: 312–321.

Lampert, W. 1978. Climatic conditions and planktonic interactions as factors controlling the regular succession of spring algal bloom and extremely clear water in Lake Constance. *Verhein Internationale Verein Limnologie* 20: 969–974.

Levine, S. 1980. Several measures of trophic structure applicable to complex food webs. *Journal of theoretical Biology* 83: 195–207.

Livingstone, D. M. 1997. An example of the simultaneous occurrence of climate-driven "sawtooth" deep-water warming/cooling episodes in several Swiss lakes. *Verhein Internationale Verein Limnologie* 26: 822–826.

Livingstone, D. M. 2000. Large-scale climatic forcing detected in historical observations of lake ice break-up. *Verhein Internationale Verein Limnologie* 27: 2775–2783.

Livingstone, D. M. 2003. Impact of secular climate change on the thermal structure of a large temperate central European lake. *Climatic Change* 57: 205–225.

Magnuson, J. J., D. M. Robertson, B. J. Benson, R. H. Wynne, D. M. Livingstone, T. Arai, R. A. Assel, R. G. Barry, V. Card, E. Kuusisto, N. G. Granin, T. D. Prowse, K. M. Stewart, and V. S. Vuglinski. 2000. Historical trends in lake and river ice cover in the Northern Hemisphere. *Science* 289: 1743–1746.

Mittelbach, G. G., A. M. Turner, D. J. Hall, J. E. Rettig, and C. W. Osenberg. 1995. Perturbation and resilience: a long-term, whole-lake study of predator extinction and reintroduction. *Ecology* 71: 2241–2254.

Müller, H., and C. Wünsch. 1999. Seasonal dynamics of cyst formation of pelagic strombidiid ciliates in a deep prealpine lake. *Aquatic Microbial Ecology* 17: 37–47.

Murdoch, W. M. 1966. "Community structure, population control, and competition"—a critique. *American Naturalist* 100: 219–226.

Nicholls, K. H. 1998. El Nino, ice cover, and Great Lakes phosphorus: implications for climate warming. *Limnology and Oceanography* 43: 715–719.

Oksanen, L., S. D. Fretwell, J. Arruda, and Niemel. 1981. Exploitation ecosystems in gradients of primary productivity. *American Naturalist* 118: 240–261.

Persson, L., A. M. De Roos, D. Claessen, P. Byström, J. Lövgren, S. Sjögren, R. Svanbäck, E. Wahlström, and E. Westman. 2003. Gigantic cannibals driving a whole-lake trophic cascade. *Proceedings of the National Academy of Sciences, USA* 100: 4035–4039.

Persson, L., S. Diehl, L. Johansson, G. Andersson, and S. F. Hamrin. 1992. Trophic interactions in temperate lake ecosystems: a test of food chain theory. *American Naturalist* 140: 59–84.

Platt, T., C. Fuentes-Yaco, and K. T. Frank. 2003. Spring algal bloom and larval fish survival. *Nature* 423: 398–399.

Post, J. R., and D. O. Evans. 1989. Size-dependent overwinter mortality of young-of-the-year yellow perch (*Perca flavescens*)—Laboratory, *in Situ* enclosure, and field experiments. *Canadian Journal of Fisheries and Aquatic Sciences* 46: 1958–1968.

Post, D. M., S. R. Carpenter, D. L. Christensen, K. L. Cottingham, J. F. Kitchell, D. E. Schindler, and J. R. Hodgson. 1997. Seasonal effects of variable recruitment of a dominant piscivore on pelagic food-web structure. *Limnology and Oceanography* 42: 722–729.

Reynolds, C. S. 1997. Vegetation Processes in the Pelagic: A Model for Ecosystem Theory. Excellence in Ecology Series. Vol 9, Ecology Institute, Oldendorf-Luhe, 371 pp.

Reynolds, C. S., and E. G. Bellinger. 1992. Patterns of abundance and dominance of the phytoplankton of Rostherne Mere, England: evidence from an 18-year data set. *Aquatic Science* 54: 10–36.

Rodionov, S., and R. A. Assel. 2003. Winter severity in the Great Lakes region: a tale of two oscillations. *Climatic Research* 24: 19–31.

Sanderson, B. L., T. R. Hrabik, J. J. Magnuson, and D. M. Post. 1999. Cyclic dynamics of a yellow perch (*Perca flavescens*) population in an oligotrophic lake: evidence for the role of intraspecific interactions.

Canadian Journal of Fisheries and Aquatic Sciences **56**: 1534–1542.

Sanford, E. 1999. Regulation of keystone predation by small changes in ocean temperature. *Science* **283**: 2095–2097.

Scheffer, M., S. Carpenter, C. Foley, and B. Walker. 2001*a*. Catastrophic shifts in ecosystems. *Nature* **413**: 591–596.

Scheffer, M., S. H. Hosper, M. L. Meijer, B. Moss, and E. Jeppesen. 1993. Alternative equilibria in shallow lakes. *TREE* **8**: 275–279.

Scheffer, M., D. Straile, E. H. van Nes, and H. Hosper. 2001*b*. Climatic warming causes regime shifts in lake food webs. *Limnology and Oceanography* **46**: 1780–1783.

Schindler, D. W. 1997. Widespread effects of climatic warming on freshwater ecosystems in North America. *Hydrological Processes* **11**: 1043–1067.

Schindler, D. W., K. G. Beaty, E. J. Fee, D. R. Cruikshank, E. R. DeBruyn, D. L. Findlay, G. A. Linsey, J. A. Shearer, M. P. Stainton, and M. A. Turner. 1990. Effects of climatic warming on lakes of the central boreal forest. *Science* **250**: 967–970.

Schindler, D. W., P. J. Curtis, B. R. Parker, and M. P. Stainton. 1996. Consequences of climate warming and lake acidification for UV-B penetration in North American boreal lakes. *Nature* **379**: 705–708.

Schmitz, O. J., E. Post, C. E. Burns, and K. M. Johnston. 2003. Ecosystem responses to global climate change: moving beyond color mapping. *Bioscience* **53**: 1199–1205.

Shapiro, J., and D. I. Wright. 1984. Lake restoration by biomanipulation—Round Lake, Minnesota, the 1st 2 Years. *Freshwater Biology* **14**: 371–383.

Sommer, U. 1989. *Plankton Ecology.* Springer.

Sommer, U., Z. M. Gliwicz, W. Lampert, and A. Duncan. 1986. The PEG-model of seasonal succession of planktonic events in fresh waters. *Archives in Hydrobiology* **106**: 433–471.

Stich, H. B., and W. Lampert. 1981. Predator evasion as an explanation of diurnal vertical migration by zooplankton. *Nature* **293**: 396–398.

Straile, D. 1995. Die saisonale Entwicklung des Kohlenstoffkreislaufes im pelagischen Nahrungsnetz des Bodensees—Eine Analyse von massenbilanzierten Flußdiagrammen mit Hilfe der Netzwerktheorie. PhD Thesis, University of Konstanz. Hartung-Gorre Verlag, Konstanz.

Straile, D. 1998. Biomass allocation and carbon flow in the pelagic food web of Lake Constance. *Archives in Hydrobiology Special Issues in Advances in Limnology* **53**: 545–563.

Straile, D. 2000. Meteorological forcing of plankton dynamics in a large and deep continental European lake. *Oecologia* **122**: 44–50.

Straile, D. 2002. North Atlantic Oscillation synchronizes food-web interactions in central European lakes. *Proceedings of the Royal Society of London B* **269**: 391–395.

Straile, D. 2004. A fresh (water) perspective on NAO ecological impacts in the North Atlantic. In: N. C. Stenseth, G. Ottersen, J. W. Hurrell, and A. Belgrano (eds), *Ecological Effects of Climate Variations in the North Atlantic.*

Straile, D., and R. Adrian. 2000. The North Atlantic Oscillation and plankton dynamics in two European lakes—two variations on a general theme. *Global Change Biology* **6**: 663–670.

Straile, D. and W. Geller. 1998. Crustacean zooplankton in Lake Constance from 1920. to 1995: response to eutrophication and reoligotrophication. *Archives in Hydrobiology. Special Issues in Advances Limnology* **53**: 255–274.

Straile, D. and A. Hälbich. 2000. Life history and multiple anti-predator defenses of an invertebrate predator: *Bythotrephes longimanus,* in a large and deep lake. *Ecology* **81**: 150–163.

Straile, D., K. Joehnk, and H. Rossknecht. 2003. Complex effects of winter warming on the physico-chemical characteristics of a deep lake. *Limnology and Oceanography* **48**: 1432–1438.

Straile, D., D. M. Livingstone, G. A. Weyhenmeyer, and D. G. George. 2003. The response of freshwater ecosystems to climate variability associated with the North Atlantic Oscillation. In: J. W. Hurrell, Y. Kushnir, G. Ottersen, and M. Visbeck (eds), *The North Atlantic Oscillation: Climatic significance and Environmental Impact,* pp. 263–279.

Tonn, W. M. and C. A. Paszkowski. 1986. Size-limited predation, winterkill, and the organization of *Umbra—Perca* fish assemblages. *Canadian Journal of Fisheries and Aquatic Sciences* **43**: 194–202.

Ulanowicz, R. E. 1986. *Growth and Development: Ecosystems Phenomenology.* Springer.

Urabe, J., J. Togari, and J. J. Elser. 2003. Stoichiometric impacts of increased carbon dioxide on a planktonic herbivore. *Global Change Biology* **9**: 818–825.

van Donk, E., L. Santamaria, and W. M. Mooij. 2003. Climate change causes regime shifts in lake food webs: a reassessment. *Limnology and Oceanography* **48**: 1350–1353.

Visser, P. M., B. W. Ibelings, B. van der Veer, J. Koedood, and L. R. Mur. 1996. Artificial mixing prevents nuisance blooms of the cyanobacterium *Microcystis* in Lake Nieuwe Meer, the Netherlands. *Freshwater Biology* **36**: 435–450.

Walther, G.-R., E. Post, P. Convey, A. Menzel, C. Parmesan, T. J. C. Beebee, J. M. Fromentin, O. Hoegh-Guldberg, and F. Bairlein. 2002.

Ecological responses to recent climate change. *Nature* **416**: 389–395.

Weyhenmeyer, G. A., T. Blenckner, and K. Pettersson. 1999. Changes of the plankton spring outburst related to the North Atlantic Oscillation. *Limnology Oceanography* **44**: 1788–1792.

Williams, R. J., and N. D. Martinez. 2004. Limits to trophic levels and omnivory in complex food webs: theory and data. *American Naturalist* **163**: E458–E468.

Williamson, C. E., G. Grad, H. J. De Lange, S. Gilroy, and D. M. Karapelou. 2002. Temperature-dependent ultraviolet responses in zooplankton: implications of climate change. *Limnology and Oceanography* **47**: 1844–1848.

Chapter 5

Allan, J. D. 1982. Feeding habits and prey consumption of three setipalpian stoneflies (Plecoptera) in a mountain stream. *Ecology* **63**: 26–34.

Allan, J. D. 1995. *Stream Ecology: Structure and Function of Running Waters*. Chapman and Hall, London.

Allen, K. R. 1951. The Horokiwi Stream: A Study of a Trout Population. Fisheries Bulletin 10. New Zealand Marine Department, Wellington, New Zealand.

Benke, A. C., and J. B. Wallace. 1997. Trophic basis of production among riverine caddisflies: implications for food web analysis. *Ecology* **78**: 1132–1145.

Benke, A. C., J. B. Wallace, J. W. Harrison, and J. W. Koebel. 2001. Food web quantification using secondary production analysis: predaceous invertebrates of the snag habitat in a subtropical river. *Freshwater Biology* **46**: 329–346.

Berlow, E. L., A. M. Neutel, J. E. Cohen, P. de Ruiter, B. Ebenman, M. Emmerson, J. W. Fox, V. A. A. Jansen, J. I. Jones, G. D. Kokkoris, D. O. Logofet, A. J. McKane, J. M. Montoya, and O. Petchey. 2004. Interaction strengths in food webs: issues and opportunities. *Journal of Animal Ecology* **73**: 585–598.

Bird, G. A., and W. Schwartz. 1996. Effect of microbes on the uptake of Co-60, Sr-85, Tc-95M, I-131 and Cs-134 by decomposing elm leaves in aquatic microcosms. *Hydrobiologia* **333**: 57–62.

Bradley, D. C., and S. J. Ormerod. 2001. Community persistence among stream invertebrates tracks the Northern Atlantic Oscillation. *Journal of Animal Ecology* **70**: 987–996.

Bunn, E. B., Davies, P. M., and Winning, M. 2003. Sources of organic carbon supporting the food web of an arid zone floodplain river. *Freshwater Biology* **48**: 619–635.

Charlebois, P. M., and Lamberti, G. A. 1996. Invading crayfish in a Michigan stream: direct and indirect

effects on periphyton and macroinvertebrates. *Journal of The North American Benthological Society* **15**: 551–563.

Clements, W. H., and M. C. Newman. 2002. *Community Ecotoxicology*. Wiley, Chichester, 336 pp.

Closs, G. P. 1996. Effects of a predatory fish (*Galaxias olidus*) on the structure of intermittent stream pool communities in southeast Australia. *Australian Journal of Ecology* **21**: 217–223.

Closs, G. P., and P. S. Lake. 1994. Spatial and temporal variation in the structure of an intermittent-stream food-web. *Ecological Monographs* **64**: 1–21.

Cohen, J. E. 1978. *Food Webs and Niche Space*. Monographs in Population Biology, No. 11. Princeton University Press, Princeton, NJ.

Cohen, J. E. and C. M. Newman. 1988. Dynamic basis of food web organization. *Ecology* **69**: 1655–1664.

Cohen, J. E., R. A. Beaver, S. H. Cousins, D. A. DeAngelis, L. Goldwasser, K. L. Heong, R. D. Holt, A. J. Kohn, J. H. Lawton, N. Martinez, R. O'Malley, L. M. Page, B. C. Patten, S. L. Pimm, G. A. Polis, M. Rejmánek, T. W. Schoener, K. Schoenly, W. G. Sprules, J. M. Teal, R. E. Ulanowicz, P. H. Warren, H. M., Wilbur, and P. Yodzis. 1993*b*. Improving food webs. *Ecology* **74**: 252–258.

Cohen, J. E., T. Jonsson, and S. R. Carpenter. 2003. Ecological community description using food web, species abundance, and body-size. *Proceedings of the National Academy of Sciences USA* **100**: 1781–1786.

Cohen, J. E., S. L. Pimm, P. Yodzis, and J. Saldaña. 1993*a*. Body sizes of animal predators and animal prey in food webs. *Journal of Animal Ecology* **62**: 67–78.

Collier, K. J., S. Bury, and Max Gibbs. 2002. A stable isotope study of linkages between stream and terrestrial food webs through spider predation. *Freshwater Biology* **47**: 1651–1659.

Cousins, S. H. 1996. Food webs: from the Lindeman paradigm to a taxonomic general theory of ecology. In: G. A. Polis, and K. O. Winemiller (eds), *Food Webs: Integration of Patterns and Dynamics*. Chapman and Hall, chapter 23, pp. 243–254.

Cross, W. F., J. P. Benstead, A. D. Rosemond, and J. B. Wallace. 2003. Consumer-resource stoichiometry in detritus-based streams. *Ecology Letters* **6**: 721–732.

Dangles, O., and B. Malmqvist. 2004. Species richness-decomposition relationships depend on species dominance. *Ecology Letters* **7**: 395–402.

Dangles, O., M. O. Gessner, F. Guerold, and E. Chauvet. 2004. Impacts of stream acidification on litter breakdown: implications for assessing ecosystem functioning. *Journal of Applied Ecology* **41**: 365–378.

Death, R. G., and M. R. Winterbourn. 1995. Diversity patterns in stream benthic invertebrate communities: the influence of habitat stability. *Ecology* **76**: 1446–1460.

De Ruiter, P. C., A.-M. Neutel, and J. C. Moore. 1995. Energetics, patterns of interaction strengths, and stability in real ecosystems. *Science* **269**: 1257–1260.

Dick, J. T. A., W. I. Montgomery, and R. W. Elwood. 1993. Replacement of the indigenous amphipod *Gammarus duebeni celticus* by the introduced *G. pulex*: differential cannibalism and mutual predation. *Journal of Animal Ecology* **62**: 79–88.

Dodds, W. K., M. A. Evans-White, N. M. Gerlanc, L. Gray, D. A. Gudder, M. J. Kemp, A. L. Lopez, D. Stagliano, E. A. Strauss, J. L. Tank, M. R. Whiles, and W. M. Wollheim. 2000. Quantification of the nitrogen cycle in a prairie stream. *Ecosystems* **3**: 574–589.

Downing, A. L., and M. Leibold. 2002. Ecosystem consequences of species richness and composition in pond food webs. *Nature* **416**: 837–841.

Edling, H., and L. R. Tranvik. 1996. Effects of pH on β-glucosidase activity and availability of DOC to bacteria in lakes. *Archiv für Hydrobiologie/Advances in Limnology* **48**: 123–132.

Eggert, S., and J. B. Wallace. 2003. Litter breakdown and invertebrate detritivores in a resource-depleted Appalachian stream. *Archiv für Hydrobiologie* **156**: 315–338.

Elser, J. J., and J. Urabe. 1999. The stoichiometry of consumer-driven nutrient recycling: theory, observations and consequences. *Ecology* **80**: 735–751.

Elton, C. S. 1927. *Animal Ecology*. Sedgwick and Jackson, London.

Emmerson, M. C., and D. G. Raffaelli. 2004. Predator–prey body size, interaction strength and the stability of a real food web. *Journal of Animal Ecology* **73**: 399–409.

Emmerson, M. C., M. Solan, C. Emes, D. M. Paterson, and D. Raffaelli. 2001. Consistent patterns and the idiosyncratic effects of biodiversity in marine ecosystems. *Nature* **411**: 73–77.

Fahey, B., and R. Jackson. 1997. Hydrological impacts of converting native forests and grasslands to pine plantations, South Island, New Zealand. *Agricultural and Forest Meteorology* **84**: 69–82.

Flecker, A. S. 1996. Ecosystem engineering by a dominant detritivore in a diverse tropical stream. *Ecology* **77**: 1845–1854.

Flecker, A. S., and C. R. Townsend. 1994. Community-wide consequences of trout introduction in New Zealand streams. *Ecological Applications* **4**: 798–807.

Friberg, N., A. Rebsdorf, and S. E. Larsen. 1998. Effects of afforestation on acidity and invertebrates in Danish streams and implications for freshwater communities in Denmark. *Water, Air, and Soil Pollution* **101**: 235–256.

Gessner, M. O., and E. Chauvet. 1994. Importance of stream microfungi in controlling breakdown rates of leaf-litter. *Ecology* **75**: 1807–1817.

Hall, R. O., J. B. Wallace, and S. L. Eggert. 2000. Organic matter flow in stream food webs with reduced detrital resource base. *Ecology* **81**: 3445–3463.

Hall, R. O., G. E. Likens, and H. M. Malcolm. 2001. Trophic basis of invertebrate production in two streams at the Hubbard Brook Experimental Forest. *Journal of the North American Benthological Society* **20**: 432–447.

Hall, S. J., and D. Raffaelli. 1993. Food webs: theory and reality. *Advances in Ecological Research* **24**: 187–239.

Hämäläinen, H., and P. Huttunen. 1996. Inferring the minimum pH of streams from macroinvertebrates using weighted averaging regression and calibration. *Freshwater Biology* **36**: 697–709.

Hieber, M., and M. O. Gessner. 2002. Contribution of stream detrivores, fungi, and bacteria to leaf breakdown based on biomass estimates. *Ecology* **83**: 1026–1038

Hildrew, A. G. 1992. Food webs and species interactions. In: P. Calow, and G. E. Petts (eds), *The Rivers Handbook* Blackwell Sciences, Oxford, pp. 309–330.

Hildrew, A. G., and S. J. Ormerod. 1995. Acidification: causes, consequences and solutions. In D. M. Harper, and A. J. D. Ferguson (eds), *The Ecological Basis for River Management*. John Wiley and Sons Ltd., London, pp. 147–160.

Hildrew, A. G., and C. R. Townsend. 1977. The influence of substrate on the functional response of *Plectrocnemia conspersa* (Curtis) larvae (Trichoptera: Polycentropodidae). *Oecologia* **31**: 21–26.

Hildrew, A. G., and C. R. Townsend. 1982. Predators and prey in a patchy environment: a freshwater study. *Journal of Animal Ecology* **51**: 797–815.

Hildrew, A. G., C. R. Townsed, and J. E. Francis. 1984. Community structure in some southern English streams: the influence of species interactions. *Freshwater Biology* **14**: 297–310.

Hildrew, A. G., C. R. Townsend, and A. Hasham. 1985. The predatory Chironomidae of an iron-rich stream: feeding ecology and food web structure. *Ecological Entomology* **10**: 403–413.

Hildrew, A. G., G. Woodward, J. H. Winterbottom, and S. Orton. 2004. Strong density-dependence in a predatory insect: larger scale experiments in a stream. *Journal of Animal Ecology* **73**: 448–458.

Hillebrand, H. 2002. Top-down versus bottom-up control of autotrophic biomass—a meta-analysis on experiments with periphyton. *Journal of the North American Benthological Society* **21**: 349–369.

Hogg, I. D., and D. D. Williams. 1996. Response of stream invertebrates to a global-warming thermal regime: an ecosystem-level manipulation. *Ecology* **77**: 395–407.

Holt, R. D. 1977. Predation, apparent competition, and the structure of prey communities. *Theoretical Population Biology* **12**: 197–229.

Huryn, A. D. 1996. An appraisal of the Allen Paradox in a New Zealand trout stream. *Limnology and Oceanography* **41**: 243–252.

Huryn, A. D. 1998. Ecosystem-level evidence for top-down and bottom-up control of production in a grassland stream system. *Oecologia* **115**: 173–183.

Hynes, H. B. N. 1975. The stream and its valley. *Verhandlungen der Internationalen Vereinigung für Theoretische und Angewandte Limnologie* **19**: 1–15.

Irons, J. G., M. W. Oswood, R. J. Stout, and C. M. Pringle. 1994. Latitudinal patterns in leaf-litter breakdown—is temperature really important? *Freshwater Biology* **32**: 401–411.

Jenkins, B., R. L. Kitching, and S. L. Pimm. 1992. Productivity, disturbance and food web structure at a local spatial scale in experimental container habitats. *Oikos* **65**: 249–255.

Jonsson, M., and B. Malmqvist. 2003. Importance of species identity and number for process rates within stream functional feeding groups. *Journal of Animal Ecology* **72**: 453–459.

Jonsson, M., O. Dangles, B. Malmqvist, and F. Guerold. 2002. Simulating species loss following perturbation: assessing the effects on process rates. *Proceedings of the Royal Society of London B* **269**: 1047–1052.

Jonsson, M., B. Malmqvist, and P. O. Hoffsten. 2001. Leaf litter breakdown rates in boreal streams: does shredder species richness matter? *Freshwater Biology* **46**: 161–171.

Junk, W. J., P. B. Bayley, and R. E. Sparks. 1989. The Flood Pulse Concept in river–floodplain systems. In: D. P. Dodge (ed.) *Canadian Fisheries and Aquatic Sciences* **106**: (Spec. Publ.) 110–127.

Keleher, C. J., and F. J. Rahel. 1996. Thermal; limits to salmonid distributions in the rocky mountain region and potential habitat loss due to global warming: a geographic information system (GIS) approach. *Transactions of the American Fisheries Society* **125**: 1–13.

Lancaster, J. 1996. Scaling the effects of predation and disturbance in a patchy environment. *Oecologia* **107**: 321–331.

Lancaster, J., and A. L. Robertson. 1995. Microcrustacean prey and macroinvertebrate predators in a stream food web. *Freshwater Biology* **34**: 123–134.

Lancaster, J., M. Real, S. Juggins, D. T. Monteith, R. J. Flower, and W. R. C. Beaumont. 1996. Monitoring temporal changes in the biology of acid waters. *Freshwater Biology* **36**: 179–201.

Ledger, M. E., and Hildrew, A. G. 2000a. Herbivory in an acid stream. *Freshwater Biology* **43**: 545–556.

Ledger, M. E. and A. G. Hildrew. 2000b. Resource depression by a trophic generalist in an acid stream. *Oikos* **90**: 271–278.

Ledger, M., and A. G. Hildrew. 2001. Growth of an acid-tolerant stonefly on biofilm from streams of contrasting pH. *Freshwater Biology* **46**: 1457–1470.

Leeks, G. J. L. 1992. Impact of plantation forestry on sediment transport processes. In: P. Billi, R. D. Hey, C. R. Thorne, and P. Tacconi (eds), *Dynamics of Gravel-Bed Rivers.* John Wiley and Sons, Chichester, pp. 651–670.

Lehman, C. L., and D. Tilman. 2000. Biodiversity, stability, and productivity in competitive communities. *American Naturalist* **156**: 534–552.

Lindeman, R. L. 1942. The trophic-dynamic aspect of ecology. *Ecology* **23**: 399–417

MacArthur, R. H. 1955. Fluctuations of animal populations and a measure of community stability. *Ecology* **36**: 533–536.

Malmqvist, B., S. D. Rundle, A. P. Covich, A. G. Hildrew, C. T. Robinson, and C. R. Townsend. *Prospects for Streams and rivers: an ecological perspective* (in press).

McCann, K. S. 2000. The diversity–stability debate. *Nature* **405**: 228–233.

McCann, K. S., A. Hastings, and G. R. Huxel. 1998. Weak trophic interactions and the balance of nature. *Nature* **395**: 794–798.

McIntosh, A. R., and C. R. Townsend. 1995. Impacts of an introduced predatory fish on mayfly grazing in New Zealand streams. *Limnology and Oceanography* **40**: 1508–1512.

Matthews, W. J., and E. G. Zimmerman. 1990. Potential effects of global warming on native fishes of the southern Great Plains and the Southwest. *Fisheries* **15**: 26–32.

May, R. M. 1972. Will a large complex system be stable? *Nature* **238**: 413–414.

May, R. M. 1973. *Stability and Complexity in Model Ecosystems.* Princeton University Press, Princeton, NJ.

Mihuc, T. B. 1997. The functional trophic role of lotic primary consumers: generalist versus specialist strategies. *Freshwater Biology* **37**: 455–462.

Milner, A. M., E. E. Knudssen, C. Soiseth, A. L. Robertson, D. Schell, L. T. Phillips, and K. Magnusson. 2000. Colonization and development of stream communities across a 200-year gradient in Glacier Bay National Park, Alaska, USA. *Canadian Journal of Fisheries and Aquatic Sciences* **57**: 2319–2355.

Mithen, S. J., and J. H. Lawton. 1986. Food-web models that generate constant predator–prey ratios. *Oecologia* **69**: 542–550.

Mohseni, O., H. G. Stefan, and J. G. Eaton. 2003. Global warming and potential changes in fish habitat in U.S. streams. *Climatic Change* **15**: 389–409.

Molander, S. H., H. Blanck, and M. Söderström. 1990. Toxicity assessment by pollution-induced community tolerance (PICT) and identification of metabolites in periphyton communities after exposure to 4,5,6-trichloroguaiacol. *Aquatic Toxicology* 18: 115–136.

Mulholland, P. J., A. V. Palumbo, J. W. Elwood, and A. D. Rosemond. 1987. Effect of acidification on leaf decomposition in streams. *Journal of the North American Benthological Society* 6: 147–158.

Nakano, S., H. Miyasaka, and N. Kuhara. 1999. Terrestrial–aquatic linkages: Riparian arthropod inputs alter trophic cascades in a stream food web. *Ecology* 80: 2435–2441.

Naeem, S. 1998. Species redundancy and ecosystem reliability. *Conservation Biology* 12: 39–45.

Neutel, A. -M., J. A. P. Heesterbeek, and P. C. de Ruiter. 2002. Stability in real food webs: weak links in long loops. *Science* 296: 1120–1123.

Ormerod, S. J., and S. J. Tyler. 1991. Exoloitaion of prey by a river bird, the dipper *Cinclus cinclus*, along acidic and circumneutral streams in upland Wales. *Freshwater Biology* 25: 105–116.

Paine, R. T. 2002. Trophic control of production in a rocky intertidal community. *Science* 296: 736–739.

Petchey, O. L., A. L. Downing, G. G. Mittelbach, L. Persson, C. F. Steiner, P. H. Warren, and G. Woodward. 2004. Species loss and the structure and functioning of multitrophic aquatic systems. *Oikos* 104: 467–478.

Poff, N. L., and J. D. Allan. 1995. Functional organization of fish assemblages in relation to hydrological variability. *Ecology* 76: 606–627.

Polis, G. A. 1998. Stability is woven by complex webs. *Nature* 395: 744–745.

Post, D. M., M. L. Pace, and N. G. Hairston. 2000. Ecosystem size determines food-chain length in lakes. *Nature* 405: 1047–1049.

Power, M. E. 1990. Effects of fish in river food webs. *Science* 250: 811–814.

Puckridge, J. T., K. F. Walker, and J. F. Costelloe 2000. Hydrological persistence and the ecology of dryland rivers. *Regulated Rivers—Research and Management* 16: 385–402.

Rosemond, A. D., C. M. Pringle, A. Ramirez, and M. J. Paul. 2001. A test of top-down and bottom-up control in a detritus-based food web. *Ecology* 82: 2279–2293.

Rosi-Marshall, E. J., and J. B. Wallace. 2002. Invertebrate food webs along a stream resource gradient. *Freshwater Biology* 47: 129–141.

Rounick, J. S., and B. J. Hicks. 1985. The stable carbon isotope ratios of fish and their invertebrate prey in four New Zealand rivers. *Freshwater Biology* 15: 207–214.

Schmid, P. E., and J. M. Schmid-Araya. 2000. Trophic relationships: integrating meiofauna into a realistic benthic food web. *Freshwater Biology* 44: 149–163.

Schmid-Araya, J. M., A. G. Hildrew, A. Robertson, P. E. Schmid, and J. Winterbottom. 2002a. The importance of meiofauna in food webs: evidence from an acid stream. *Ecology* 83: 1271–1285.

Schmid-Araya, J. M., P. E. Schmid, A. Robertson, J. Winterbottom, C. Gjerlov, and A. G. Hildrew. 2002b. Connectance in stream food webs. *Journal of Animal Ecology* 71: 1056–1062.

Schmid, P. E., M. Tokeshi, and J. M. Schmid-Araya. 2000. Relation between population density and body size in stream communities. *Science* 289: 557–1560.

Schreiber, E. S. G., G. P. Quinn, and P. S. Lake. 2003. Distribution of an alien aquatic snail in relation to flow variability, human activities and water quality. *Freshwater Biology* 48: 951–961.

Simon, K. S., E. F. Benfield, and S. A. Macko. 2003. Food web structure and the role of epilithic biofilms in cave streams. *Ecology* 84: 2395–2406.

Speirs, D. C., W. S. C. Gurney, J. H. Winterbottom, and A. G. Hildrew. 2000. Long-term demographic balance in the Broadstone Stream insect community. *Journal of Animal Ecology* 69: 45–58.

Stead, T. K., J. M. Schmid-Araya, and A. G. Hildrew. All creatures great and small: patterns in the stream benthos across the whole range of metazoan body size. *Freshwater Biology* 48: 532–547.

Steinmetz, J., S. L. Kohler, and D. A. Soluk. 2003. Birds are overlooked top predators in aquatic food webs. *Ecology* 84: 1324–1328.

Sterner, R. W. 1995. Elemental stoichiometry of species in ecosystems. In: C. Jones and J. Lawton (eds), *Linking Species and Ecosystems.* Chapman and Hall, New York.

Stone, M. K. and J. B. Wallace. 1998. Long term recovery of a mountain stream from clear-cut logging: the effects of forest succession on benthic invertebrate community structure. *Hydrobiologia* 353: 107–119.

Sutcliffe, D. W., and A. G. Hildrew. 1989. Invertebrate communities in acid streams. In: R. Morris, E. W. Taylor, D. J. A. Brown, and J. A. Brown (eds), *Acid Toxicity and Aquatic Animals.* Seminar Series of the Society for Experimental Biology, Cambridge University Press, pp. 13–29.

Tavares-Cromar, A. F., and D. D. Williams. 1996. The importance of temporal resolution in food web analysis: Evidence from a detritus-based stream. *Ecological Monographs* 66: 91–113.

Thompson, R. M., and C. R. Townsend. 1999. The effect of seasonal variation on the community structure and food-web attributes of two streams: implications for food-web science. *Oikos* 87: 75–88.

Thompson, R. M., and C. R. Townsend. 2003. Impacts on stream food webs of native and exotic forest: an intercontinental comparison. *Ecology* 84: 145–161.

Thompson, R. M., and C. R. Townsend. Energy availability, spatial heterogeneity and ecosystem size predict food-web structure in streams. *Oikos* (in press).

Tokeshi, M. 1999. *Species Coexistence: Ecological and Evolutionary Perspectives*. Blackwell Science.

Townsend, C. R. 1989. The patch dynamics concept of stream community ecology. *Journal of the North American Benthological Society* **8**: 36–50.

Townsend, C. R. 1996. Invasion biology and ecological impacts of brown trout (*Salmo trutta*) in New Zealand. *Biological Conservation* **78**: 13–22.

Townsend, C. R. 2003. Individual, population, community, and ecosystem consequences of a fish invader in New Zealand streams. *Conservation Biology* **17**: 38–47.

Townsend, C. R., and A. G. Hildrew. 1979. Resource partitioning by two freshwater invertebrate predators with contrasting foraging strategies. *Journal of Animal Ecology* **48**: 909–920.

Townsend, C. R., S. Dolédec, and M. R. Scarsbrook. 1997. Species traits in relation to temporal and spatial heterogeneity in streams: a test of habitat templet theory. *Freshwater Biology* **37**: 367–387.

Townsend, C. R., R. M. Thompson, A. R. McIntosh, C. Kilroy, E. Edwards, and M. R. Scarsbrook. 1998. Disturbance, resource supply and food-web architecture in streams. *Ecology Letters* **1**: 200–209.

Tuchman, N. C., K. A. Wahtera, R. G. Wetzel, and J. A. Teeri. 2003. Elevated atmospheric CO_2 alters leaf litter quality for stream ecosystems: an *in situ* leaf decomposition study. *Hydrobiologia* **495**: 203–211.

Ulanowicz, R. E., and C. J. Puccia. 1990. Mixed trophic impacts in ecosystems. *Coenoses* **5**: 7–16.

Usio, N. 2000. Effects of crayfish on leaf processing and invertebrate colonisation of leaves in a headwater stream: decoupling of a trophic cascade. *Oecologia* **124**: 608–614.

Usio, N. and C. R. Townsend. 2001. The significance of the crayfish *Paranephrops zealandicus* as shredders in a New Zealand headwater stream. *Journal of Crustacean Biology* **21**: 354–359.

Vanni, M. J. 2002. Nutrient cycling by animals in freshwater ecosystems. *Annual Review of Ecology and Systematics* **33**: 341–370.

Vannote, R. L., G. W. Minshall, K. W. Cummins, J. R. Sedell, and C. E. Cushing. 1980. The river continuum concept. *Canadian Journal of Fisheries and Aquatic Sciences* **37**: 130–137.

Walker, B. H. 1992. Biodiversity and ecological redundancy. *Conservation Biology* **6**: 18–23.

Ward, J. V., G. Bretschko, M. Brunke, D. Danielopol, G. Gibert, T. Gonser, and A. G. Hildrew. 1988. A metazoan perspective of riparian/groundwater ecology. *Freshwater Biology* **40**: 531–569.

Warren, P. H. 1996. Structural constraints on food web assembly. In: M. E. Hochberg, J. Clobert, and R. Barbault (eds), *Aspects of the Genesis and Maintenance of Biological Diversity*. Oxford University Press, Oxford, pp. 142–161.

Webb, B. W. 1996. Trends in stream and river temperature. *Hydrological Processes* **10**: 205–226.

Wikramanayake, E. D., and P. B. Moyle. 1989. Ecological structure of tropical fish assemblages in wet-zone streams of Sri Lanka. *Journal of Zoology (London)* **218**: 503–526.

Williams, R. J., and N. D. Martinez. 2000. Simple rules yield complex food webs. *Nature* **404**: 180–183.

Winemiller, K. O. 1990. Spatial and temporal variation in tropical fish trophic networks. *Ecological Monographs* **60**: 331–367.

Winemiller, K. O. 1996. Dynamic diversity: fish communities of tropical rivers. In: M. L. Cody and J. A. Smallwood (eds), *Long-term Studies of Vertebrate Communities*. Academic Press, pp. 99–134.

Winner, R. W., M. W. Boesel, and M. P. Farrell. 1980. Insect community structure as an index of heavy-metal pollution in lotic ecosystems. *Canadian Journal of Fisheries and Aquatic Sciences* **37**: 647–655.

Wipfli, M. S., and J. Caouette. 1998. Influence of salmon carcasses on stream productivity: response of biofilm and benthic macroinvertebrates in southeastern Alaska, USA. *Canadian Journal of Fisheries and Aquatic Sciences* **55**: 1503–1511.

Woodward, G., and A. G. Hildrew. 2001. Invasion of a stream food web by a new top predator. *Journal of Animal Ecology* **70**: 273–288.

Woodward, G., and A. G. Hildrew, 2002*a*. Body-size determinants of niche overlap and intraguild predation within a complex food web. *Journal of Animal Ecology* **71**: 1063–1074.

Woodward, G., and A. G. Hildrew. 2002*b*. Differential vulnerability of prey to an invading top predator: integrating field surveys and laboratory experiments. *Ecological Entomology* **27**: 732–744.

Woodward, G., and A. G. Hildrew. 2002*c*. Food web structure in riverine landscapes. *Freshwater Biology* **47**: 777–798.

Woodward, G., and A. G. Hildrew. 2002*d*. The impact of a sit-and-wait predator: separating consumption and prey emigration. *Oikos* **99**: 409–418.

Woodward, G., J. I. Jones, and A. G. Hildrew. 2002. Community persistence in Broadstone Stream (U.K.) over three decades. *Freshwater Biology* **47**: 1419–1435.

Woodward, G., D. C. Speirs, and A. G. Hildrew. Quantification and resolution of a complex, size-structured food web. *Advances in Ecological Research* (in press).

Wootton, J. T. 1997. Estimates and test of per capita interaction strength: diet, abundance and impact of intertidally foraging birds. *Ecological Monographs* **67**: 45–64.

Wootton, J. T., M. S. Parker, and M. E. Power. 1996. The effect of disturbance on river food webs. *Science* **273**: 1558–1561.

Wright, J. F., D. W. Sutcliffe, and M. T. Furse. eds. 2000. *Assessing the Quality of Fresh Waters: Rivpacs and Other Techniques*. Freshwater Biological Association, Ambleside, 373 pp.

Young, R. G., and A. D. Huryn. 1997. Longitudinal patterns of organic matter transport and turnover along a New Zealand grassland river. *Freshwater Biology* **38**: 93–107.

Chapter 6

Azzalini, A. 1996. *Statistical Inference*. Chapman and Hall, London.

Camacho, J., R. Guimerà, and L. A. Nunes Amaral. 2002. Robust patterns in food web structure. *Physical Review Letters* **88**: 1–4.

Cohen, J. E. 1990. A stochastic theory of community food webs. VI. Heterogeneous alternatives to the cascade model. *Theoretical Population Biology* **37**: 55–90.

Cohen, J. E., and C. Newman. 1985. A stochastic theory of community food webs. I. Models and aggregated data. *Proceedings of the Royal Society* **B224**: 421–448.

Cohen, J. E., F. Briand, and C. Newman. 1990. *Community Food Webs: Data and Theory*. Springer, Berlin.

Cohen, J. E., S. L. Pimm, P. Yodzis, and J. Saldaña. 1993. Body size of animal predators and animal prey in food webs. *Journal of Animal Ecology* **62**: 67–78.

Forbes, S. A. 1887. The lake as a microcosm. *Bulletin of the Peoria Scientific Association* 537–550.

Link, J. 2002. Does food web theory work for marine ecosystems? *Marine Ecology Progress Series* **230**: 1–9.

Martinez, N. D. 1992. Constant connectance in community food webs. *The American Naturalist* **139**: 1208–1218.

Martinez, N. D. 1993. Effects of resolution on food web structure. *Oikos* **60**: 403–412.

May, R. M. 1973. *Stability and Complexity in Model Ecosystems*. Princeton University Press, Princeton, NJ.

Murtaugh, P. A. 1994. Statistical analysis of food webs. *Biometrics* **50**: 1199–1202.

Murtaugh, P. A., and D. R. Derryberry. 1998. Models of connectance in food webs. *Biometrics* **54**: 754–761.

Murtaugh, P. A., and J. P. Kollath. 1997. Variation in trophic fractions and connectance in food webs. *Ecology* **78**: 1382–1387.

Neubert, M. G., S. C. Blumenshine, D. E. Duplisea, T. Jonsson, and B. Rashleigh. 2000. Body size and food web structure: testing the equiprobability assumption of the cascade model. *Oecologia* **123**: 241–251.

Pimm, S. L. 1982. *Food Webs*. Chapman and Hall, London.

Pruesse, G., and F. Ruskey. 1994. Generating linear extensions fast. *SIAM Journal of Computing* **23**: 373–386.

Schmid-Araya, J. M., P. E. Schmid, A. Robertson, J. Winterbottom, C. Gjerlov, and A. G. Hildrew. 2002. Connectance in stream food webs. *Journal of Animal Ecology* **71**: 1056–1061.

Solow, A. R. 1996. On the goodness of fit of the cascade model. *Ecology* **77**: 1294–1297.

Solow, A. R., and A. R. Beet. 1998. On lumping species in food webs. *Ecology* **79**: 2013–2018.

Varol, Y., and D. Rotem. 1981. An algorithm to generate all topological sorting arrangements. *Computer Journal* **24**: 83–84.

Williams, D. A. 1982. Extrabinomial variation in logistic linear models. *Applied Statistics* **31**: 144–148.

Williams, R. J., and N. D. Martinez. 2000. Simple rules yield complex food webs. *Nature* **404**: 180–183.

Winemiller, K. O., and G. A. Polis. 1996. Food webs: what do they tell us about the world? In: G. A. Polis and K.O. Winemiller (eds), *Food webs: Integration of Pattern and Dynamics*. Chapman and Hall, London, pp. 1–22.

Chapter 7

Asaeda, T., and L. H. Nam. 2002. Effects of rhizome age on the decomposition rate of *Phragmites australis* rhizomes. *Hydrobiologia* **485**: 205–208.

Chapin, F. S., E. S. Zaveleta, V. T. Eviner, R. L. Naylor, P. M. Vitousek, H. L. Reynolds, D. U. Hoooper, S. Lavorel, O. E. Sala, S. E. Hobbie, M. C. Mack, and S. Díaz. 2000. Consequences of changing biodiversity. *Nature* **405**: 234–242.

Christian, R. R., and J. J. Luczkovich. 1999. Organizing and understanding a winter's seagrass foodweb network through effective trophic levels. *Ecological Modelling* **117**: 99–124.

Cohen, J. E., and 23 others. 1993. Improving food webs. *Ecology* **74**: 252–258.

Gordon, M. S., G. A. Bartholomew, A. D. Grinnell, C. B. Jorgensen, and F. N. White. 1972. *Animal Physiology*. Macmillan Co., New York, 592 pp.

MacArthur, R. H. 1957. On the relative abundance of bird species. Proceedings of the National Academy of Sciences, USA **43**: 293–295.

Lambers, H. 1985. Respiration in intact plants and tissues: its regulation and dependence on environmental factors, metabolism and invaded organisms. In: R. Douce and D. A. Day (eds), *Encyclopedia of Plant Physiology*, Vol. 18. Springer Verlag. Berlin, pp. 418–473.

Landsberg, J. J. 1986. *Physiological Ecology of Forest Production*. Academic Press, London, 198 pp.

Martinez, N. 1991. Artifacts or attributes? Effects of resolution on the Little Roke Lake food web. *Ecological Monographs* **61**: 367–392.

May, R. 1973. *Stability and Complexity in Model Ecosystems*. Princeton University Press, Princeton, NJ.

McCann, K., A. Hastings, and G. R. Huxel. 1998. Weak trophic interactions and the balance of nature. *Nature* **395**: 794–798.

McGrady-Steed, J., P. M. Harris, and P. J. Morin. 2001. Biodiversity regulates ecosystem predictability. *Nature* **412**: 34–36.

Milo, R., S. Itzkovitz, N. Kashtan, R. Levitt, S. Shen-Orr, I. Ayzenshtat, M. Sheffer, and U. Alon. 2004. Superfamilies of evolved and designed networks. *Science* **303**: 1538–1542.

Morris, J. T., and K. Lajtha. 1986. Decomposition and nutrient dynamics of litter from four species of fresh water emergent macrophytes. *Hydrobiology* **131**: 215–223.

Odum, E. P. 1969. The strategy of ecosystem development. *Science* **164**: 262–270.

Qualls, R. G., and C. J. Richardson. 2003. Factors controlling concentration, export, and decomposition of dissolved organic nutrients in the everglades of Florida. *Biogeochemistry* **62**(2): 197–229.

Paine, R. T. 1988. Food webs: road maps of interactions or grist for theoretical development. *Ecology* **69**: 1648–1654.

Pimm, S. L. 1982. *Food Webs*. Chapman and Hall, London, 219 pp.

Salovius, S., and E. Bonsdorff. 2004. Effects of depth, sediment and grazers on the degradation of drifting filamentous algae (*Cladophora glomerata* and *Pilayella littoralis*). *Journal of Experimental Marine Biology and Ecology* **298**: 93–109.

Schomberg, H. H., and J. L. Steiner. 1997. Estimating crop residue decomposition coefficients using substrate-induced respiration. *Soil Biology and Biochemistry* **29**: 1089–1097.

Ulanowicz, R. E. 1986*a*. *Growth and Development: Ecosystems Phenomenology*. ToExcel Press, 203 pp.

Ulanowicz, R. E. 1986*b*. A phenomenological perspective of ecological development. In: T. M. Poston and R. Purdy (eds), *Aquatic Toxicology and Environmental Fate, Vol. 9*. ASTM STP 921. American Society for Testing and Materials, Philadelphia, PA, pp. 73–8l.

Ulanowicz, R. E. 1997. *Ecology, the Ascendent Perspective*. Columbia University Press, 201 pp.

Ulanowicz, R. E. 2002. The balance between adaptability and adaptation. *BioSystems* **64**: 13–22.

Ulanowicz, R. E., and J. Norden. 1990. Symmetrical overhead in flow networks. *International Journal of Systems Sciences* **21**: 429–437.

Ulanowicz, R. E., C. Bondavalli, and M. S. Egnotovich. 1998. Network Analysis of throphic dynamics in South Florida Ecosystem FY97: The Florida Bay Ecosystem. Annual Report to the United States Geological Service Biological Resource Division, University of Miami, Coral Gablel, FL.

Van Santvoort, P. J., G. J. De Lange, J. Thomson, S. Colley, F. J. Meysman, and C. P. Slomp. 2002. Oxidation and origin of organic matter in surficial eastern mediterranean hemipelagic sediments. *Aquatic Geochemistry* **8**: 153–175.

Zorach, A. C., and R. E. Ulanowicz. 2003. Quantifying the complexity of flow networks: how many roles are there? *Complexity* **8**: 68–76.

Chapter 8

Beddington, J. R., and Basson, M. 1994. The limits to exploitation on land and sea. *Philosophical Transactions of the Royal Society* **B343**: 87–92.

Belgrano, A., A. P. Allen, B. J. Enquist. and J. F. Gillooly. 2002. Allometric scaling of maximum population density: a common rule for marine phytoplankton and terrestrial plants. *Ecology Letters* **5**: 611–613.

Benoit, E., and M. J. Rochet. 2004. A continuous model of biomass size spectra governed by predation and the effects of fishing on them. *Journal of Theoretical Biology* **226**: 9–21.

Beverton, R. J. H., and S. J. Holt. 1959. A review of the lifespan and mortality rates of fish in nature and their relationship to growth and other physiological characteristics. *Ciba Foundation Colloquim on Ageing* **5**: 142–180.

Blazka, P., T. Backiel, and F. B. Taub. 1980. Trophic relationships and efficiencies. In: E. D. Le Cren and R. H. Lowe-McConnell (eds), *The Functioning of Freshwater Ecosystems*. Cambridge University Press, Cambridge, pp. 393–410.

Boudreau, P. R., and L. M. Dickie. 1989. Biological model of fisheries production based on physiological and ecological scalings of body size. *Canadian Journal of Fisheries and Aquatic Sciences* **46**: 614–623.

Boudreau, P. R., and L. M. Dickie. 1992. Biomass spectra of aquatic ecosystems in relation to fisheries yield. *Canadian Journal of Fisheries and Aquatic Sciences* **49**: 1528–1538.

Boyle, P. R., and S. V. Bolettzky. 1996. Cephalopod populations: definition and dynamics. *Philosophical Transactions of the Royal Society* **351**: 985–1002.

Brey, T. 1990. Estimating productivity of macrobenthic invertebrates from biomass and mean individual weight. *Meeresforschung* **32**: 329–343.

Brey, T. 1999. Growth performance and mortality in aquatic macrobenthic invertebrates. *Advances in Marine Biology* **35**: 153–223.

Brown, J. H. 1995. *Macroecology.* University of Chicago, Press, Chicago, IL.

Brown, J. H., and J. F. Gillooly. 2003. Ecological food webs: high-quality data facilitate theoretical unification. *Proceedings of the National Academy of Sciences, USA* **100**: 1467–1468.

Brown, J. H., and G. B. West. 2000. *Scaling in Biology.* Oxford University Press, Oxford.

Charnov, E. L. 1993. *Life History Invariants: Some Explorations of Symmetry in Evolutionary Ecology.* Oxford University Press, Oxford.

Chase, J. M. 2000. Are there real differences among aquatic and terrestrial food webs? *Trends in Ecology and Evolution* **15**: 408–412.

Cohen, J. E., T. Jonsson, and S. R. Carpenter, 2003. Ecological community description using the food web, species abundance and body size. *Proceedings of the National Academy of Sciences, USA* **100**: 1781–1786.

Cohen, J. E., S. L. Pimm, P. Yodzis, and J. Saldaña. 1993. Body sizes of animal predators and animal prey in food webs. *Journal of Animal Ecology* **62**: 67–78.

Cushing, D. H. 1975. *Marine Ecology and Fisheries.* Cambridge University Press, Cambridge.

Cyr, H. 2000. Individual energy use and the allometry of population density. In: J. H. Brown and G. B. West (eds), *Scaling in Biology.* Oxford University Press, Oxford, pp. 267–295.

Damuth, J. 1981. Population density and body size in mammals. *Nature* **290**: 699–700.

Denney, N. H., S. Jennings, and J. D. Reynolds. 2002. Life history correlates of maximum population growth rates in marine fishes. *Proceedings of the Royal Society: Biological Sciences* **269**: 2229–2237.

Dickie, L. M. 1976. Predation, yield and ecological efficiency in aquatic food chains. *Journal of the Fisheries Research Board of Canada* **33**: 313–316.

Dickie, L. M., S. R. Kerr, and P. R. Boudreau. 1987. Size-dependent processes underlying regularities in ecosystem structure. *Ecological Monographs* **57**: 233–250.

Dinmore, T. A., and S. Jennings. 2004. Predicting body size distribution in benthic infaunal communities. *Marine Ecology Progress Series* **276**: 289–292.

Duplisea, D. E. 2000. Benthic organism size-spectra in the Baltic Sea in relation to the sediment environment. *Limnology and Oceanography* **45**: 558–568.

Duplisea, D. E., and S. R. Kerr. 1995. Application of a biomass size spectrum model to demersal fish data from the Scotian shelf. *Journal of Theoretical Biology* **177**: 263–269.

Fenchel, T. 1974. Intrinsic rate of natural increase: the relationship with body size. *Oecologia* **14**: 317–326.

France, R., M. Chandler, and R. Peters. 1998. Mapping trophic continua of benthic food webs: body size d15N relationships. *Marine Ecology Progress Series* **174**: 301–306.

Fry, B., and R. B. Quinones. 1994. Biomass spectra and stable-isotope indicators of trophic level in zooplankton of the northwest Atlantic. *Marine Ecology Progress Series* **112**: 201–204.

Gaedke, U. 1993. Ecosystem analysis based on biomass size distributions: a case study of a planktonic community in a large lake. *Limnology and Oceanography* **38**: 112–127.

Gaston, K. J., and T. M. Blackburn. 2000. *Pattern and Process in Macroecology.* Blackwell Science, Oxford.

Gislason, H., and J. C. Rice. 1998. Modelling the response of size and diversity spectra of fish assemblages to changes in exploitation. *ICES Journal of Marine Sciences* **55**: 362–370.

Gunderson, D. R., and P. H. Dygert. 1988. Reproductive effort as a predictor of natural mortality rate. *Journal du Conseil, Conseil International pour l'Exploration de la Mer* **44**: 200–209.

Hall, S. J., and D. G. Raffaelli. 1993. Food webs: theory and reality. *Advances in Ecological Research* **24**: 187–239.

Harvey, P. H., and M. D. Pagel. 1991. *The Comparative Method in Evolutionary Biology.* Oxford University Press, Oxford.

Jennings, S., and J. L. Blanchard. 2004. Fish abundance with no fishing: predictions based on macroecological theory. *Journal of Animal Ecology* **73**: 632–642.

Jennings, S., and S. Mackinson. 2003. Abundance–body mass relationships in size structured food webs. *Ecology Letters* **6**: 971–974.

Jennings, S., and K. J. Warr. 2003. Smaller predator–prey body size ratios in longer food chains. *Proceedings of the Royal Society Biological Sciences* **270**: 1413–1417.

Jennings, S., J. K. Pinnegar, N. V. C. Polunin, and T. Boon. 2001. Weak cross-species relationships between body size and trophic level belie powerful size-based trophic structuring in fish communities. *Journal of Animal Ecology* **70**: 934–944.

Jennings, S., J. K. Pinnegar, N. V. C. Polunin, and K. J. Warr. 2002a. Linking size-based and trophic analyses of benthic community structure. *Marine Ecology Progress Series* **226**: 77–85.

Jennings, S., K. J. Warr, and Mackinson, S. 2002b. Use of size-based production and stable isotope analyses to predict trophic transfer efficiencies and predator–prey body mass ratios in food webs. *Marine Ecology Progress Series* **240**: 11–20.

Kerr, S. R. 1974. Theory of size distribution in ecological communities. *Journal of the Fisheries Research Board of Canada* **31**: 1859–1862.

Kerr, S. R., and L. M. Dickie. 2001. *The Biomass Spectrum: A Predator–Prey Theory of Aquatic Production*. Columbia University Press, New York.

Li, W. K. W. 2002. Macroecological patterns of phytoplankton in the northwestern Atlantic Ocean. *Nature* **419**: 154–157.

Murawski, S. A., and J. S. Idoine. 1992. Multispecies size composition: a conservative property of exploited fishery systems. *Journal of Northwest Atlantic Fishery Science* **14**: 79–85.

Nee, S., A. F. Read, J. J. D. Greenwood, and P. H. Harvey. 1991. The relationship between abundance and body size in British birds. *Nature* **351**: 312–313.

Olive, P. J. W., J. K. Pinnegar, N. V. C. Polunin, G. Richards, and R. Welch. 2003. Isotope trophic-step fractionation: a dynamic equilibrium model. *Journal of Animal Ecology* **72**: 608–617.

Pauly, D. 1980. On the interrelationships between natural mortality, growth parameters and mean environmental temperature in 175 fish stocks. *Journal du Conseil, Conseil International pour l'Exploration de la Mer* **39**: 175–192.

Peters, R. H. 1983. *The Ecological Implications of Body Size*. Cambridge University Press, Cambridge.

Pimm, S. L. 1991. *The balance of nature? Ecological Issues in the Conservation of Species and Communities*. University of Chicago Press, Chicago, IL.

Polunin, N. V. C., and J. K. Pinnegar. 2002. Ecology of fishes in marine food-webs. In: P. J. Hart and J. D. Reynolds (eds), *Handbook of Fish and Fisheries*. Blackwell, Oxford.

Pope, J. G., J. G. Shepherd, and J. Webb. 1994. Successful surf-riding on size spectra: the secret of survival in the sea. *Philosophical Transactions of the Royal Society* **343**: 41–49.

Pope, J. G., T. K. Stokes, S. A. Murawski, and S. I. Iodoine. 1988. A comparison of fish size composition in the North Sea and on Georges Bank. In: W. Wolff, C. J. Soeder, and F. R. Drepper (eds), *Ecodynamics: Contributions to Theoretical Ecology*. Springer Verlag, Berlin, pp. 146–152.

Post, D. M. 2002*a*. The long and short of food chain length. *Trends in Ecology and Evolution* **17**: 269–277.

Post, D. M. 2002*b*. Using stable isotopes to estimate trophic position: models, methods and assumptions. *Ecology* **83**: 703–718.

Rice, J., and H. Gislason. 1996. Patterns of change in the size spectra of numbers and diversity of the North Sea fish assemblage, as reflected in surveys and models. *ICES Journal of Marine Sciences* **53**: 1214–1225.

Rice, J. C., N. Daan, J. G. Pope, and H. Gislason. 1991. The stability of estimates of suitabilities in MSVPA over four years of data from predator stomachs. *ICES Marine Science Symposia* **193**: 34–45.

Schwinghamer, P. 1981. Characteristic size distributions of integral benthic communities. *Canadian Journal of Fisheries and Aquatic Sciences* **38**: 1255–1263.

Schwinghamer, P., B. Hargrave, D. Peer, and C. M. Hawkins. 1986. Partitioning of production and respiration among size groups of organisms in an intertidal benthic community. *Marine Ecology Progress Series* **31**: 131–142.

Sheldon, R. W., and S. R. Kerr. 1972. The population density of monsters in Loch Ness. *Limnology and Oceanography* **17**: 769–797.

Sheldon, R. W., A. Prakash, and W. H. Sutcliffe. 1972. The size distribution of particles in the Ocean. *Limnology and Oceanography* **17**: 327–340.

Slobodkin, L. B. 1960. Ecological energy relationships at the population level. *American Naturalist* **94**: 213–236.

Sprules, W. G., and J. D. Stockwell. 1995. Size based biomass and production models in the St Lawrence Great Lakes. *ICES Journal of Marine Sciences* **52**: 705–710.

Stephens, D. W., and J. R. Krebs. 1986. *Foraging Theory*. Princeton University Press, Princeton, NJ.

Thiebaux, M. L., and L. M. Dickie. 1992. Models of aquatic biomass size spectra and the common structure of their solutions. *Journal of Theoretical Biology* **159**: 147–161.

Thiebaux, M. L., and L. M. Dickie. 1993. Structure of the body size spectrum of the biomass in aquatic ecosystems: a consequence of allometry in predator–prey interactions. *Canadian Journal of Fisheries and Aquatic Sciences* **50**: 1308–1317.

Ursin, E. 1973. On the prey size preferences of cod and dab. *Meddelelser fra Danmarks Fiskeri-og Havundersogelser* **7**: 85–98.

Ware, D. M. 2000. Aquatic ecosystems: properties and models. In: P. J. Harrison and T. R. Parsons (eds), *Fisheries Oceanography: An Integrative Approach to Fisheries Ecology and Management*. Blackwell Science, Oxford, pp. 161–194.

Chapter 9

Abarca-Arenas, L. G., and R. E. Ulanowicz. 2002. The effects of taxonomic aggregation on network analysis. *Ecological Modelling* **149**: 285–296.

Abarca-Arenas, L. G., and E. Valero-Pacheco. 1993. Toward a trophic model of Tamiahua, a coastal lagoon in Mexico. In: V. Christensen and D. Pauly (eds), *Trophic Models of Aquatic Ecosystems*. ICLARM Conference Proceedings 26, pp. 181–185.

Ainley, D. G., W. R. Fraser, W. O. Smith, T. L. Hopkins, and J. J. Torres. 1991. The structure of upper level pelagic food webs in the Antarctic: effect of phytoplankton on distribution. *Journal of Marine Systems* 2: 111–122.

Ainsworth, C., B. Ferriss, E. Leblond, and S. Guenette. 2001. The Bay of Biscay France; 1998 and 1970 models. *University of British Columbia, Fisheries Centre Research Reports* 9(4): 271–313.

Aliño, P. M., L. T. McManus, J. W. McManus, C. Nañola, M. D. Fortes, G. C. Trono, and G. S. Jacinto. 1993. Initial parameter estimations of a coral reef flat ecosystem in Bolinao, Pangasinan, Northwestern Philippines. In: V. Christensen and D. Pauly (eds), *Trophic Models of Aquatic Ecosystems*. ICLARM Conference Proceedings 26, pp. 252–258.

Andersen, K. P., and E. Ursin. 1977. A multispecies extension to the Beverton and Holt theory, with accounts of phosphorus circulation and primary production. *Meddelelser fra Danmarks Fiskeri-og Havundersogelser, NS* 7: 319–435.

Arias-Gonzalez, J. E., B. Delesalle, B. Slavat, and R. Galzin. 1997. Trophic functioning of the Tiahura reef sector, Moorea Island, French Polynesia. *Coral Reefs* 16: 231–246.

Arreguin-Sanchez, F., and L. E. Calderon-Aguilera. 2002. Evaluating harvesting strategies for fisheries in the central Gulf of California ecosystem. *University of British Columbia, Fisheries Centre Research Reports* 10(2): 135–141.

Arreguin-Sanchez F., J. C. Seijo, and E. Valero-Pacheco. 1993. An application of ECOPATH II to the north central shelf ecosystem of Yucatan, Mexico. In: V. Christensen and D. Pauly (eds), *Trophic Models of Aquatic Ecosystems*. ICLARM Conference Proceedings 26, pp. 269–278.

Atwell, L., K. A. Hobson, and H. E. Welch. 1998. Biomagnification and bioaccumulation of mercury in an arctic marine food web: insights from stable nitrogen isotope analysis. *Canadian Journal of Fisheries and Aquatic Sciences* 55: 1114–1121.

Aydin, K. Y., V. V. Lapko, V. I. Radchenko, and P. A. Livingston. 2002. A comparison of the eastern and western Bering Sea shelf and slope ecosystems through the use of mass-balance food web models. NOAA Technical Memorandum NMFS-AFSC-130.

Azam, F., T. Fenchel, J. G. Field, J. S. Gray, L. A. Meyer-Reil, and F. Thingstad. 1983. The ecological role of water-column microbes in the sea. *Marine Ecology Progress Series* 10: 257–263.

Baird, D., J. M. Glade, and R. E. Ulanowicz. 1991. The comparative ecology of six marine ecosystems. *Philosophical Transactions of the Royal Society of London B* 333: 15–29.

Barrera-Oro, E. 2002. The role of fish in the Antarctic marine food web: differences between inshore and offshore waters in the southern Scotia Arc and west Antarctic Peninsula. *Antarctic Science* 14: 293–309.

Beattie, A., and M. Vaconcellos. 2002. Previous ecosystem models of Northern BC. *University of British Columbia, Fisheries Centre Research Reports* 10(1): 78–80.

Bengtsson, J. 1994. Confounding variables and independent observations in comparative analyses of food webs. *Ecology* 75: 1282–1288.

Bersier, L. F., C. Banasek-Richter, and M. F. Cattin. 2002. Quantitative descriptors of food-web matrices. *Ecology* 83: 2394–2407.

Bjornsson, H. 1998. Calculating capelin consumption by Icelandic cod using a spatially disaggregated simulation model. In: *Fishery Stock Assessment Models*. University of Alaska Sea Grant, AK-SG-98-01: 703–718.

Bogstad, B., K. H. Hauge, and O. Ulltang. 1997. MULTSPEC—a multispecies model for fish and marine mammals in the Barents Sea. *Journal of Northwest Atlantic Fishery Science* 22: 317–341.

Borga, K., G. W. Gabrielsen, and J. U. Skaare. 2001. Biomagnification of organochlorines along a Barents Sea food chain. *Environmental Pollution* 113: 187–198.

Bradford-Grieve, J., and S. Hanchet. 2002. Understanding ecosystems: a key to managing fisheries? *Water and Atmosphere* 10: 16–17.

Bradford-Grieve, J. M., P. K. Probert, S. D. Nodder, D. Thompson, J. Hall, S. Hanchet, P. Boyd, J. Zeldis, A. N. Baker, H. A. Best, N. Broekhuizen, S. Childerhouse, M. Clark, M. Hadfield, K. Safi, and I. Wilkinson. 2003. Pilot trophic model for subantarctic water over the Southern Plateau, New Zealand: a low biomass, high transfer efficiency system. *Journal of Experimental Marine Biology and Ecology* 289: 223–262.

Briand, F., and J. E. Cohen. 1984. Community food webs have scale-invariant structure. *Nature* 307: 264–267.

Briand, F., and J. E. Cohen. 1987. Environmental correlates of food chain length. *Science* 238: 956–960.

Brodeur, R. D., and W. G. Pearcy. 1992. Effects of environmental variability on trophic interactions and food web structure in a pelagic upwelling ecosystem. *Marine Ecology Progress Series* 84: 101–119.

Browder, J. A. 1993. A pilot model of the Gulf of Mexico continental shelf. In: V. Christensen, and D. Pauly (eds), *Trophic Models of Aquatic Ecosystems*. ICLARM Conference Proceedings 26, pp. 279–284.

Buchary, E. 2001. Preliminary reconstruction of the Icelandic marine ecosystem in 1950. *University of British Columbia, Fisheries Centre Research Reports* **9**(4): 198–206.

Buchary, E., J. Alder, S. Nurhakim, and T. Wagey. 2002. The use of ecosystem-based modelling to investigate multi-species management strategies for capture fisheries in the Bali Strait, Indonesia. *University of British Columbia, Fisheries Centre Research Reports* **10**(2): 24–32.

Bulman, C., F. Althaus, X. He, N. J. Bax, and A. Williams. 2001. Diets and trophic guilds of demersal fishes of the south-eastern Australian shelf. *Marine and Freshwater Research* **52**: 537–548.

Bundy, A. 2001. Fishing on ecosystems: the interplay of fishing and predation in Newfoundland–Labrador. *Canadian Journal of Fisheries and Aquatic Sciences* **58**: 1153–1167.

Bundy, A., and D. Pauly. 2000. Selective harvesting by small-scale fisheries: ecosystem analysis of San Miguel Bay, Philippines. *Fisheries Research* **53**: 263–281

Bundy, A., G. R. Lilly, and P. A. Shelton. 2000. A Mass Balance Model of the Newfoundland-Labrador Shelf. Canadian Technical Report of Fisheries and Aquatic Sciences, No. 2310, 157 p.

Carr, M. H., T. W. Anderson, and M. A. Hixon. 2002. Biodiversity, population regulation, and the stability of coral-reef fish communities. *Proceedings of the National Academy of Sciences, USA* **99**: 11241–11245.

Carr, M. H., J. E. Neigel, J. A. Estes, S. Andelman, R. R. Warner, and J. L. Largier. 2003. Comparing marine and terrestrial ecosystems: implications for the design of coastal marine reserves. *Ecological Applications* **13**: S90-S107.

Cheung, W.-L., R. Watson, and T. Pitcher. 2002. Policy simulation of fisheries in the Hong Kong marine ecosystem. *University of British Columbia, Fisheries Centre Research Reports* **10**(2): 46–53.

Cho, B. C., and F. Azam. 1988. Major role of bacteria in biogeochemical fluxes in the ocean's interior. *Nature* **332**: 441–443.

Christensen, V. 1995. A model of trophic interactions in the North Sea in 1981, the year of the stomach. *Dana* **11**: 1–28.

Christensen, V. 1998. Fishery-induced changes in a marine ecosystem: insight from models of the Gulf of Thailand. *Journal of Fish Biology* **53(A)**: 128–142.

Christensen, V., and D. Pauly. 1992. ECOPATH II—A software for balancing steady-state models and calculating network characteristics. *Ecological Modelling* **61**: 169–185.

Clarke, G. L. 1946. Dynamics of production in a marine area. *Ecological Monographs* **16**: 323–335.

Clarke, T. A., A. O. Flechsig, and R. W. Grigg. 1967. Ecological studies during Project Sealab II. *Science* **157**: 1381–1389.

Closs, G. P., S. R. Balcombe, and M. J. Shirley. 1999. Generalist predators, interaction strength and food-web stability. *Advances in Ecological Research* **28**: 93–126.

Cohen, J. E. 1989. Food webs and community structure. In: J. Roughgarden, R. M. May, and S. A. Levin (eds), *Perspectives in Ecological Theory*. Princeton University Press, Princeton, NJ, pp. 181–202.

Cohen, J. E. 1994. Marine and continental food webs: three paradoxes? *Philosophical Transactions of the Royal Society of London B* **343**: 57–69.

Cohen, J. E., and F. Briand. 1984. Trophics links of community food web. *Proceedings of the National Academy of Sciences, USA* **81**: 4105–4109.

Cohen, J. E., and C. M. Newman. 1988. Dynamic basis of food web organization. *Ecology* **69**: 1655–1664.

Cohen, J. E., F. Briand, and C. M. Newman. 1990. *Community Food Webs: Data and Theory*. Springer-Verlag, New York.

Cohen, J. E., R. A. Beaver, S. H. Cousins, D. L. DeAngelis, L. Goldwasser, K. L. Heong, R. D. Holt, A. J. Kohn, J. H. Lawton, N. D. Martinez, R. O'Malley, L. M. Page, B. C. Patten, S. L. Pimm, G. A. Polis, M. Rejmanek, T. W. Schoener, K. Schoenly, W. G. Sprules, J. M. Teal, R. E. Ulanowicz, P. H. Warren, H. M. Wilbur, and P. Yodzis. 1993. Improving food webs. *Ecology* **74**: 252–258.

Constable, A. J. 2001. The ecosystem approach to managing fisheries: achieving conservation objectives for predators of fished species. *CCAMLR Science* **8**: 37–64.

Constable, A. J., W. K. de la Mare, D. J. Agnew, I. Everson, and D. Miller. 2000. Managing fisheries to conserve the Antarctic marine ecosystem: practical implementation of the Convention on the Conservation of Antarctic Marine Living Resources. *ICES Journal of Marine Sciences* **57**: 778–791.

Cox, S. P., T. E. Essington, J. F. Kitchell, S. J. D. Martell, C. J. Walters, C. Boggs, and I. Kaplan. 2002. Reconstructing ecosystem dynamics in the central pacific Ocean, 1952–1998. II. A preliminary assessment of the trophic impacts of fishing and effects on tuna dynamics. *Canadian Journal of Fisheries and Aquatic Sciences* **59**: 1736–1747.

Cousins, S. H. 1996. Food webs: from the Lindeman paradigm to a taxonomic general theory of ecology. In: G. A. Polis and K. O. Winemiller (eds), *Food Webs: Integration of Patterns and Dynamics*. Chapman and Hall, pp. 243–254.

Crowder, L. B., D. P. Reagan, and D. W. Freckman. 1996. Food web dynamics and applied problems. In: G. A. Polis and K. O. Winemiller (eds), *Food Webs:*

Integration of Patterns and Dynamics. Chapman and Hall, New York, pp. 327–336.

de Ruiter, C. P., A. Neutal, and J. C. Moore. 1995. Energetics, patterns of interaction strengths, and stability in real ecosystems. *Science* **269**: 1257–1260.

Dommasnes, A., V. Christensen, B. Ellertsen, C. Kvamme, W. Melle, L. Nottestad, T. Pedersen, S. Tjelmeland, and D. Zeller. 2001. Ecosystem model of the Norwegian and Barents Seas. *University of British Columbia, Fisheries Centre Research Reports* **9**(4): 213–240.

Dunne, J. A., R. J. Williams, and N. D. Martinez. 2002*a*. Network structure and biodiversity loss in food webs: robustness increases with connectance. *Ecology Letters* **5**: 558–567.

Dunne, J. A., R. J. Williams, and N. D. Martinez. 2002*b*. Food-web structure and network theory: the role of connectance and size. *Proceedings of the National Academy of Sciences, USA* **99**: 12917–12922.

Dunne, J. A., R. J. Williams, and N. D. Martinez. 2004. Network structure and robustness of marine food webs. *Marine Ecology Progress Series* **273**: 291–302.

Elton, C. S. 1958. *The Ecology of Invasions by Animals and Plants*. Methuen, London. 181 p.

Emery, A. R. 1978. The basis for fish community structure: marine and freshwater comparisions. *Environmental Biology of Fishes* **3**: 33–47.

Essington, T. E., J. F. Kitchell, C. Boggs, D. E. Schindler, R. J. Olson, and R. Hilborn. 2002. Alternative fisheries and the predation rate of yellowfin tuna in the eastern Pacific Ocean. *Ecological Applications* **12**: 724–734.

Etter, R. J., and L. S. Mullineaux. 2001. Deep sea communities. In: M. D. Bertness, S. D. Gaines, M. E. Hay, and M. A. Sunderland (eds), *Marine Community Ecology*. Sinauer Associates, Inc., pp. 724–734.

Field, J. G., R. J. M. Crawfod, P. A. Wickens, C. L. Moloney, K. L. Cochrane, and C. A. Villacastin-Herrero. 1991. Network analysis of Benguela pelagic food webs. Benguela Ecology Programme, Workshop on Seal–Fishery Biological Interactions. University of Cape Town, 16–20 September, 1991, SW91/M5a.

Fisk, A. T., K. A. Hobson, and R. J. Norstrom. 2001. Influence of chemical and biological factors on trophic transfer of persistent organic pollutants in the northwater polynya marine food web. *Environmental Science and Technology* **35**: 732–738.

Fulton, B., and T. Smith. 2002. ECOSIM case study: Port Phillip Bay, Australia. *University of British Columbia, Fisheries Centre Research Reports* **10**(2): 83–93.

Gardner, M. R., and W. R. Ashby. 1970. Connectance of large dynamical (cybernetic) systems: critical values for stability. *Nature* **228**: 784.

Godinot, O., and V. Allain. 2003. A preliminary ECOPATH model of the warm pool pelagic ecosystem. 16th Meeting of the Standing Committee on Tuna and Billfish, Mooloolaba, Queensland, Australia, 9–16 July 2003. URL, available at: http://spc.org.nc/OceanFish/Html/SCTB/SCTB16/BBRG5.pdf.

Goldwasser, L., and J. Roughgarden. 1993. Construction and analysis of a large Caribbean food web. *Ecology* **74**: 1216–1233.

Goldwasser, L., and J. Roughgarden. 1997. Sampling effects and the estimation of food-web properties. *Ecology* **78**: 41–54.

Gomes, M. D. C. 1993. Predictions Under Uncertainty: Fish Assemblages and Food Webs on the Grand Banks of Newfoundland. Social and Economic Studies No. 51, ISER, Memorial University of Newfoundland, St John's, Nfld.

Greenstreet, S. P. R., A. D. Bryant, N. Broekhuizen, S. J. Hall, and M. R. Heath. 1997. Seasonal variation in the consumption of food by fish in the North Sea and implications for food web dynamics. *ICES Journal of Marine Sciences* **54**: 243–266.

Gribble, N. A. 2001. A model of the ecosystem and associated penaeid prawn community, in the far northern Great Barrier Reef. In: E. Wolanski (ed.), *Oceanographic Process and Coral Reefs, Physical and Biological Links in the Great Barrier Reef*. CRC Press, New York, pp. 189–207.

Guenette, S., and T. Morato. 2001. The Azores Archipelago in 1997. *University of British Columbia, Fisheries Centre Research Reports* **9**(4): 241–270.

Hall, S. J., and D. Raffaelli. 1991. Food-web patterns: lessons from a species-rich web. *Journal of Animal Ecology* **60**: 823–842.

Hall, S. J., and D. G. Raffaelli. 1993. Food webs: theory and reality. *Advances in Ecological Research* **24**: 187–236.

Hardy, A. C., 1924. The herring in relation to its animate environment, Part 1. The food and feeding habits of the herring. *Fisheries Investigations London, Series II* **7**: 1–53.

Harvey, C. J., S. P. Cox, T. E. Essington, S. Hansson, and J. F. Kitchell. 2003. An ecosystem model of food web and fisheries interactions in the Baltic Sea. *ICES Journal of Marine Sciences* **60**: 939–950.

Havens, K., 1992. Scale and structure in natural food webs. *Science* **257**: 1107–1109.

Haydon, D., 1994. Pivotal assumptions determining the relationship between stability and complexity: an analytical synthesis of the stability–complexity debate. *American Naturalist* **144**: 14–29.

Heymans, J., and D. Baird. 2000. Network analysis of the northern Benguela ecosystem by means of NETWRK and ECOPATH. *Ecological Modelling* **131**: 97–118.

Heymans, J. J., L. J. Shannon, and A. Jarre. 2004. Changes in the northern Benguela ecosystem over three decades: 1970s, 1980s, 1990s. *Ecological Modelling* **172**: 179–195.

Hobson, K. A., and H. E. Welch. 1992. Determination of trophic relationships within a high Arctic marine food web using ^{13}C and ^{15}N analysis. *Marine Ecology Progress Series* **84**: 9–18.

Hobson, K. A., W. G. Ambrose, and P. E. Renaud. 1995. Sources of primary production, benthic–pelagic coupling, and trophic relationships within the Northeast Water Polynya: insights from ^{13}C and ^{15}N analysis. *Marine Ecology Progress Series* **128**: 1–10.

Hogetsu, K. 1979. Biological productivity of some regions of Japan. In: M. J. Dunbar, (ed), *Marine Production Mechanisms*. Cambridge University Press, Cambridge, pp. 71–87.

Hollowed, A. B., N. Bax, R. Beamish, J. Collie, M. Fogarty, P. Livingston, J. Pope, and J. C. Rice. 2000. Application of multispecies models in assessment of impacts of commercial fishing. Symposium proceedings on the ecosystem effects of fishing, *ICES Journal of Marine Sciences* **57**: 707–719.

Hop, H., K. Borga, G. W. Gbrielsen, L. Kleivane, and J. U. Skaare. 2002. Food web magnification of persistent organic pollutants in poikilotherms and homeotherms from the Barents Sea. *Environmental Science and Technology* **36**: 2589–2597.

Jackson, J. B. C., M. X. Kirby, W. H. Berger, K. A. Bjorndal, L. W. Botsford, B. J. Bourque, R. H. Bradbury, R. Cooke, J. Erlandson, J. A. Estes, T. P. Hughes, S. Kidwell, C. B. Lange, H. S. Lenihan, J. M. Pandolfi, C. H. Peterson, R. S. Steneck, M. J. Tegner, and R. R. Warner. 2001. Historical overfishing and the recent collapse of coastal ecosystems. *Science* **293**: 629–638.

Jarre, A., P. Muck, and D. Pauly. 1991. Two approaches for modelling fish stock interactions in the Peruvian upwelling ecosystem. *ICES Marine Science Symposia* **193**: 178–184.

Jarre-Teichmann, A., and D. Pauly. 1993. Seasonal changes in the Peruvian Upwelling ecosystem. In: V. Christensen and D. Pauly (eds), *Trophic Models of Aquatic Ecosystems*. ICLARM Conference Proceedings 26, pp. 307–314.

Jarre-Teichmann, A., L. Shannon, C. Moloney, and P. Wickens. 1998. Comparing trophic flows in the southern Benguela to those in other upwelling ecosystems. *South African Journal of Marine Science* **19**: 391–414.

Kaehler, S., E. A. Pakhomov, and C. D. McQuaid. 2000. Trophic structure of the marine food web at the Prince Edward Islands (Southern Ocean) determined by ^{13}C and ^{15}N analysis. *Marine Ecology Progress Series* **208**: 13–20.

Kauzinger, C. M. K., and P. J. Morin. 1998. Productivity controls food-chain properties in microbial communities. *Nature* **395**: 495–497.

Kenny, D., and C. Loehle. 1991. Are food webs randomly connected? *Ecology* **72**: 1794–1797.

Kitchell, J. F., C. H. Boggs, X. He, and C. J. Walters. 1999. Keystone predators in the Central Pacific. In: *Ecosystem Approaches for Fisheries Management*. University of Alaska Sea Grant, AK-SG-99-01: 665–684.

Kitchell, J. F., T. E. Essington, C. H. Boggs, D. E. Schindler, and C. J. Walters. 2002. The role of sharks and longline fisheries in a Pelagic ecosystem of the Central Pacific. *Ecosystems* **5**: 202–216.

Krause, A. E., K. A. Frank, D. M. Mason, R. E. Ulanowicz, and W. M. Taylor. 2003. Compartments revealed in food-web structure. *Nature* **426**: 282–285.

Lindeman, R. L. 1942. The trophic-dynamic aspect of ecology. *Ecology* **23**: 399–418.

Link, J. 2002. Does food web theory work for marine ecosystems? *Marine Ecology Progress Series* **230**: 1–9.

Lotka, A. J. 1925. *Elements of Physical Biology*. Williams and Watkins, Baltimore, MD.

Lyons, K. G., and M. W. Schwartz. 2001. Rare species loss alters ecosystem function—invasion resistance. *Ecology Letters* **4**: 358–365.

MacArthur, R. H. 1955. Fluctuation of animal populations and a measure of community stability. *Ecology* **36**: 533–536.

Manickchand-Heileman, S., F. Arreguin-Sanchez, A. Lara-Dominguez, and L. A. Soto. 1998*b*. Energy flow and network analysis of Terminos Lagoon, SW Gulf of Mexico. *Journal of Fish Biology* **53(A)**: 179–197.

Manickchand-Heileman, S., L. A. Soto, and E. Escobar. 1998*a*. A preliminary trophic model of the continental shelf, South-western Gulf of Mexico. *Estuarine, Coastal and Shelf Science* **46**: 885–899.

Mann, K. H., and J. R. N. Lazier. 1991. *Dynamics of Marine Ecosystems: Biological–Physical Interactions in the Oceans*. Blackwell Scientific Publishers, Oxford.

Martinez, N. D. 1991. Artifacts or attributes? Effects of resolution on the Little Rock Lake food web. *Ecological Monographs* **61**: 367–392.

Martinez, N. D. 1992. Constant connectance in community food webs. *American Naturalist* **139**: 1208–1218.

Martinez, N. D. 1993. Effects of resolution of food web structure. *Oikos* **60**: 403–412.

Martinez, N. D. 1994. Scale-dependent constraints on food-web structure. *American Naturalist* **144**: 935–953.

Martinez, N. D., B. A. Hawkins, H. A. Dawah, and B. P. Feifarek. 1999. Effects of sampling effort on characterization of food-web structure. *Ecology* **80**: 1044–1055.

May, R. M. 1972. Will a large complex system be stable? *Nature* **238**: 413–414.

May, R. M. 1973. *Stability and Complexity in Model Ecosystems.* Princeton University Press, Princeton, NJ.

McCann, K., A. Hastings, and G. R. Huxel. 1998. Weak trophic interactions and the balance of nature. *Nature* **395**: 794–798.

Mendoza, J. J. 1993. A preliminary biomass budget for the northeastern Venezuela shelf. In: V. Christensen and D. Pauly (eds), *Trophic Models of Aquatic Ecosystems.* ICLARM Conference Proceedings 26, pp. 285–297.

Mendy, A. N., and E. Buchary. 2001. Constructing the Icelandic marine ecosystem model for 1997 using a mass balance modeling approach. *University of British Columbia, Fisheries Centre Research Reports* **9**(4): 182–197.

Mohammed, E. 2001. A preliminary model for the Lancaster Sound Region in the 1980s. *University of British Columbia, Fisheries Centre Research Reports* **9**(4): 99–110.

Murphy, E. J. 1995. Spatial structure of the Southern Ocean ecosystem: predator-prey linkages in Southern Ocean food webs. *Journal of Animal Ecology* **64**: 333–347.

National Marine Fisheries Service (NMFS). 1999. Ecosystem-Based Fishery Management. A Report to Congress by the Ecosystems Principles Advisory Panel. US DOC, Silver Spring, MD.

Niquil, N., J. E. Arias-Gonzalez, B. Delesalle, and R. E. Ulanowicz. 1999. Characterization of the planktonic food web of Takapoto Atoll lagoon, using network analysis. *Oecologia* **118**: 232–241.

Odum, E. P. 1960. Organic production and turnover in old field succession. *Ecology* **41**: 34–49.

Odum, H. T. 1957. Trophic structure and productivity of Silver Springs, Florida. *Ecological Monographs* **27**: 55–112.

Okey, T. A. 2001. A "straw-man" Ecopath model of the Middle Atlantic Bight continental shelf, United States. *University of British Columbia, Fisheries Centre Research Reports* **9**(4): 151–166.

Okey, T. A., and D. Pauly. 1999. A mass-balanced model of trophic flows in Prince William Sound: de-compartmentalizing ecosystem knowledge. In: *Ecosystem Approaches for Fisheries Management.* University of Alaska Sea Grant, AK-SG-99-01, pp. 621–635.

Okey, T. A., and R. Pugilese 2001. A preliminary Ecopath model of the Atlantic continental shelf adjacent to the southeastern United States. *University of British Columbia, Fisheries Centre Research Reports* **9**(4): 167–181.

Olivieri, R. A., A. Cohen, and F. P. Chavez. 1993. An ecosystem model of Monterey Bay, California. In: V. Christensen and D. Pauly (eds), *Trophic Models of Aquatic Ecosystems.* ICLARM Conference Proceedings 26, pp. 315–322.

Olson, R. J., and G. M. Watters. 2003. A model of the pelagic ecosystem in the eastern Tropical Pacific Ocean. *Inter-American Tropical Tuna Commission Bulletin* **22**(3): 135–218.

Opitz, S. 1993. A quantitative model of the trophic interactions. in a Caribbean coral reef ecosystem. In: V. Christensen and D. Pauly (eds), *Trophic Models of Aquatic Ecosystems.* ICLARM Conference Proceedings 26, pp. 269–278.

Ortiz, M., and M. Wolff. 2002. Dynamical simulation of mass-balance trophic models for benthic communities of north-central Chile: assessment of resilience time under alternative management scenarios. *Ecological Modelling* **148**: 277–291.

Paine, R. T. 1966. Food web complexity and species diversity. *American Naturalist* **100**: 65–75.

Paine, R. T. 1988. Food webs: road maps of interactions or grist for theoretical development? *Ecology* **69**: 1648–1654.

Pauly, D., and V. Christensen. 1993. Stratified models of large marine ecosystems: a general approach and an application to the South China Sea. In: K. Sherman, L. M. Alexander, and B. D. Gold (eds), *Large Marine Ecosystems: Stress, Mitigation, and Sustainability.* AAAS Press, Washington, DC, pp. 148–174.

Pauly, D., V. Christensen, J. Dalsgaard, R. Froese, and F. Torres. 1998. Fishing down marine food webs. *Science* **279**: 860–863.

Pedersen, S. A. 1994. Multispecies interactions on the offshore West Greenland shrimp grounds. *Journal of Marine Sciences, ICES C.M. Documents* **1994/P**: 2.

Pedersen, S., and D. Zeller. 2001. Multispecies interactions in the West Greenland ecosystem: importance of the shrimp fisheries. *University of British Columbia, Fisheries Centre Research Reports* **9**(4): 111–127.

Petersen, C. G. J. 1918. The sea bottom and its production of fish food. *Reports of the Danish Biological Station* **25**: 1–62.

Pimm, S. L. 1982. *Food Webs.* Chapman and Hall, London.

Pimm, S. L., and J. C. Rice. 1987. The Dynamics of Multispecies, Multi-Life-Stage Models of Aquatic Food webs. *Theoretical Population Biology* **32**: 303–325.

Pimm, S. L., J. H. Lawton, and J. E. Cohen. 1991. Food web patterns and their consequences. *Nature* **350**: 669–674.

Polis, G. A. 1991. Complex trophic interactions in deserts: an empirical critique of food-web theory. *American Naturalist* **138**: 123–155.

Polis, G. A., and D. R. Strong. 1996. Food web complexity and community dynamics. *American Naturalist* **147**: 813–846.

Polovina, J. J. 1984. Model of a coral reef ecosystem. I. The ECOPATH model and its application to French Frigate Shoals. *Coral Reefs* **3**: 1–11.

Pomeroy, L. R. 2001. Caught in the food web: complexity made simple? *Scientia Marina* **65**: 31–40.

Power, M. E., and L. S. Mills. 1995. The keystone cops meet in Hilo. *Trends in Ecology and Evolution* **10**: 182–184.

Purcell, J. E. 1986. Predation on fish eggs and larvae by pelagic cnidarians and ctenophores. *Bulletin of Marine Science* **37**: 739–755.

Qiyong, Z., L. Qiumian, L. Youtong, and Z. Yueping. 1981. Food web of fishes in Minnan-Taiwanchientan fishing ground. *Acta Oceanologica Sinica* **3**: 275–290.

Raffaelli, D. 2000. Trends in research on shallow water food webs. *Journal of Experimental Marine Biology and Ecology* **250**: 223–232.

Raffaelli, D. G., and S. J. Hall 1996. Assessing the relative importance of trophic links in food webs. In: G. A. Polis and K. O. Winemiller (eds), *Food Webs: Integration of Patterns and Dynamics*. Chapman and Hall, New York, pp. 185–191.

Reagan, D. P., G. R. Camilo, and R. B. Waide. 1996. The community food web: major properties and patterns of organization. In: D. P. Reagan and R. B. Waide (eds), *The Food Web of a Tropical Rain Forest*. University of Chicago Press, Chicago, IL, pp. 461–510.

Reid, K., and J. P. Croxall. 2001. Environmental response of upper trophic-level predators reveals a system change in an Antarctic marine ecosystem. *Proceedings of the Royal Society of London B* **268**: 377–384.

Rysgaard, S., T. G. Nielsen, and B. W. Hansen. 1999. Seasonal variation in nutrients, pelagic primary production and grazing in a high-Arctic coastal marine ecosystem, Young Sound, Northeast Greenland. *Marine Ecology Progress Series* **179**: 13–25.

Sandberg, J., R. Elmgren, and F. Wulff. 2000. Carbon flows in Baltic Sea food webs—a re-evaluation using a mass balance approach. *Journal of Marine Systems* **25**: 249–260.

Schlesinger, W. H. 1997. *Biogeochemistry, An Analysis of Global Change*. Academic Press, San Diego.

Schoener, T. W. 1989. Food webs from the small to the large. *Ecology* **70**: 1559–1589.

Shannon, L. J., and A. Jarre-Teichmann. 1999. A model of trophic flows in the northern Bengula upwelling system during the 1980s. *South African Journal of Marine Science* **21**: 349–366.

Sherr, E. B., B. F. Sherr, and G. A. Paffenhoeffer. 1986. Phagotrophic protozoa as food for metazoans: A "missing" trophic link in marine pelagic food webs? *Marine Microbial Food Webs* **1**: 61–80.

Silvestre, G., S. Selvanathan, and A. H. M. Salleh. 1993. Preliminary trophic model of the coastal fishery resources of Brunei Darussalam, South China Sea. In: V. Christensen and D. Pauly (eds), *Trophic Models of*

Aquatic Ecosystems. ICLARM Conference Proceedings 26, pp. 300–306.

Solow, A., and A. Beet. 1998. On lumping species in food webs. *Ecology* **79**: 2013–2018.

Stanford, R., K. Lunn, and S. Guenette. 2001. A preliminary ecosystem model for the Atlantic coast of Morocco in the mid-1980s. *University of British Columbia, Fisheries Centre Research Reports* **9**(4): 314–344.

Steele, J. H. 1985. A comparison of terrestrial and marine ecological systems. *Nature* **313**: 355–358.

Steele, J. H. 1991. Can ecological theory cross the land–sea boundary? *Journal of Theoretical Biolgoy* **153**: 425–436.

Sugihara, G., K. Schoenly, and A. Trombla. 1989. Scale invariance in food web properties. *Science* **245**: 48–52.

Tittlemier, S. A., A. T. Fisk, K. A. Hobson, and R. J. Norstrom. 2002. Examination of the bioaccumulation of halogenated dimethyl bipyrroles in an Arctic marine food web using stable nitrogen isotope analysis. *Environmental Pollution* **116**: 85–93.

Tjelmeland, S., and B. Bogstad. 1998. MULTSPEC—a review of a multispecies modelling project for the Barents Sea. *Fisheries Research* **37**: 127–142.

Trites, A. W., P. A. Livingston, S. Mackinson, M. C. Vasconcellos, A. M. Springer, and D. Pauly. 1999. Ecosystem change and the decline of marine mammals in the eastern Bering Sea: testing the ecosystem shift and commercial whaling hypotheses. *University of British Columbia, Fisheries Centre Research Reports* **7**(1): 106 p.

Vega-Cendejas, M. E., F. Arreguin-Sanchez, and M. Hernandez. 1993. Trophic fluxes on the Campeche Bank, Mexico. In: V. Christensen and D. Pauly (eds), *Trophic Models of Aquatic Ecosystems*. ICLARM Conference Proceedings 26, pp. 206–213.

Venier, J. M., and D. Pauly. 1997. Trophic dynamics of a Florida Keys coral reef ecosystem. *Proceedings of the 8th International Coral Reef Symposium* **1**: 915–920.

Verity, P. G. and V. Smetacek. 1996. Organism life cycles, predation, and the structure of marine pelagic ecosystems. *Marine Ecology Progress Series* **130**: 277–293.

Volterra, V. 1926. Fluctuations in the abundance of a species considered mathematically. *Nature* **118**: 558–560.

Walters, C., V. Christensen, and D. Pauly. 1997. Structuring dynamic models of exploited ecosystems from trophic mass-balance assessments. *Reviews in Fish Biology and Fisheries* **7**: 139–172.

Warren, P. H. 1994. Making connections in food webs. *Trends in Ecology and Evolution* **9**: 136–141.

Welch, H. E., M. A. Bergmann, T. D. Siferd, K. A. Martin, M. F. Curtis, R. E. Crawford, R. J. Conover, and H. Hop. 1992. Energy flow through the marine ecosystem of the Lancaster Sound region, arctic Canada. *Arctic* **45**: 343–357.

Whipple, S., J. S. Link, L. P. Garrison, and M. J. Fogarty. 2000. Models of predation and fishing mortality in aquatic ecosystems. *Fish and Fisheries* **1**: 22–40.

Williams, R. J., and N. D. Martinez. 2000. Simple rules yield complex food webs. *Nature* **6774**: 180–183.

Winemiller, K. O. 1990. Spatial and temporal variation in tropical fish trophic networks. *Ecological Monographs* **60**: 27–55.

Winemiller, K. O., and G. A. Polis. 1996. Food webs: what can they tell us about the world? In: G. A. Polis and K. O. Winemiller (eds), *Food Webs: Integration of Patterns and Dynamics*. Kluwer Academic Publishers, Boston, pp. 1–22.

Wolff, M. 1994. A trophic model for Tongoy Bay—a system exposed to suspended scallop culture (Northern Chile). *Journal of Experimental Marine Biology and Ecology* **182**: 149–168.

Yodzis, P. 1998. Local trohpodynamics and the interaction of marine mammals and fisheries in the Benguela ecosystem. *Journal of Animal Ecology* **67**: 635–658.

Yodzis, P. 2000. Diffuse effects in food webs. *Ecology* **81**: 261–266.

Zeller, D., and K. Freire. 2001. A preliminary North-east Atlantic marine ecosystem model: Faroe islands and ICES area Vb. *University of British Columbia, Fisheries Centre Research Reports* **9**(4): 207–212.

Chapter 10

Albert, R., J. Jeong, and A.-L. Barabasi. 2000. Error and attack tolerance of complex networks.

Baird, D., and R. E. Ulanowicz. 1989. The seasonal dynamics of the Chesapeake Bay ecosystem. *Ecological Monographs* **59**: 329–364.

Berlow, E. 1999. Strong effects of weak interactions in ecological communities. *Nature* **398**: 330–334.

Brose, U., N. D. Martinez, and R. J. Williams. 2003a. Estimating species richness: sensitivity to sample coverage and insensitivity to spatial patterns. *Ecology* **84**: 2364–2377.

Brose, U., R. J. Williams, and N. D. Martinez. 2003b. Comment on "Foraging adaptation and the relationship between food-web complexity and stability." *Science* **301**: 918b.

Caldarelli, G., P. G. Higgs, and A. J. McKane. 1998. Modelling coevolution in multispecies communities. *Journal of Theoretical Biology* **193**: 345–358.

Camacho, J., R. Guimera, and L. A. N. Amaral. 2002. Analytical solution of a model for complex food webs. *Physical Review E* **65**: 030901(R).

Christensen, V., and C. J. Walters. 2004. Ecopath with Ecosim: methods, capabilities and limitations. *Ecological Modelling* **172**: 109–139.

Cohen, J. E., F. Briand, and C. M. Newman. 1990. *Community Food Webs: Data and Theory*. Springer-Verlag, New York.

Daily, G. C. (ed.). 1997. *Nature's Services: Societal Dependence on Natural Ecosystems*. Island Press, Washington DC.

De Angelis, D. L. 1975. Stability and connectance in food web models. *Ecology* **56**: 238–243.

de Ruiter, P. C., A.-M. Neutel, and J. C. Moore. 1995. Energetics, patterns of interaction strengths, and stability in real ecosystems. *Science* **269**: 1257–1260.

Diehl, S., and M. Feissel. 2001. Intraguild prey suffer from enrichment of their resources: a microcosm experiment with ciliates. *Ecology* **82**: 2977–2983.

Drossel, B., P. G. Higgs, and A. J. McKane. 2001. The influence of predator–prey population dynamics on the long-term evolution of food web structure. *Journal of Theoretical Biology* **208**: 91–107.

Dunne, J. A., R. J. Williams, and N. D. Martinez. 2002a. Food web structure and network theory: the role of connectance and size. *Proceedings of the National Academy of Sciences, USA* **99**: 12917–12922.

Dunne, J. A., R. J. Williams, and N. D. Martinez. 2002b. Network structure and biodiversity loss in food webs: robustness increases with connectance. *Ecology Letters* **5**: 558–567.

Dunne, J. A., R. J. Williams, and N. D. Martinez. 2004. Network structure and robustness of marine food webs. *Marine Ecology Progress Series* **273**: 291–302.

Elton, C. S. 1927. *Animal Ecology*. Sidgwick and Jackson, London.

Elton, C. S. 1958. *Ecology of Invasions by Animals and Plants*. Chapman and Hall, London.

Field, J. G., R. J. M. Crawford, P. A. Wickens, C. L. Moloney, K. L. Cochrane, and C. A. Villacastic-Herraro. 1991. Network analysis of Benguela pelagic food webs. Benguela Ecology Programme, Workshop on Seal-Fishery Biological Interactions. University of Cape Town, 16–20 September, 1991, BEP/SW91/M5a.

Fussmann, G. F., and G. Heber. 2002. Food web complexity and chaotic population dynamics. *Ecology Letters* **5**: 394–401.

Garlaschelli, D., G. Caldarelli, and L. Pietronero. 2003. Universal scaling relations in food webs. *Nature* **423**: 165–168.

Girvan, M., and M. E. J. Newman. 2002. Community structure in social and biological networks. *Proceedings of the National Academy of Sciences, USA* **99**: 8271–8276.

Hairston, N. G., J. D. Allen, R. K. Colwell, D. J. Futuyma, J. Howell, M. D. Lubin, J. Mathias, and

J. H. Vandermeer. 1968. The relationship between species diversity and stability: and experimental approach with protozoa and bacteria. *Ecology* **49**: 1091–1101.

Hassell, M. P. 1978. *The Dynamics of Arthropod Predator–Prey Systems*. Princeton University Press, Princeton, NJ.

Hastings, A., and T. Powell. 1991. Chaos in a three-species food chain. *Ecology* **72**: 896–903.

Hastings, A., C. L. Hom, S. Ellner, P. Turchin, and H. C. J. Godfray. 1993. Chaos in ecology: is mother nature a strange attractor. *Annual Review of Ecology and Systematics* **24**: 1–33.

Hector, A. et al. (34 authors). 1999. Plant diversity and productivity experiments in European grasslands. *Science* **286**: 1123–1127.

Holling, C. S. 1959. Some characteristics of simple types of predation and parasitism. *Canadian Entomology* **91**: 385–399.

Holling, C. S. 1966. The strategy of building models of complex ecological systems. In: K. E. F. Watt (ed.), *Systems Analysis in Ecology*. Academic Press, New York, pp. 195–214.

Hutchinson, G. E. 1959. Homage to Santa Rosalia, or why are there so many kinds of animals? *American Naturalist* **93**: 145–159.

Kondoh, M. 2003a. Foraging adaptation and the relationship between food-web complexity and stability. *Science* **299**: 1388–1391.

Kondoh, M. 2003b. Response to Comment on "Foraging adaptation and the relationship between food-web complexity and stability." *Science* **301**: 918c.

Krause, A. E., K. A. Frank, D. M. Mason, R. E. Ulanowicz, and W. W. Taylor. 2003. Compartments revealed in food-web structure. *Nature* **426**: 282–285.

Lawlor, L. R. 1978. A comment on randomly constructed model ecosystems. *American Naturalist* **112**: 445–447.

Lindeman, R. L. 1942. The tropic-dynamic aspect of ecology. *Ecology* **23**: 399–418.

MacArthur, R. H. 1955. Fluctuations of animal populations and a measure of community stability. *Ecology* **36**: 533–536.

Martinez, N. D. 1991. Artifacts or attributes? Effects of resolution on the Little Rock Lake food web. *Ecological Monographs* **61**: 367–392.

Martinez, N. D. 1992. Constant connectance in community food webs. *American Naturalist* **140**: 1208–1218.

Martinez, N. D., B. A. Hawkins, H. A. Dawah, and B. Feifarek. 1999. Effects of sample effort on characterization of food-web structure. *Ecology* **80**: 1044–1055.

May, R. M. 1972. Will a large complex system be stable? *Nature* **238**: 413–414.

May, R. M. 1973. *Stability and Complexity in Model Ecosystems*, 2nd edn. Princeton University Press, Princeton, NJ.

May, R. M. 1974. Biological populations with non-overlapping generations: stable points, stable cycles, and chaos. *Science* **186**: 645–647.

May, R. M. 1976. Simple mathematical models with very complicated dynamics. *Nature* **261**: 459–467.

May, R. M. 2001. *Stability and Complexity in Model Ecosystems*. Princeton Landmarks in Biology Edition. Princeton University Press, Princeton, NJ.

McCann, K. 2000. The diversity–stability debate. *Nature* **405**: 228–233.

McCann, K., and A. Hastings. 1997. Re-evaluating the omnivory–stability relationship in food webs. *Proceedings of the Royal Society London B* **264**: 1249–1254.

McCann, K., and P. Yodzis. 1994a. Biological conditions for chaos in a three-species food chain. *Ecology* **75**: 561–564.

McCann, K., and P. Yodzis. 1994b. Nonlinear dynamics and population disappearances. *American Naturalist* **144**: 873–879.

McCann, K., A. Hastings, and G. R. Huxel. 1998. Weak trophic interactions and the balance of nature. *Nature* **395**: 794–798.

Moore, J. C., and H. W. Hunt. 1988. Resource compartmentation and the stability of real ecosystems. *Nature* **333**: 261–263.

Murdoch, W. W., and A. Oaten. 1975. Predation and population stability. *Advances in Ecological Research* **9**: 1–131.

Naeem, S., L. J. Thompson, S. P. Lawlor, J. H. Lawton, and R. M. Woodfin. 1994. Declining biodiversity can alter the performance of ecosystems. *Nature* **368**: 734–737.

Neutel, A.-M., J. A. P. Heesterbeek, and P. C. deRuiter. 2002. Stability in real food webs: weak links in long loops. *Science* **296**: 1120–1123.

Odum, E. 1953. *Fundamentals of Ecology*. Saunders, Philadelphia, PA.

Paine, R. T. 1966. Food web complexity and species diversity. *American Naturalist* **100**: 65–75.

Pimm, S. L., and J. H. Lawton. 1978. On feeding on more than one trophic level. *Nature* **275**: 542–544.

Pimm, S. L., and J. H. Lawton. 1977. Number of trophic levels in ecological communities. *Nature* **268**: 329–331.

Polis, G. A. 1991. Complex desert food webs: an empirical critique of food web theory. *American Naturalist* **138**: 123–155.

Post, D. M., M. E. Conners, and D. S. Goldberg. 2000. Prey preference by a top predator and the stability of linked food chains. *Ecology* **81**: 8–14.

Skalski, G. T., and J. F. Gilliam. 2001. Functional responses with predator interference: viable alternatives to the Holling type II model. *Ecology* **82**: 3083–3092.

Strogatz, S. H. 2001. Exploring complex networks. *Nature* **410**: 268–276.

Tilman, D., P. B. Reich, J. Knops, D. Wedin, T. Mielke, and C. Lehman. 2001. Diversity and productivity in a long-term grassland experiment. *Science* **294**: 843–845.

Ulanowicz, R. E. 2002. The balance between adaptability and adaptation. *BioSystems* **64**: 13–22.

Watts, D. J., and S. H. Strogatz. 1998. Collective dynamics of 'small-world' networks. *Nature* **393**: 440–442.

Williams, R. J., and N. D. Martinez. 2000. Simple rules yield complex food webs. *Nature* **404**: 180–183.

Williams, R. J., and N. D. Martinez. 2004*a*. Trophic levels in complex food webs: theory and data. *American Naturalist* **163**: 458–468.

Williams, R. J., and N. D. Martinez. 2004*b*. Stabilization of chaotic and non-permanent food web dynamics. *European Physics Journal B* **38**: 297–303.

Williams, R. J., and N. D. Martinez. 2004*c*. Diversity, complexity, and persistence in large model ecosystems. Santa Fe Institute Working Paper 04–07–022.

Yodzis, P. 1981. The stability of real ecosystems. *Nature* **289**: 674–676.

Yodzis, P. 1998. Local trophodynamics and the interaction of marine mammals and fisheries in the Benguela ecosystem. *Journal of Animal Ecology* **67**: 635–658.

Yodzis, P. 2000. Diffuse effects in food webs. *Ecology* **81**: 261–266.

Yodzis, P. 2001. Must top predators be culled for the sake of fisheries? *Trends in Ecology and Evolution* **16**: 78–84.

Yodzis, P., and S. Innes. 1992. Body-size and consumer–resource dynamics. *American Naturalist* **139**: 1151–1173.

Chapter 11

Abrams, P. A. 1982. Functional responses of optimal foragers. *The American Naturalist* **120**: 382–390.

Abrams, P. A. 1984. Foraging time optimization and interactions in food webs. *American Naturalist* **124**: 80–96.

Abrams, P. A. 1992. Predators that benefit prey and prey that harm predators: unusual effects of interacting foraging adaptations. *American Naturalist* **140**: 573–600.

Abrams, P. A. 2000. The evolution of predator–prey interactions: theory and evidence. *Annual Review of Ecological Systems* **31**: 79–105.

Abrams, P. A., and H. Matsuda. 1996. Positive indirect effects between prey species that share predators. *Ecology* **77**: 610–616.

Abrams, P. A., B. A. Menge, G. G. Mittelbach, D. Spiller, and P. Yodzis. 1996. The role of indirect effects in food webs. In: G. A. Polis and K. O. Winemiller (eds), *Food Webs: Integration of Patterns and Dynamics*. Chapman and Hall, New York, pp. 371–395.

Bolker, B., M. Holyoak, V. Krivan, L. Rowe, and O. Schmitz. 2003. Connecting theoretical and empirical studies of trait-mediated interactions. *Ecology* **84**: 1101–1114.

Brose, U., R. J. Williams, and N. D. Martinez. 2003. Comments on "foraging adaptation and the relationship between food-web complexity and stability." *Science* **301**: 917c.

Chen, X., and J. E. Cohen. 2001*a*. Transient dynamics and food-web complexity in the Lotka-Volterra cascade model. *Proceedings of The Royal Society of London, Series B* **268**: 869–877.

Chen, X., and J. E. Cohen. 2001*b*. Global stability, local stability and permanence in model food webs. *Journal of Theoretical Biology* **212**: 223–235.

DeAngelis, D. L. 1975. Stability and connectance in food web models. *Ecology* **56**: 238–243.

De Ruiter, P. C., A. Neutel, and J. C. Moore. 1995. Energetics, patterns of interaction strengths, and stability in real ecosystems. *Science* **269**: 1257–1260.

Dunne, J. A., R. J. Williams, and N. D. Martinez. 2002. Network structure and biodiversity loss in food webs: robustness increases with connectance. *Ecology Letters* **5**: 558–567.

Emlen, J. M. 1966. The role of time and energy in food preference. *American Naturalist* **100**: 611–617.

Emmerson, M. C., and D. G. Raffaelli. 2004. Predator–prey body size, interaction strength and the stability of a real food web. *Journal of Animal Ecology* **73**: 399–409.

Endler, J. A. 1986. Defence against predators. In: M. E. Feder and G. V. Lauder (eds), *Predator–Prey Relationships*. University of Chicago Press, Chicago, IL, pp. 109–134.

Fryxell, L. M., and P. Lundberg, 1997. *Individual Behavior and Community Dynamics*. Chapman and Hall, New York.

Gilpin, M. E. 1975. Stability of feasible predator–prey systems. *Nature* **254**: 137–139.

Goldwasser, L., and J. Roughgarden. 1993. Construction of a large Caribbean food web. *Ecology* **74**: 1216–1233.

Haydon, D. T. 2000. Maxmully stable model ecosystems can be highly connected. *Ecology* **81**: 2631–2636.

Holling, C. S. 1959. Some characteristics of simple types of predation and parasitism. *Canadian Entomologist* **91**: 385–398.

Holt, R. D. 1977. Predation, apparent competition, and the structure of prey communities. *Theoretical Population Biology* **12**: 197–229.

Holt, R. D. 1983. Optimal foraging and the form of the predator isocline. *American Naturalist* **122**: 521–541.

Holt, R. D., and G. A. Polis. 1997. A theoretical framework for intraguild predation. *American Naturalist* **149**: 745–754.

Hughes, R. N. 1990. *Behavioural Mechanisms of Food Selection,* NATO-ASI Series, Vol. G20. Springler-Verlag, Berlin.

Ives, A. R., and A. P. Dobson. 1987. Antipredator behavior and the population dynamics of simple predator–prey systems. *American Naturalist* **130**: 431–437.

Keitt, T. H. 1997. Stability and complexity on a lattice: coexistence of species in an individual-based food web model. *Ecological Modelling* **102**: 243–258.

Kondoh, M. 2003*a*. Foraging adaptation and the relationship between food-web complexity and stability. *Science* **299**: 1388–1391.

Kondoh, M. 2003*b*. Response to comment on "foraging adaptation and the relationship between food-web complexity and stability." *Science* **301**: 918c.

Křivan, V. 1998. Effects of optimal antipredator behavior of prey on predator–prey dynamics: the role of refuges. *Theoretical Population Biology* **53**: 131–142.

Křivan, V. 2000. Optimal intraguild foraging and population stability. *Theoretical Population Biology* **58**: 79–94.

Law, R., and R. D. Morton. 1996. Permanence and the assembly of ecological communities. *Ecology* **77**: 762–775.

Lawlor, L. R. 1978. A comment on randomly constructed model ecosystems. *American Naturalist* **112**: 445–447.

Lima, S. L. 1992. Life in a multi-predator environment: some considerations for antipredatory vigilance. *Annals of Zoology—Fennica* **29**: 217–226.

Lima, S. L., and L. M. Dill. 1996. Behavioral decisions made under the risk of predation: a review and prospectus. *Candian Journal of Zoology* **68**: 619–640.

MacArthur, R. H., and E. R. Pianka. 1966. On optimal use of a patchy environment. *American Naturalist* **100**: 603–609.

MacNamara, J. M., and A. I. Houston. 1987. Starvation and predation as factors limiting population size. *Ecology* **68**: 1515–1519.

Matsuda, H., P. A. Abrams, and M. Hori. 1993. The effect of adaptive antipredator behavior on exploitative competition and mutualism between predators. *Oikos* **68**: 549–559.

Matsuda, H., M. Hori, and P. A. Abrams. 1994. Effects of predator-specific defense on predator persistence and community complexity. *Evolutionary Ecology* **8**: 628–639.

Matsuda, H., M. Hori, and P. A. Abrams. 1996. Effects of predator-specific defense on biodiversity and community complexity in two-trophic-level communities. *Evolutionary Ecology* **10**: 13–28.

May, R. M. 1972. Will a large complex system be stable? *Nature* **238**: 413–414.

May, R. M. 1973. *Stability and Complexity in Model Ecosystems.* Princeton University Press, Princeton, NJ.

Martinez, N. D. 1991. Artifacts or attributes? Effects of resolution on the Little Rock Lake food web. *Ecological Monographs* **61**: 367–392.

McCann, K. and A. Hastings. 1997. Re-evaluating the omnivory-stability relationship in food webs. *Proc. R. Soc. Lond. B.* **264**: 1249–1254.

McCann, K., A. Hastings, and G. R. Huxel. 1998. Weak trophic interactions and the balance of nature. *Nature* **395**: 794–798.

Menge, B. A. 1994. Indirect effects in marine rocky intertidal interaction webs: pattern and importance. *Ecological Monographs* **65**: 21–74.

Neutel, A., J. A. P. Heesterbeek, and P. C. de Ruiter. 2002. Stability in real food webs: weak links in long loops. *Science* **296**: 1120–1123.

Nunney, L. 1980. The stability of complex model ecosystems. *American Naturalist* **115**: 639–649.

Paine, R. T. 1988. Food webs: road maps of interactions or grist for theoretical development? *Ecology* **69**: 1648–1654.

Pelletier, J. D. 2000. Are large complex ecosystems more unstable? A theoretical reassessment with predator switching. *Mathematical Biosciences* **163**: 91–96.

Petchey, O. L. 2000. Prey diversity, prey composition, and predator population dynamics in experimental microcosms. *Journal of Animal Ecology* **69**: 874–882.

Pimm, S. L. 1991. *The Balance of Nature?: Ecological Issues in the Conservation of Species and Communities.* University of Chicago Press, Chicago, IL.

Pimm, S. L., and J. H. Lawton. 1977. On the number of trophic levels. *Nature* **268**: 329–331.

Pimm, S. L., and J. H. Lawton, 1978. On feeding on more than one trophic level. *Nature* **275**: 542–544.

Polis, G. A. 1991. Complex desert food webs: an empirical critique of food web theory *American Naturalist* **138**: 123–155.

Post, W. M., and S. L. Pimm. 1983. Community assembly and food web stability. *Mathematical Bioscience* **64**: 169–192.

Post, D. M., M. E. Conners, and D. S. Goldberg. 2000. Prey preference by a top predator and the stability of linked food chains. *Ecology* **81**: 8–14.

Roxburgh, S. H. and J. B. Wilson. 2000. Stability and coexistence in a lawn community: mathematical prediction of stability using a Community matrix with parameters derived from competition experiments. *Oikos* **88**: 395–408.

Schmitz, O. J., and G. Booth. 1997. Modelling food web complexity: The consequences of individual-based, spatially explicit behavioural ecology on trophic interactions. *Evolutionary Ecology* **11**: 379–398.

Schoener, T. W. 1993. On the relative importance of direct versus indirect effects in ecological communities. In: H. Kawanabe, J. E. Cohen, and K. Iwasaki (eds), *Mutualism and Community Organization: Behavioral, Theoretical and Food Web Approaches*. Oxford University Press, Oxford, pp. 366–411.

Sih, A. 1984. Optimal behavior and density-dependent predation. *American Naturalist* **123**: 314–326.

Sole, R. V., and J. M. Montoya. 2001. Complexity and fragility in ecological networks. *Proceedings of The Royal Society of London, Series B* **268**: 2039–2045.

Sutherl, W. 1996. *From Individual Behaviour to Population Ecology*. Oxford University Press, Oxford.

Stephens, D. W., and J. R. Krebs. 1987. *Foraging Theory*. Princeton University Press, Princeton, NJ.

Sterner, R. W., A. Bajpai, and T. Adams. 1997. The enigma of food chain length: absence of theoretical evidence for dynamic constraints. *Ecology* **78**: 2258–2262.

Tansky, M. 1978. Switching effect in prey–predator system. *Journal of Theoretical Biology* **70**: 263–271.

Teramoto, E., K. Kawasaki, and N. Shigesada. 1979. Switching effect of predation on competitive prey species. *Journal of Theoretical Biology* **79**: 305–315.

Vos, M., S. M. Berrocal, F. Karamaouna, L. Hemerik, and L. E. M. Vet. 2001. Plant-mediated indirect effects and the persistence of parasitoid–herbivore communities. *Ecology Letters* **4**: 38–45.

Warren, P. H. 1989. Spatial and temporal variation in tropical fish trophic networks. *Oikos* **55**: 299–311.

Wilmers, C. C., S. Sinha, and M. Brede. 2002. Examining the effects of species richness on community stability: an assembly model approach. *Oikos* **99**: 364–368.

Wilson, D. S., and J. Yoshimura. 1994. On the coexistence of specialists and generalists. *American Naturalist* **144**: 692–707.

Winemiller, K. O. 1990. Spatial and temporal variation in tropical fish trophic networks. *Ecological Monographs* **60**: 331–367.

Yodzis, P. 1981. The stability of real ecosystems. *Nature* **289**: 674–676.

Yodzis, P. 2000. Diffuse effects in food webs. *Ecology* **81**: 261–266.

Chapter 12

Alaska Sea Grant. 1993. Is it food? Addressing Marine Mammals and Seabirds Declines. Alaska Sea Grant Report No. 93–01. University of Alaska Fairbanks, Fairbanks.

Anderson, P. J. 2000. Pandalid shrimp as indicators of ecosystem regime shift. *Journal of Northwest Atlantic Fisheries Science* **27**: 1–10.

Anderson, P. J., J. F. Piatt. 1999. Community reorganization in the Gulf of Alaska following ocean climate regime shift. *Marine Ecology Progress Series* **189**: 117–123.

Angliss, R. P., and K. L. Lodge. 2002. Alaska marine mammals stock assessment, 2002. U.S. Department Commerce, NOAA Technical Memorandum NMFS-AFSC-133, 224 p.

Anker-Nilssen, T., R. T. Barret, and J. V. Krasnov. 1997. Long- and short-term responses of seabirds in the Norwegian and Barents Seas to changes in stocks of prey fish. In: *Forage Fishes in Marine Ecosystems*. Alaska Sea Grant College Program Publication AK-SG-97-01. University of Alaska, Fairbanks, pp. 683–698.

Anthony, J. A., D. D. Roby, and K. R. Turco. 2000. Lipid content and energy density of forage fishes from the northern Gulf of Alaska. *Journal of Experimental Marine Biology and Ecology* **248**: 53–78

Bailey, K. M. 2000. Shifting control of recruitment of walleye pollock *Theragra chalcogramma* after a major climatic and ecosystem change. *Marine Ecology Progress Series* **198**: 215–224.

Bailey, K. M., and S. A. Macklin. 1994. Analysis of patterns in larval Walleye pollock theragra-chalcogramma survival and wind mixing events in Shelikof Strait, Gulf of Alaska. *Marine Ecology Progress Series* **113**: 1–12

Bailey, K. M., and S. J. Picquelle. 2002. Larval distribution of offshore spawning flatfish in the Gulf of Alaska: potential transport pathways and enhanced onshore transport during ENSO events. *Marine Ecology Progress Series* **236**: 205–217

Ballance, L. T., and R. L. Pitman. 1999. Foraging ecology of tropical seabirds. In: N. J. Adams, and R. H. Slotow (eds), *Proceedings of the 22nd International Ornithology Congress*, Durban, 2057–2071. BirdLife South Africa, Johannesburg.

Behrenfeld, M. J., and Z. S. Kolber, 1999. Widespread iron limitation of phytoplankton in the South Pacific Ocean. *Science* **283**: 840–843.

Benson, A. J., and A. W. Trites. 2002. Ecological effects of regime shifts in the Bering Sea and North Pacific Ocean. *Fish and Fisheries* **3**: 95–113.

Berenboim, B. I., A. V. Dolgov, V. A. Korzhev, and N. A. Yaragina. 2000. The impact of cod on the dynamics of Barents Sea shrimp (*Pandalus borealis*) as determined by multispecies models. *Journal of Northwest Atlantic Fisheries Science* **27**: 69–75

Bidigare, R. R., and M. E. Ondrusek. 1996. Spatial and temporal variability of phytoplankton pigment distribution in the central equatorial Pacific Ocean. *Deep-Sea Research II* **43**: 809–833.

Blindheim, J., V. Borovkov, B. Hansen, S-A. Malmberg, W. R. Turrell, and S. Østerhus. 2000. Upper layer cooling and freshening in the Norwegian Sea in relation to atmospheric forcing. *Deep-Sea Research I* **47**: 655–680.

Bogstad B., T. Haug, and S. Mehl. 2000. Who eats whom in the Barents Sea? *NAMMCO Scientific Publication* **2**: 98–119.

Bond, N. A., J. E. Overland, M. Spillane, and P. Stabeno. 2004. Recent shifts in the state of the North Pacific. *Geophysical Reasearch Letters* **30**: 2183–2186.

Boyd, P. W. et al. 2000. A mesoscale phytoplankton bloom in the polar Southern Ocean stimulated by iron fertilization. *Nature* **407**: 695–702.

Brodeur, R. D. 1998. *In situ* observations of the associations between juvenile fishes and schyphomedusae in the Bering Sea. *Marine Ecology Progress Series* **163**: 11–20.

Brodeur, R. D., and D. M. Ware. 1992. Long-term variability in zooplankton biomass in the sub-arctic Pacific Ocean. *Fisheries Oceanography* **1**: 32–38.

Brodeur, R. D., B. W. Frost, S. R. Hare, R. C. Francis, and W. J. Jr Ingraham. 1999. Interannual variations in zooplankton biomass in the Gulf of Alaska, and covariation with California Current zooplankton biomass. In: K. Sherman and Q. Tang (eds), *Large Marine Ecosystems of the Pacific Rim: Assessment, Sustainability, and Management*. Blackwell Science Inc., Malden, MA, pp. 106–137.

Burek, K. A., F. M. D. Gulland, G. Sheffield, D. Calkins, E. Keyes, T. R. Spraker, A. W. Smith, D. E. Skilling, J. Evermann, J. L. Stott, and A. W. Trites. 2003. Disease agents in Steller sea lions in Alaska: a review and analysis of serology data from 1975–2000. *Fisheries Centre Reports* **11**(4): 26.

Carpenter, S. R. 1990. Large-scale perturbations: opportunities for innovation. *Ecology* **71**: 2038–2043.

Carpenter, S. R., and J. F. Kitchell. 1993. The trophic cascade in lakes. Cambridge University Press, Cambridge.

Chavez, F. P., J. Ryan, S. E. Lluch-Cota, and M. Niquen. 2003. From anchovies to sardines and back: Multidecadal change in the Pacific Ocean. *Science* **299**: 217–221.

Choi, S. J., K. T. Frank, W. C. Legget, and K. Drinkwater. 2004. Transition to an alternate state in a continental shelf ecosystem. *Canadian Journal of fisheries and Aquatic Sciences* **61**: 505–510.

Ciannelli, L., K. S. Chan, K. M. Bailey, and N. C. Stenseth. 2004. Non-additive effects of the environment on the survival of a large marine fish population. *Ecology* **85**(12).

Coale, K. H. et al. 1996. A massive phytoplankton bloom induced by an ecosystem-scale iron fertilization experiment in the Equatorial Pacific Ocean. *Nature* **383**: 495–501.

Cohen, J. E., F. Briand, and C. M. Newman. 1990. *Community Food Webs*. Springer-Verlag, 308 pp.

Compagno, L. 1984. Sharks of the world. *FAO Species Catalogue* **4**: 655.

Cury P., A. Bakun, R. J. M. Crawford, A. Jarre, R. A. Quinones, L. J. Shannon, and H. M. Verheye. 2000. Small pelagics in upwelling systems: patterns of interaction and structural changes in "wasp-waist" ecosystems. *ICES Journal of Marine Sciences* **57**: 603–618.

Cushing, D. H. 1982. *Climate and Fisheries*. Academic Press, London.

Cushing, D. H. 1990. Plankton production and year-class strength in fish populations: An update of the match/mismatch hypothesis. *Advances in Marine Biology* **26**: 249–294.

Dalpadado, P., and H. R. Skjoldal. 1996. Abundance, maturity and growth of krill species *Thysanoessa inermis* and *T. longicaudata* in the Barents Sea. *Marine Ecology Progress Series* **144**: 175–183.

Dalpadado, P., N. Borkner, B. Bogstad, and S. Mehl. 2001. Distribution of *Themisto* spp (Amphipoda) in the Barents Sea and predator–prey interactions. *ICES Journal of Marine Sciences* **58**: 876–895.

Dalpadado, P., B. Bogstad, H. Gjøsæter, S. Mehl, and H. R. Skjoldal. 2002. Zooplankton–fish interactions in the Barents Sea. In: K. Sherman and H. R. Skjoldal (eds), *Large Marine Ecosystems of The North Atlantic*. Elsevier, Amsterdam, pp. 449 + xiv.

Dickson, R. R., T. J. Osborn, J. W. Hurrell, J. Meincke, J. Blindheim, B. Ådlandsvik, T. Vinje, G. Alekseev, and W. Maslowski. 2000. The Arctic Ocean response to the North Atlantic Oscillation. *Journal of Climate* **13**: 2671–2696.

Dolgov, A. V. 2002. The role of capelin (*Mallotus villosus*) in the food web of the Barents Sea. *ICES Journal of Marine Sciences* **59**: 1034–1045

Dragesund, O., J. Hamre, and O. Ulltang. 1980. Biology and population dynamics of the Norwegian spring-spawning herring. *Rapports et Procès-verbaux des Réunions Conseil international pour l'Exploration de la Mer* **177**: 43–71.

Dragoo, D. E., G. V. Byrd, and D. B. Irons. 2000. Breeding status and population trends of seabirds in Alaska in 1999. U.S. Fish and Wildlife Service Report AMNWR 2000/02.

Ellertsen, B., P. Fossum, P. Solemdal, and S. Sundby. 1989. Relation between temperature and survival of eggs and first-feeding larvae of northeast Arctic cod (*Gadus morhua* L.). *Rapports et Procés-Verbaux des Réunions Conseil international pour l'Exploration de la Mer* **191**: 209–219.

Elton, C. 1958. *The Ecology of Invasions by Animals and Plants*. Methuen and Co., London.

Estes, J. A., M. T. Tinker, T. M. Williams, and D. F. Doak. 1998. Killer whale predation on sea otters linking oceanic and nearshore ecosystems. *Science* **282**: 473–476.

Food and Agricultural Organization (FAO). 2002. The state of world fisheries and aquaculture. Available at: www.fao.org.

Fossum, P. 1996. A study of first-feeding herring (*Clupea harengus* L) larvae during the period 1985–1993. *ICES Journal of Marine Science* **53**: 51–59.

Francis, R. C., S. R. Hare, A. B. Hollowed, and W. S. Wooster. 1998. Effects of interdecadal climate variability on the oceanic ecosystems of the NE Pacific. *Fisheries Oceanography* **7**: 1–21.

Furevik, T. 2001. Annual and interannual variability of Atlantic Water temperatures in the Norwegian and Barents Seas: 1980–1996. *Deep Sea Research I* **48**: 383–404.

Gillett, N. P., H. F. Graf, and T. J. Osborn. 2003. Climate change and the North Atlantic Oscillation. In: J. W. Hurrell, Y. Kushnir, G. Ottersen, and M. Visbeck (eds), *The North Atlantic Oscillation: Climatic Significance and Environmental Impact*. American Geophysical Union, Washington DC, pp. 193–209.

Gjøsæter, H., and B. Bogstad. 1998. Effects of the presence of herring (*Clupea harengus*) on the stock-recruitment relationship of Barents Sea capelin (*Mallotus villosus*). *Fisheries Research* **38**: 57–71.

Gjøsæter, H., P. Dalpadado, and A. Hassell. 2002. Growth of Barents Sea capelin (*Mallotus villosus*) in relation to zooplankton abundance. *ICES Journal of Marine Sciences* **59**: 959–967.

Grandperrin R., 1975. Structures trophiques aboutissant aux thons de longues lignes dans le Pacifique sud-ouest tropical. Thèse de Doctorat, Université de Marseille. 296 pp.

Gueredrat, J. A. 1971. Evolution d'une population de copepodes dans le systeme des courants equatoriaux de l'ocean Pacifique. Zoogeographie, ecologie et diversite specifique. *Marine Biology* **9**: 300–314.

Hampton, J., and D. A. Fournier. 2001. A spatially dis-aggregated, length-based, age-structured population model of yellowfin tuna (*Thunnus albacares*) in the western and central Pacific Ocean. *Marine and Freshwater Research* **52**: 937–963.

Hamre, J. 1994. Biodiversity and exploitation of the main fish stocks in the Norwegian–Barents Sea ecosystem. *Biodiversity and Conservation* **3**: 473–492

Hamre, J. 2000. Effects of Climate and Stocks Interactions on the Yield of North-east Arctic Cod. Results from Multi-species Model Run. Report No. ICES CM 2000/V:04.

Hamre, J. 2003. Capelin and herring as key species for the yield of north-east arctic cod. Results from multi-species model runs. *Scientia Marina* **67**: 315–323.

Hassel, A., H. R. Skjoldal, H. Gjøsæter, H. Loeng, and L. Omli. 1991. Impact of grazing from capelin (*Mallotus villosus*) on zooplankton—a case study in the Northern Barents Sea in August 1985. *Polar Research* **10**: 371–388.

Hasting, A. 1988. Food web theory and stability. *Ecology* **69(6)**: 1665–1668.

Haug, T., and K. T. Nilssen. 1995. Ecological mplications of harp seal *Phoca groenlandica* invasions in northern Norway. In: A. S. Blix, L. Walløe, and Ø. Ulltang (eds), *Whales, Seals, Fish and Man. Developments in Marine Biology* 4. Elsevier, Amsterdam, pp. 545–556.

Hjermann, D. Ø., N. C. Stenseth, and G. Ottersen. 2004. Competition among fisherman and fish causes the collapse of the Barents Sea capelin. *Proceedings of the National Academy of Sciences USA* **101(32)**: 11679–11684.

Hoerling, M. P., J. W. Hurrell, and T. Y. Xu. 2001. Tropical origins for recent North Atlantic climate change. *Science* **292**: 90–92.

Hollowed, A. B., S. R. Hare, and W. S. Wooster. 2001. Pacific Basin climate variability and patterns of Northeast Pacific marine fish production. *Progress in Oceanography* **49**: 257–282.

Hunt, G. L., P. Stabeno, G. Walters, E. Sinclair, R. D. Brodeur., Napp, J. M., and N. A. Bond, 2002. Climate change and control of the southeastern Bering Sea pelagic ecosystem. *Deep-Sea Research II* **49**: 5821–5853.

Hurrell, J. W. 1995. Decadal trends in the North Atlantic Oscillation: regional temperatures and precipitation. *Science* **169**: 676–679.

Hurrell, J. W., Y. Kushnir, G. Ottersen, and M. Visbeck. 2003. An overview of the North Atlantic Oscillation. In: J. W. Hurrell, Y. Kushnir, G. Ottersen, and M. Visbeck (eds), *The North Atlantic Oscillation: Climatic Significance and Environmental Impact*. Geophysical Monograph Series. American Geophysical Union, pp. 1–35. Washington DC,

Huse, G., and R. Toresen. 1996. A comparative study of the feeding habits of herring (*Clupea harengus*, Clupeidae, L.) and capelin (*Mallotus villosus*, Osmeridae, Müller) in the Barents Sea. *Sarsia* **81**: 143–153.

Huse, G., and R. Toresen. 2000. Juvenile herring prey on Barents Sea capelin larvae. *Sarsia* **85**: 385–391.

Ingvaldsen, R., H. Loeng, G. Ottersen, and B. Ådlandsvik. 2003. Climate variability in the Barents Sea during the 20th century with focus on the 1990s. *ICES Marine Sciences Symposium* **219**: 160–168.

Jackson, J. B. C., M. X. Kirby, W. H. Berger, K. A. Bjorndal, L. W. Botsford, B. J. Bourque, R. H. Bradbury, R. Cooke, J. Erlandson, J. A. Estes, T. P. Hughes, S. Kidwell, C. B. Lange, H. S. Lenihan, J. M. Pandolfi, C. H. Peterson,

R. S. Steneck, M. J. Tegner, and R. R. Warner. 2001. Historical overfishing and the recent collapse of coastal ecosystems. *Science* **293**: 629–638.

Jorgensen, T. 1990. Long-term changes in age at sexual maturity of Northeast Arctic cod (*Gadus morhua* L). *Journal Du Conseil* **46**: 235–248.

Kendall, A. W., J. D. Schumacher, and S. Kim. 1996. Walleye pollock recruitment in Shelikof Strait: applied fisheries oceanography. *Fisheries Oceanography* **5**: 4–18.

Krasnov, Y. V., and R. T. Barrett. 1995. Large-scale interactions between seabirds, their prey and man in the southern Barents Sea. In: H. R. Skjoldal, C. C. E. Hopkins, K. E. Erikstad, and H. P. Leinaas (eds), *Ecology of Fjords and Coastal Waters*. Elsevier Science, Amsterdam, 623, pp. 443–456.

Krause, A. E., K. A. Frank, D. M. Mason, R. E. Ulanowicz, and W. W. Taylor. 2003. Compartments revealed in food-web structure. *Nature* **426**: 282–285.

Landry, M. R., and D. L. Kirchman. 2002. Microbial community structure and variability in the tropical Pacific. *Deep-Sea Research II* **49**: 2669–2693.

Landry, M. R., J. Constantinou, and J. Kirshtein. 1995. Microzooplankton grazing in the central equatorial Pacific during February and August, 1992. *Deep-Sea Research* **42**: 657–671.

Law, R. 2000. Fishing, selection, and phenotypic evolution. *ICES Journal of Marine Sciences* **57**: 659–668.

Le Borgne, R., and M. Rodier. 1997. Net zooplankton and the biological pump: a comparison between the oligotrophic and mesotrophic equatorial Pacific. *Deep-Sea Research* **44**: 2003–2024.

Legand, M., P. Rourret, P. Fourmanoir, R. Grandperrin, J. A. Gueredrat, A. Mitchel, P. Rancurel, R. Repelin, and C. Roger. 1972. Relations trophiques et distributions verticales en milieu pélagique dans l'ocean Pacifique intertropical, centre ORSTOM de Noumea, New Caledonia.

Lehodey, P. 2001. The pelagic ecosystem of the tropical Pacific Ocean: dynamic spatial modelling and biological consequences of ENSO. *Progress in Oceanography* **49**: 439–468.

Lehodey, P., M. Bertignac, J. Hampton, A. Lewis, and J. Picaut. 1997. El Nino Southern Oscillation and tuna in the western Pacific. *Nature* **389**: 715–718.

Lehodey, P., F. Chai, and J. Hampton. 2003. Modelling climate-related variability of tuna populations from a coupled ocean–biogeochemical–populations dynamics model. *Fisheries Oceanography* **12**: 483–494.

Lima, M., N. C. Stenseth, and F. M. Jaksic. 2002. Food web structure and climate effects on the dynamics of small mammals and owls in semi-arid Chile. *Ecology Letters* **5**: 273–284.

Loeng, H. 1989. The influence of temperature on some fish population parameters in the Barents Sea. *Journal of Northwest Atlantic Fisheries Science* **9**: 103–113.

Loeng, H. 1991. Features of the physical oceanographic conditions of the Barents Sea. *Polar Research* **10**: 5–18.

Loeng, H., J. Blindheim B., Ådlandsvik and G. Ottersen. 1992. Climatic variability in the Norwegian and Barents Seas. *ICES Marine Sciience Symposium* **195**: 52–61.

MacArthur, R. 1955. Fluctuations of animal populations and a measure of community stability. *Ecology* **36**: 533–536.

Mackas, D. L., R. E. Thomson, and M. Galbraith. 2001. Changes in the zooplankton community of the British Columbia continental margin, 1985–1999, and their covariation with oceanographic conditions. *Canadian Journal of Fisheries and Aquatic Sciences* **58**: 685–702.

Mann, K. H. 1993. Physical Oceanography, food-chains, and fish stocks—a review. *ICES Journal of Marine Science* **50**: 105–119.

Mantua, N. J., and S. R. Hare. 2002. The Pacific decadal oscillation. *Journal of Oceanography* **58**: 35–44.

Mantua, N. J., S. R. Hare, Y. Zhang, J. M. Wallace, and R. C. Francis. 1997. A Pacific interdecadal climate oscillation with impacts on salmon production. *Bulletin of the American Meteorological Society* **78**: 1069–1079.

Mauchline, J. 1998. The biology of calanoid copepods. *Advances in Marine Biology*. Academic Press, London:.

May, R. M. 1973. *Stability and Complexity in Model Ecosystems*. Princeton University Press, Princeton, NJ.

Maynard Smith, J. 1974. *Models in Ecology*. Cambridge University Press.

McFarlane, G. A., J. R. King, and R. J. Beamish. 2000. Have there been recent changes in climate? Ask the fish. *Progress in Oceanography* **47**: 147–169.

McGowan, J. A., D. R. Cayan, and L. M. Dorman. 1998. Climate-Ocean variability and ecosystem response in the Northeast Pacific. *Science* **281**: 210–217.

McPhaden, M. J., and J. Picaut. 1990. El Niño-Southern Oscillation index displacements of the Western Equatorial Pacific Warm Pool. *Science* **50**: 1385–1388.

Mehl, S. 1991. The Northeast Arctic cod stock's place in the Barents Sea ecosystem in the 1980s: an overview. *Polar Research* **10**: 525–534.

Mehlum, F., and Gabrielsen, G. W. 1995. Energy expenditure and food consumption by seabird populations in the Barents Sea region. In: H. R. Skjoldal, C. Hopkins, K. E. Erikstad, and H. P. Leinaas (eds), *Ecology of Fjords and Coastal Waters*. Elsevier Science, Amsterdam, pp. 457–470.

Melle, W., and H. R. Skjoldal. 1998. Reproduction, life cycles and distributions of *Calanus finmarchicus* and *C. hyperboreus* in relation to environmental conditions in the Barents Sea. Dr. Scient Department of Fisheries and Marine Biology, University of Bergen, Bergen, Norway.

Merati, N., and R. D. Brodeur. 1996. Feeding Habits and Daily Ration of Juvenile Walleye Pollock, *Theragra chalcogramma*, in the Western Gulf of Alaska. NOAA Technical Report NMFS **126**: 65–79.

Mueter, F. J., and B. L. Norcross. 2000. Changes in species composition of the demersal fish community in near-shore waters of Kodiak Island, Alaska. *Canadian Journal of Fisheries and Aquatic Sciences* **57**: 1169–1180.

Murtugudde, R. G., S. R. Signorini, J. R. Christian, A. J. Busalacchi, C. R. McClain, and J. Picaut 1999. Ocean color variability of the tropical Indo-Pacific Basin observed by SeaWiFS during 1977–1998. *Journal of Geophysical Research* **104**: 18351–18366.

Nesterova, V. N. 1990. *Plankton Biomass Along the Drift Route of Cod Larve* (in Russian). PINRO Press, Murmansk, 64 pp.

National Research Council. 2003. *Decline of Steller Sea lion in Alaskan Waters*. National Academy Press, Washington DC.

Nilssen, K. T., O. P. Pedersen, L. P. Folkow, and T. Haug. 2000. Food consumption estimates of Barents Sea harp seals. *NAMMCO Scientific Publications* **2**: 9–28.

Odum, E. P. 1963. *Ecology*. Holt, Rinehart and Winston, New York.

Odum, E. P. 1985. Trends expected in stressed ecosystems. *BioScience* **35**: 419–422.

Olsen, A., L. G. Anderson, and T. Johannessen. 2002. The impact of climate variations on fluxes of oxygen in the Barents Sea. *Continental Shelf Research* **22**: 1117–1128.

Orensanz, J. M., J. Armstrong, D. Armstrong, and R. Hilborn. 1998. Crustaceans resources are vulnerable to serial depletion—the multifaceted decline of crab and shrimps fisheries in the Greater Gulf of Alaska. *Reviews in Fish Biology and Fisheries* **8**: 117–176.

Ottersen, G., and H. Loeng. 2000. Covariability in early growth and year-class strength of Barents Sea cod, haddock, and herring: the environmental link. *ICES Journal of Marine Sciences* **57**: 339–348.

Ottersen, G., and N. C. Stenseth. 2001. Atlantic climate governs oceanographic and ecological variability in the Barents Sea. *Limnology and Oceanography* **46**: 1774–1780.

Paine, R. T. 1980. Food webs—linkage, interaction strength and community infrastructure—the 3rd Tansley Lecture. *Journal of Animal Ecology* **49**: 667–685.

Panasenko, L. D. 1984. Feeding of the Barents Sea capelin. *Journal of Marine Sucuess, ICES CM Documents* **1984/H**: 6, 16.

Parsons, L. S., and W. H. Lear, 2001. Climate variability and marine ecosystem impacts: a North Atlantic perspective. *Progress in Oceanography* **49**: 167–188.

Parsons, T. R. 1986. Ecological relations. In: *The Gulf of Alaska: Physical Environment and Biological Resources*. D. W. Hood and S. T. Zimmerman (eds), pp. 561–570. Minerals Management Service, Springfield, VA.

Parsons, T. R. 1996. The impact of industrial fisheries on the trophic structure of marine ecosystems. In: G. A. Polis, and K. O. Winemiller (eds), *Food Webs*. Chapman and Hall, New York, pp. 352–357.

Pauly D., V. Christensen, J. Dalsgaard, R. Froese, and F. Torres. 1998. Fishing down marine food webs. *Science* **279**: 860–863.

Philander, S. G. H. 1990. *El Niño, La Niña, and the Southern Oscillation.*, Vol., Academic Press, New York.

Piatt, J. P., and P. Anderson. 1996. Response of common murres to the Exxon Valdez oil spill and long-term changes in the Gulf of Alaska marine ecosystem. In: S. D. Rice, R. B. Spies, D. A. Wolfe, and B. A. Wright (eds), *Proceedings of Exxon Valdez Oil Spill Symposium*. American Fisheries Society Symposium 18, Bethesda, MD, pp. 720–737.

Pimm, S. L. 1982. *Food Webs*. Chapman and Hall, 219 pp.

Polovina, J. J., G. T. Mitchum, and G. T. Evans. 1995. Decadal and basin-scale variation in mixed layer depth and the impact on biological production in the central and North Pacific, 1960–1988. *Deep Sea Research* **42**: 1201–1716.

Pozo-Vazquez, D., M. J. Esteban-Parra, F. S. Rodrigo, and Y. Castro-Diez. 2000. An analysis of the variability of the North Atlantic Oscillation in the time and the frequency domains. *International Journal of Climatology* **20**: 1675–1692.

Raffaelli, D. 2000. Trends in research on shallow water fiid webs. *Journal of Experimental Biology and Ecology* **250**: 223–232.

Reed, R. K., and J. D. Schumacher. 1987. Physical oceanography. In: D. W. Z. S. Hood, (ed.), *The Gulf of Alaska: Physical Environment and Biological Resources*. US Department of Commerce, NOAA, p. 655.

Rey F., H. R. Skjoldal, and D. Slagstad. 1987. Primary production in relation to seasonal changes in the Barents Sea. In: H. Loeng (ed.), *The Effect of Oceanographic Conditions on Distribution and Population Dynamics of Commercial Fish Stocks in the Barents Sea*. Institute of Marine Research, Bergen, Norway, pp. 29–46.

Rogers, J. C. 1984. The assosiation between the North Atlantic Oscillation and the Southern Oscillation in the northern hemisphere. *Monsoon and Weather Reviews* **112**: 1999–2015.

Roman, M. R., H. G. Dam, R. Le Borgne, and X. Zhang. 2002. Latitudinal comparisons of equatorial Pacific zooplankton. *Deep-Sea Research Part II-Topical Studies in Oceanography* **49**: 2695–2711.

Stock Assessment and Fishery Evaluation (SAFE). 2003. Report for the Ground Fish Resources of the Gulf of Alaska. North Pacific Fishery Management Council, Anchorage, Alaska. Available at: http://afsc.noaa.gov/refm/stocks/assessments.htm.

Sakshaug, E. 1997. Biomass and productivity distributions and their variability in the Barents Sea. *ICES Journal of Marine Sciences* 54: 341–350.

Sakshaug, E., A. Bjørge, B. Gulliksen, H. Loeng and F. Mehlum (eds), 1992. Økosystem Barentshavet. Lillehammer.

Sambrotto, R. N., and C. J. Lorenzen. 1987. Phytoplankton and primary production. In: D. W. Z. S. Hood (ed.), *The Gulf of Alaska: Physical Environment and Biological Resources.* US Department of Commerce, NOAA, p. 655.

Schindler, D. W. 1985. Detecting ecosystem responses to anthropogenic stress. *Canadian Journal of Fisheries and Aquatic Sciences.* 44(Suppl. 1): 6–25.

Sekino, T., and N. Yamamura. 1999. Diel vertical migration of zooplankton: optimum migrating schedule based on energy accumulation. *Evolutionary Ecology* 13: 267–282.

Sinclair, E. H., and T. K. Zeppelin. 2003. Seasonal and spatial differences in diet in the western stock of Steller sea lions (*Eumetopias jubatus*). *Journal of Mammalogy* 83(4): 973–990.

Springer, A. M. 1993. Report of the seabird working group. In: *Is it food?: Addressing Marine Mammal and Seabird Declines* Workshop Summary. University of Alaska, Alaska Sea Grant Report 93–01, Fairbanks, AK, pp. 14–29.

Springer, A. M., J. A. Estes, G. B. van Vliet, T. M. Williams, D. F. Doak, E. M. Danner, K. A. Forney, and B. Pfister 2003. Sequential megafaunal collapse in the North Pacific Ocean: An ongoing legacy of industrial whaling? *Proceedings of the National Academy of Sciences, USA* 100: 12223–12228.

Stabeno, P., N. A. Bond, A. J. Hermann, C. W. Mordy, and J. E. Overland. 2004. Meteorology and oceanography of the northern Gulf of Alaska. *Continetal Shelf Research* 24: 859–897.

Steele, J. H. 1974. *The Structure of Marine Ecosystems.* Harvard University Press, Cambridge, MA, 128 pp.

Steele, J. H., and M. Schumacher. 2000. Ecosystem structure before fishing. *Fisheries Research* 44: 201–205.

Stenseth, N. C., W. Falck, O. N. Bjornstad, and C. J. Krebs. 1997. Population regulation in snowshoe hare and Canadian lynx: Asymmetric food web configurations between hare and lynx. *Proceedings of the National Academy of Sciences, USA* 94: 5147–5152.

Stenseth, N. C., A. Mysterrud, G. Ottersen, J. W. Hurrel, K. S. Chan, and M. Lima. 2002. Ecological effects of climate fluctuations. *Science* 297: 1292–1296.

Stenseth, N. C., G. Ottersen, J. W. Hurrell, A. Mysterud, M. Lima, K. S. Chan, N. G. Yoccoz, and B. Ådlandsvik 2003. Studying climate effects on ecology through the use of climate indices: the North Atlantic Oscillation, El Niño Southern Oscillation and beyond. *Proceedings of the Royal Society of London Series B* 270: 2087–2096.

Strasburg, D. W. 1958. Distribution, abundance, and habits of pelagic sharks in the central Pacific Ocean. *Fisheries Bulletin* 58: 335–361.

Sturdevant, M. V. 1999. Forage Fish Diet Overlap, 1994–1996. EXXON VALDEZ Oil Spill Restoration Project. Final Report (Restoration Project 97163C).

Toresen, R., and O. J. Østvedt. 2000. Variation in the abundance of Norwegian spring-spawning herring (*Clupea harengus*, Clupeidae) throughout the 20th century and the influence of climatic fluctuations. *Fish and Fisheries* 1: 231–251.

Trenberth, K. E., and J. W. Hurrell, 1994. Decadal Atmosphere-Ocean variations in the Pacific. *Climate Dynamics* 9: 303–319.

Tricas, T. C. 1979. Relationships of the blue shark, *Prionace glauca*, and its prey species near Santa Catalina Island, California. *Fisheries Bulletin* 77: 175–182.

Trites, A. W., and L. P. Donnelly, 2003. The decline of Steller sea lions in Alaska: A review of the nutritional stress hypothesis. *Mammal Review* 33: 3–28.

Ushakov, N. G., and D. V. Prozorkevich. 2002. The Barents Sea capelin—a review of trophic interrelations and fisheries. *ICES Journal of Marine Sciences* 59: 1046–1052.

Verity, P. G., V. Smetacek, and T. J. Smayda. 2002. Status, trends and the future of the marine pelagic ecosystem. *Environmental Conservation* 29: 207–237.

Vinogradov, M. E. 1981. Ecosystems of equatorial upwellings. In: A. R. Longhurst (ed.), *Analysis of Marine Ecosystems.* Academic Press, pp. 69–93.

Watters, G. M., R. J. Olson, R. C. Francis, P. C. Fiedler, J. J. Polovina, S. B. Reilly, K. Y. Aydin, C. H. Boggs, T. E. Essington, C. J. Walters, and J. F. Kitchell. 2003. Physical forcing and the dynamics of the pelagic ecosystem in the eastern tropical Pacific: simulations with ENSO-scale and global-warming climate drivers. *Canadian Journal of Fisheries and Aquatic Sciences* 60: 1161–1175.

Worm, B., and R. A. Myers. 2003. Meta-analysis of cod-shrimp interactions reveals top-down control in oceanic food webs. *Ecology* 84: 162–173.

Yang, M. S., and M. W. Nelson. 2000. Food Habits of the Commercially Important Groundfishes in the Gulf of Alaska in 1990, 1993, and 1996. US Department of Commerce, NOAA Technical Memornadum, NMFS-AFSC-112, 174 pp.

Yang, M. 2004. Diet changes of Pacific cod (Gadus macrocephalus) in Pavlof Bay associated with climate changes in the Gulf of Alaska between 1980 and 1995. *Fishery Bulletin*.

Zenkevitch, L. 1963. *Biology of the Seas of the USSR*. George Allen and Unwin, London.

Zhang, Y., J. M. Wallace, and D. S. Battisti. 1997. ENSO-like interdecadal variability. *Journal of Climate* **10**: 1004–1020.

Ådlandsvik, B., and H. Loeng. 1991. A study of the climatic system in the Barents Sea. *Polar Research* **10**: 45–49.

Chapter 13

Babcock, R. C., S. Kelly, N. T. Shears, J. W. Walker, and T. J. Willis. 1999. Changes in community structure in temperate marine reserves. *Marine Ecology-Progress Series* **189**: 125–134.

Ballesteros, E. 1991. Els vegetals i la zonació litoral: espècies, comunitats i factors que influeixen en la seva distribució. Arxius de la Secció de Ciències C. I., Institut d'Estudis Catalans, Barcelona.

Bascompte, J., C. Melian, and E. Sala. The structure of species interaction strength and the overfishing of marine food webs (submitted).

Berlow, E. L. 1999. Strong effects of weak interactions in ecological communities. *Nature* **6725**: 330–334.

Berlow, E. L., S. A. Navarrete, C. J. Briggs, M. E. Power, and B. A. Menge. 1999. Quantifying variation in the strengths of species interactions. *Ecology* **80**: 2206–2224.

Boudouresque, C. F., and M. Verlaque. 2002. Biological pollution in the Mediterranean Sea: invasive versus introduced macrophytes. *Marine Pollution Bulletin* **44**: 32–38.

Carpenter, S. R., and J. F. Kitchell. 1974. The trophic cascade in lakes. Cambridge University Press, Cambridge.

Dayton, P. K. 1971. Competition, disturbance, and community organization: the provision and subsequent utilization of space in a rocky intertidal community. *Ecological Monographs* **41**: 351–389.

Dayton, P. K., V. Currie, T. Gerrodette, B. D. Keller, R. Rosenthal, and D. Ven Tresca. 1984. Patch dynamics and stability of some California kelp communities. *Ecological Monographs* **54**: 253–289.

Dayton, P. K., M. J. Tegner, P. B. Edwards, and K. L. Riser. 1998. Sliding baselines, ghosts, and reduced expectations in kelp forest communities. *Ecological Applications* **8**: 309–322.

Dayton, P. K., M. J. Tegner, P. E. Parnell, and P. B. Edwards. 1992. Temporal and spatial patterns of disturbance and recovery in a kelp forest community. *Ecological Monographs* **62**: 421–445.

Dulvy, N. K., R. P. Freckleton, and N. V. C. Polunin. 2004. Coral reef cascades and the indirect effects of predator removal by exploitation. *Ecology Letters* **7**: 410–416.

Estes, J. A., and D. O. Duggins. 1995. Sea otters and kelp forests in Alaska—Generality and variation in a community ecological paradigm. *Ecological Monographs* **65**: 75–100.

Estes, J. A., M. T. Tinker, T. M. Williams, and D. F. Doak. 1998. Killer whale predation on sea otters linking oceanic and nearshore ecosystems. *Science* **282**: 473–476.

Fagan, W. F. 1997. Omnivory as a stabilizing feature of natural communities. *American Naturalist* **150**: 554–567.

Friedlander, A. M., and E. E. DeMartini. 2002. Contrasts in density, size, and biomass of reef fishes between the northwestern and the main Hawaiian islands: the effects of fishing down apex predators. *Marine Ecology Progress Series* **230**: 253–264.

García-Rubies, A. 1996. Estudi ecològic de les poblacions de peixos sobre substrat rocós a la Mediterrània Occidental: efecte de la fondària, el substrat, l'estacionalitat i la protecció. PhD Thesis, Universitat de Barcelona, Spain.

Garrabou, J., E. Ballesteros, and M. Zabala. 2002. Structure and dynamics of north-western Mediterranean rocky benthic communities along a depth gradient. *Estuarine Coastal and Shelf Science* **55**: 493–508.

Goreau, T. 1959. The ecology of Jamaican coral reefs. I. Species composition and zonation. *Ecology* **40**: 67–90.

Graham, M. H. 2000. *Planktonic Patterns and Processes in the Giant Kelp Macrocystis Pyrifera*. Ph D Thesis, University of California, San Diego

Graham, M. H. 2004. Effects of local deforestation on the diversity and structure of Southern California giant kelp forest food webs. *Ecosystems* **7**: 341–357.

Grigg, R. W. 1983. Community structure succession and development of coral reefs in Hawaii, USA. *Marine Ecology Progress Series* **11**: 1–14.

Grigg, R. W., and J. E. Maragos. 1974. Recolonization of hermatypic corals on submerged lava flows in Hawaii. *Ecology* **55**: 387–395.

Grosholz, E. D., G. M. Ruiz, C. A. Dean, K. A. Shirley, J. L. Maron, and P. G. Connors. 2000. The impacts of a nonindigenous marine predator in a California bay. *Ecology* **81**: 1206–1224.

Hairston, N. G., F. E. Smith, and L. B. Slobodkin. 1960. Community structure, population control, and competition. *American Naturalist* **94**: 421–425.

Hereu, B. 2004. The role of trophic interactions between fishes, sea urchins and algae in the northwestern

Mediterranean rocky infralittoral. PhD thesis, Universitat de Barcelona.

Hughes, T. P. 1994. Catastrophes, phase shifts, and large-scale degradation of a caribbean coral reef. *Science* **265**: 1547–1551.

Jackson, J. B. C., and E. Sala. 2001. Unnatural oceans. *Scientia Marina* **65**: 273–281.

Jackson J. B. C., M. X. Kirby, W. H. Berger, K. A. Bjorndal, L. W. Botsford, B. J. Bourque, R. H. Bradbury, R. Cooke, J. Erlandson, J. A. Estes, T. P. Hughes, S. Kidwell, C. B. Lange, H. S. Lenihan, J. M. Pandolfi, C. H. Peterson, R. S. Steneck, M. J. Tegner, and R. R. Warner. 2001. Historical overfishing and the recent collapse of coastal ecosystems. *Science* **293**: 629–638.

Jones, G. P., M. I. McCormick, M. Srinivasan, and J. V. Eagle. 2004. Coral decline threatens fish biodiversity in marine reserves. *Proceedings of the National Academy of Sciences, USA* **101**: 8251–8253.

Knowlton, N. 2001. The future of coral reefs. *Proceedings of the National Academy of Sciences, USA* **98**: 5419–5425.

Knowlton, N., and F. Rohwer 2003. Multispecies microbial mutualisms on coral reefs: The host as a habitat. *American Naturalist* **162**: S51–S62.

Kokkoris, G. D., A. Y. Troumbis, and J. H. Lawton. 1999. Patterns of species interaction strength in assembled theoretical competition communities. *Ecology Letters* **2**: 70–74.

Link, J. 2002. Does food web theory work for marine ecosystems? *Marine Ecology Progress Series* **230**: 1–9.

Margalef, R. 1991. Teoria de los sistemas ecológicos. Publicacions de la Universitat de Barcelona, Barcelona.

Margalef, R. 1997. *Our Biosphere*. Ecology Institute, Oldendorf/Luhe.

May, R. M. 1973. *Stability and Complexity in Model Ecosystems*. Princeton University Press, Princeton, NJ.

May, R. M. 1974. Biological populations with non-overlapping generations: stable points, stable cycles, and chaos. *Science* **186**: 645–647.

McCann, K., and A. Hastings. 1997. Re-evaluating the omnivory–stability relationship in food webs. *Proceedings of the Royal Society of London Series B: Biological Sciences* **264**: 1249–1254.

McCann, K., A. Hastings, and G. R. Huxel. 1998. Weak trophic interactions and the balance of nature. *Nature* **395**: 794–798.

McClanahan, T. R. 1995. A coral reef ecosystem-fisheries model—impacts of fishing intensity and catch selection on reef structure and processes. *Ecological Modelling* **80**: 1–19.

Meinesz, A. 2002. Killer Algae. University of Chicago Press, Chicago, IL.

Micheli, F., B. S. Halpern, L. W. Botsford, and R. R. Warner. Trajectories and correlates of community change in no-take marine reserves. *Ecological Applications* (in press).

Myers, R. A., and B. Worm. 2003. Rapid worldwide depletion of predatory fish communities. *Nature* **423**: 280–283.

Navarrete, S. A., and B. A. Menge. 1996. Keystone predation and interaction strength—interactive effects of predators on their main prey. *Ecological Monographs* **66**: 409–429.

Odum, E. 1969. The strategy of ecosystem development. *Science* **164**: 262–270.

Paine, R. T. 1966. Food web complexity and species diversity. *American Naturalist* **100**: 65–75.

Paine, R. T. 1971. A short term experimental investigation of resource partitioning in a New Zealand rocky intertidal habitat. *Ecology* **52**: 1096–1106.

Paine, R. T. 1980. Food webs: Linkage, interaction strength and community infrastructure. *Journal of Animal Ecology* **49**: 669–685.

Paine, R. T. 1992. Food-web analysis through field measurement of per capita interaction strength. *Nature* **355**: 73–75.

Palumbi, S. R. 2002. The evolution explosion: how humans cause rapid evolutionary change. W.W. Norton & Company.

Pandolfi, J. M., R. H. Bradbury, E. Sala, T. P. Hughes, K. A. Bjorndal, R. G. Cooke, D. McArdle, L. McClenachan, M. J. H. Newman, G. Paredes, R. R. Warner, and J. B. C. Jackson 2003. Global trajectories of the long-term decline of coral reef ecosystems. *Science* **301**: 955–958.

Pauly, D., V. Christensen, J. Dalsgaard, R. Froese, and F. Torres. 1998. Fishing down marine food webs. *Science* **279**: 860–863.

Pinnegar, J. K., N. V. C. Polunin, P. Francour, F. Badalamenti, R. Chemello M. L. Harmelin-Vivien, B. Hereu, M. Milazzo, M. Zabala, G. D'Anna, and C. Pipitone. 2000. Trophic cascades in benthic marine ecosystems: lessons for fisheries and protected-area management. *Environmental Conservation* **27**: 179–200.

Rosenzweig, M. L. 1971. Paradox of enrichment: destabilization of exploitation ecosystems in ecological time. *Science* 171: 385–387.

Russ, G. R., and A. C. Alcala. 1996. Marine reserves—rates and patterns of recovery and decline of large predatory fish. *Ecological Applications* **6**: 947–961.

Sala, E. 1997. The role of fishes in the organization of a Mediterranean sublittoral community. 2. Epifaunal communities. *Journal of Experimental Marine Biology and Ecology* **212**: 45–60.

Sala, E. 2004. The past and present topology and structure of Mediterranean subtidal rocky shore food webs. *Ecosystems* **7**: 333–340.

Sala, E., and C. F. Boudouresque. 1997. The role of fishes in the organization of a Mediterranean sublittoral community. 1. Algal communities. *Journal of Experimental Marine Biology and Ecology* **212**: 25–44.

Sala, E., C. F. Boudouresque, and M. HarmelinVivien. 1998. Fishing, trophic cascades, and the structure of algal assemblages: evaluation of an old but untested paradigm. *Oikos* **82**: 425–439.

Sala, E., and M. H. Graham. 2002. Community-wide distribution of predator–prey interaction strength in kelp forests. *Proceedings of the National Academy of Sciences, USA* **99**: 3678–3683.

Sala, E., and M. Zabala, 1996. Fish predation and the structure of the sea urchin *Paracentrotus lividus* populations in the NW Mediterranean. *Marine Ecology Progress Series* **140**: 71–81.

Sala, E., O. Aburto-Oropeza, M. Reza, G. Paredes, and L. G. Lopez-Lemus. 2004. Fishing down coastal food webs in the Gulf of California. *Fisheries* **29**: 19–25.

Shears, N. T., and R. C. Babcock. 2003. Continuing trophic cascade effects after 25 years of no-take marine reserve protection. *Marine Ecology Progress Series* **246**: 1–16.

Shiganova, T. A., and Y. V. Bulgakova. 2000. Effects of gelatinous plankton on Black Sea and Sea of Azov fish and their food resources. *ICES Journal of Marine Science* **57**: 641–648.

Springer, A. M., J. A. Estes, G. B. van Vliet, T. M. Williams, D. F. Doak, E. M. Danner, K. A. Forney, and B. Pfister. 2003. Sequential megafaunal collapse in the North Pacific Ocean: an ongoing legacy of industrial whaling? *Proceedings of the National Academy of Sciences, USA* **100**: 12223–12228.

Steneck, R. S. 1998. Human influences on coastal ecosystems: does overfishing create trophic cascades? *Trends in Ecology and Evolution* **13**: 429–430.

Steneck, R. S., and J. Carlton. 2001. Human alterations of marine communities: Students beware! In: M. Bertness, S. Gaines, and M. E. Hay (eds), *Marine Community Ecology*. Sinauer, Sunderland, MA, p 445–468.

Sugihara, G. 1982. Niche hierarchy: Structure assembly and organization in natural communities. PhD thesis, Princeton University, Princeton NS.

Sugihara G., L. F. Bersier, T. R. E. Southwood, S. L. Pimm, and R. M. May. 2003. Predicted correspondence between species abundances and dendrograms of niche similarities. *Proceedings of the National Academy of Sciences, USA* **100**: 5246–5251.

Vadas, R. L., and R. S. Steneck. 1995. Overfishing and inferences in kelp-sea urchin interactions.

In: H. R. Skjoldal, C. Hopkins, K. E. Eirkstad, and H. P. Leinaas (eds), *Ecology of Fjords and Coastal Waters*, Elsevier Science, London, pp. 509–524.

Verlaque, M. 1987. Relations entre *Paracentrotus lividus* (Lamarck) et le phytobenthos de Mediterranee occidentale. In: C. F. Boudouresque (ed.), *Colloque International sur* Paracentrotus lividus *et les oursins comestibles*. GIS Posidonie, Marseille, France, pp. 5–36.

Witman, J. D., and K. P. Sebens. 1992. Regional variation in fish predation intensity: a historical perspective in the Gulf of Maine. *Oecologia* **90**: 305–315.

Wootton, J. T. 1997. Estimates and tests of per capita interaction strength: diet, abundance, and impact of intertidally foraging birds. *Ecological Monographs* **67**: 45–64.

Yodzis, P. 2001. Must top predators be culled for the sake of fisheries? *Trends in Ecology and Evolution* **16**: 78–84.

Zabala, M., and E. Ballesteros. 1989. Surface-dependent strategies and energy flux in benthic marine communities or, why corals do not exist in the Mediterranean. *Scientia Marina* **53**: 3–17.

Chapter 14

Abrams, P. A. 1993. Effect of increased productivity on the abundance of trophic levels. *American Naturalist* **141**: 351–371.

Abrams, P. A. 1995. Monotonic or unimodal diversity–productivity gradients: what does competition theory predict? *Ecology* **76**: 2019–2027.

Abrams, P. A. 2001. The effect of density-independent mortality on the coexistence of exploitative competitors for renewing resources. *American Naturalist* **158**: 459–470.

Adrian, R., S. A. Wickham, and N. M. Butler. 2001. Trophic interactions between zooplankton and the microbial community in contrasting food webs: the epilimnion and deep chlorophyll maximum of a mesotrophic lake. *Aquatic Microbial Ecology* **24**: 83–97.

Amarasekare, P. 2003. Diversity–stability relationships in multitrophic systems: an empirical exploration. *Journal of Animal Ecology* **72**: 713–724.

Amarasekare, P., and R. M. Nisbet. 2001. Spatial heterogeneity, source–sink dynamics, and the local coexistence of competing species. *American Naturalist* **158**: 572–584.

Anderson, T. R., and D. W. Pond. 2000. Stoichiometric theory extended to micronutrients: Comparison of the roles of essential fatty acids, carbon and nitrogen in the nutrition of marine copepods. *Limnology and Oceanography* **45**: 1162–1167.

Armstrong, R. A., and R. McGehee. 1980. Competitive exclusion. *American Naturalist* **115**: 151–170.

Attayde, J. L., and L. A. Hansson. 1999. Effects of nutrient recycling by zooplankton and fish on phytoplankton communities. *Oecologia* **121**: 47–54.

Bell, T. 2002. The ecological consequences of unpalatable prey: phytoplankton response to nutrient and predator additions. *Oikos* **99**: 59–68.

Berninger, U. G., S. A. Wickham, and B. J. Finlay. 1993. Trophic coupling within the microbial food web: a study with fine temporal resolution in a eutrophic freshwater ecosystem. *Freshwater Biology* **30**: 419–432.

Bohonak, A. J., and D. G. Jenkins. 2003. Ecological and evolutionary significance of dispersal by freshwater invertebrates. *Ecology Letters* **6**: 783–796.

Borer, E. T., K. Anderson, C. A. Blanchette, B. Broitman, S. D. Cooper, and B. S. Halpern. 2002. Topological approaches to food web analyses: a few modifications may improve our insights. *Oikos* **99**: 397–401.

Borer, E. T., E. W. Seabloom, J. B. Shurin, K. E. Anderson, C. A. Blanchette, B. Broitman, S. D. Cooper, B. S. Halpern. *In press*. What determines the strength of a trophic cascade? *Ecology*.

Borrvall, C., B. Ebenman, and T. Jonsson. 2000. Biodiversity lessens the risk of cascading extinction in model food webs. *Ecology Letters* **3**: 131–136.

Brönmark, C., S. D. Rundle, and A. Erlandsson. 1991. Interactions between fresh-water snails and tadpoles—competition and facilitation. *Oecologia* **87**: 8–18.

Brose, U., A. Ostling, K. Harrison, and N. D. Martinez. 2004. Unified spatial scaling of species and their trophic interactions. *Nature* **428**: 167–171.

Cardinale, B. J., M. A. Palmer, and S. L. Collins. 2002. Species diversity enhances ecosystem functioning through interspecific facilitation. *Nature* **415**: 426–429.

Carpenter, S. R., J. F. Kitchell, J. R. Hodgson, P. A. Cochran, J. J. Elser, M. M. Elser, D. M. Lodge et al. 1987. Regulation of lake primary productivity by food web structure. *Ecology* **68**: 1863–1876.

Chase, J. M., P. A. Abrams, J. P. Grover, S. Diehl, P. Chesson, R. D. Holt, S. A. Richards et al. 2002. The interaction between predation and competition: a review and synthesis. *Ecology Letters* **5**: 302–315.

Chase, J. M., and M. A. Leibold. 2002. Spatial scale dictates the productivity–biodiversity relationship. *Nature* **416**: 427–430.

Chase, J. M., W. G. Wilson, and S. A. Richards. 2001. Foraging trade-offs and resource patchiness: theory and experiments with a freshwater snail community. *Ecology Letters* **4**: 304–312.

Chesson, P. 2000. Mechanisms of maintenance of species diversity. *Annual Review of Ecological and Systemstic* **31**: 343–366.

Chesson, P., and N. Huntly. 1997. The roles of harsh and fluctuating conditions in the dynamics of ecological communities. *American Naturalist* **150**: 519–553.

Cohen, J. E. 1978. *Food Webs and Niche Space*. Princeton University Press, Princeton, NJ.

Connell, J. H. 1978. Diversity in tropical rain forests and coral reefs. *Science* **199**: 1302–1310.

Cornell, H. V., and J. H. Lawton. 1992. Species interactions, local and regional processes, and limits to the richness of ecological communities: a theoretical perspective. *Journal of Animal Ecology* **61**: 1–12.

DeMott, W. R. 1998. Utilization of a cyanobacterium and a phosphorus-deficient green alga as complementary resources by daphnids. *Ecology* **79**: 2463–2481.

Diehl, S. 1992. Fish predation and benthic community structure: the role of omnivory and habitat complexity. *Ecology* **73**: 1646–1661.

Diehl, S. 1995. Direct and indirect effects of omnivory in a littoral lake community. *Ecology* **76**: 1727–1740.

Diehl, S., and M. Feissel. 2001. Intraguild prey suffer from enrichment of their resources: A microcosm experiment with ciliates. *Ecology* **82**: 2977–2983.

Dodson, S. I., S. E. Arnott, and K. L. Cottingham. 2000. The relationship in lake communities between primary productivity and species richness. *Ecology* **81**: 2662–2679.

Downing, A. L., and M. A. Leibold. 2002. Ecosystem consequences of species richness and composition in pond food webs. *Nature* **416**: 837–841.

Duffy, J. E. 2002. Biodiversity and ecosystem function: the consumer connection. *Oikos* **99**: 201–219.

Duffy, J. E., K. S. Macdonald, J. M. Rhode, and J. D. Parker. 2001. Grazer diversity, functional redundancy, and productivity in seagrass beds: an experimental test. *Ecology* **82**: 2417–2434.

Duffy, J. E., J. P. Richardson, and E. A. Canuel. 2003. Grazer diversity effects on ecosystem functioning in sea grass beds. *Ecology Letters* **6**: 637–645.

Dunne, J. A., R. J. Williams, and N. D. Martinez. 2002*a*. Network structure and biodiversity loss in food webs: robustness increases with connectance. *Ecological Letters* **5**: 558–567.

Dunne, J. A., R. J. Williams, and N. D. Martinez. 2002*b*. Food-web structure and network theory: the role of connectance and size. *Proceeding of the National Academy of Sciences, USA* **99**: 12917–12922.

Eklöv, P., and T. VanKooten. 2001. Facilitation among piscivorous predators: effects of prey habitat use. *Ecology* **82**: 2486–2494.

Finke, D. L. and R. F. Denno. 2004. Predator diversity dampens trophic cascades. *Nature* **429**: 407–410.

Finlay, B. J., S. C. Maberly, and J. I. Cooper. 1997. Microbial diversity and ecosystem function. *Oikos* **80**: 209–213.

Flöder, S., and U. Sommer. 1999. Diversity in planktonic communities: an experimental test of the intermediate disturbance hypothesis. *Limnology and Oceanography* **44**: 1114–1119.

Forbes, S. A. 1887. The lake as a microcosm. *Bulletin of the Peoria Scientific Association* 77–87.

Fukami, T., and P. J. Morin. 2003. Productivity–biodiversity relationships depend on the history of community assembly. *Nature* 424: 423–426.

Fussmann, G. F., S. P. Ellner, and N. G. Hairston. 2003. Evolution as a critical component of plankton dynamics. *Proceedings of the Royal Society of London B* 270: 1015–1022.

Garlaschelli, D., G. Caldarelli, and L. Pietronero. 2003. Universal scaling relations in food webs. *Nature* **423**: 165–168.

Gaston, K. J., and T. M. Blackburn. 1999. A critique for macroecology. *Oikos* **84**: 353–363.

Groner, E., and A. Novoplansky. 2003. Reconsidering diversity–productivity relationships: directness of productivity estimates matters. *Ecology Letters* 6: 695–699.

Grover, J. P. 1995. Competition, herbivory, and enrichment—nutrient-based models for edible and inedible plants. *American Naturalist* 145: 746–774.

Gurevitch, J., J. A. Morrison, and L. V. Hedges. 2000. The interaction between competition and predation: a meta-analysis of field experiments. *American Naturalist* 155: 435–453.

Haddad, N. M., D. Tilman, J. Haarstad, M. Ritchie, and J. M. H. Knops. 2001. Contrasting effects of plant richness and composition on insect communities: a field experiment. *American Naturalist* 158: 17–35.

Harrison, S. S. C., and A. G. Hildrew. 2001. Epilithic communities and habitat heterogeneity in a lake littoral. *Journal of Animal Ecology* 70: 692–707.

Havel, J. E., and J. B. Shurin. 2004. Mechanisms, effects and scales of dispersal in freshwater zooplankton. *Limnology and Oceanography* 49: 1229–1238.

Havens, K. 1992. Scale and structure in natural food webs. *Science* 257: 1107–1109.

Hawkins, B. A., and E. E. Porter. 2003. Does herbivore diversity depend on plant diversity? The case of California butterflies. *American Naturalist* 161: 40–49.

Hillebrand, H. 2003. Opposing effects of grazing and nutrients on diversity. *Oikos* **100**: 592–600.

Hillebrand, H. 2004. On the generality of the latitudinal gradient of diversity. *American Naturalist* 163: 192–211.

Hillebrand, H., and T. Blenckner. 2002. Regional and local impact on species diversity—from pattern to processes. *Oecologia* 132: 479–491.

Hillebrand, H., and B. J. Cardinale 2004. Consumer effect size declines with prey diversity. *Ecology Letters* 7: 192–201.

Hillebrand, H., B. Worm, and H. K. Lotze. 2000. Marine microbenthic community structure regulated by nitrogen loading and grazing pressure. *Marine Ecology Progress Series* **204**: 27–38.

Holt, R. D., and M. Loreau. 2001. Biodiversity and ecosystem functioning: the role of trophic interactions and the importance of system openness. In: A. P. Kinzig, S. W. Pacala, and D. Tilman (eds), *The Functional Consequences of Biodiversity*. Princeton University Press, Princeton, NJ, pp. 246–262.

Holt, R. D., J. Grover, and D. Tilman. 1994. Simple rules for interspecific dominance in systems with exploitative and apparent competition. *American Naturalist* **144**: 741–771.

Holt, R. D., J. H. Lawton, G. A. Polis, and N. D. Martinez. 1999. Trophic rank and the species–area relationship. *Ecology* 80: 1495–1505.

Horner-Devine, M. C., M. A. Leibold, V. H. Smith, and B. J. M. Bohannan. 2003. Bacterial diversity patterns along a gradient of primary productivity. *Ecology Letters* 6: 613–622.

Huisman, J., and F. J. Weissing. 1999. Biodiversity of plankton by species oscillations and chaos. *Nature* **402**: 407–410.

Huston, M. 1979. A general hypothesis of species diversity. *American Naturalist* **113**: 81–101.

Huston, M. A. 1994. *Biological Diversity: The Coexistence of Species in Changing Landscapes*. Cambridge University Press, Cambridge.

Interlandi, S. J., and S. S. Kilham. 2001. Limiting resources and the regulation of diversity in phytoplankton communities. *Ecology* 82: 1270–1282.

Irigoien, X., J. Huisman, and R. P. Harris. 2004. Global biodiversity patterns of marine phytoplankton and zooplankton. *Nature* 429: 863–867.

Jeppesen, E., J. P. Jensen, M. Sondergaard, T. Lauridsen, and F. Landkildehus. 2000. Trophic structure, species richness and biodiversity in Danish lakes: changes along a phosphorus gradient. *Freshwater Biology* 45: 201–218.

Jonsson, M., and B. Malmqvist. 2000. Ecosystem process rates increases with animal species richness: evidence from leave-eating aquatic insects. *Oikos* 89: 519–523.

Jonsson, M., and B. Malmqvist. 2003*a*. Importance of species identity and number for process rates within different stream invertebrate functional feeding groups. *Journal of Animal Ecology* 72: 453–459.

Jonsson, M., and B. Malmqvist. 2003*b*. Mechanisms behind positive diversity effects on ecosystem functioning: testing the facilitation and interference hypotheses. *Oecologia* 134: 554–559.

Jonsson, M., B. Malmqvist, and P. O. Hoffsten. 2001. Leaf litter breakdown rates in boreal streams: does shredder species richness matter? *Freshwater Biology* **46**: 161–171.

Karez, R., S. Engelbert, and U. Sommer. 2000. Co-consumption and protective coating: two new proposed effects of epiphytes on their macroalgal hosts in mesograzer–epiphyte–host interactions. *Marine Ecology Progress Series* 205: 85–93.

Keddy, P. A. 1989. *Competition*. Chapman and Hall, London.

Knapp, R. A., K. R. Matthews, and O. Sarnelle. 2001. Resistance and resilience of alpine lake fauna to fish introductions. *Ecological Monographs* 71: 401–421.

Kneitel, J. M., and T. E. Miller. 2002. Resource and top-predator regulation in the pitcher plant (*Sarracenia purpurea*) inquiline community. *Ecology* 83: 680–688.

Kondoh, M. 2001. Unifying the relationships of species richness to productivity and disturbance. *Proceedings of the Royal Society of London B.* 268: 269–271.

Koricheva, J., C. P. H. Mulder, B. Schmid, J. Joshi, and K. Huss-Danell. 2000. Numerical responses of different trophic groups of invertebrates to manipulations of plant diversity in grasslands. *Oecologia* 125: 271–282.

Krause, A. E., K. A. Frank, D. M. Mason, R. E. Ulanowicz, and W. W. Taylor. 2003. Compartments revealed in food-web structure. *Nature* 426: 282–285.

Leibold, M. A. 1989. Resource edibility and the effects of predators and productivity on the outcome of trophic interactions. *American Naturalist* 134: 922–949.

Leibold, M. A. 1996. A graphical model of keystone predators in food webs: Trophic regulation of abundance, incidence, and diversity patterns in communities. *American Naturalist* 147: 784–812.

Leibold, M. A. 1999. Biodiversity and nutrient enrichment in pond plankton communities. *Evolution and Ecological Research* 1: 73–95.

Leibold, M. A., J. M. Chase, J. B. Shurin, and A. L. Downing. 1997. Species turnover and the regulation of trophic structure. *Annual Review of Ecology and Systematics* 28: 467–494.

Leibold M. A., M. Holyoak, N. Mouquet, P. Amarasekare, J. M. Chase, M. F. Hoopes, R. D. Holt, J. B. Shurin, R. Law, D. Tilman, M. Loreau, and A. Gonzalez. 2004. The metacommunity concept: a framework for multiscale community ecology. *Ecology Letters* 7: 601–613.

Leiden, N. L., and Traunspurger, W. 2002. Nematoda. In: S. D. Rundle, A. L. Robertson, and J. M. Schmid-Araya (eds), *Freshwater Meiofauna: Biology and Ecology*. Backhuys Publishers, pp. 63–104.

Levins, R. 1979. Coexistence in a variable environment. *American Naturalist* 114: 765–783.

Lindeman, R. L. 1942. The trophic-dynamic aspect of ecology. *Ecology* 23: 399–418.

Loladze, I., Y. Kuang, J. J. Elser, and W. F. Fagan 2004. Competition and stoichiometry: coexistence of two predators on one prey. *Theoretical Population Biology* 65: 1–15.

Loreau, M. 2000. Biodiversity and ecosystem functioning: recent theoretical advances. *Oikos* 91: 3–17.

Lotze, H. K., B. Worm, and U. Sommer. 2000. Propagule banks, herbivory, and nutrient supply control population development and dominance patterns in macroalgal blooms. *Oikos* 89: 46–58.

Lubchenco, J. 1978. Plant species diversity in a marine intertidal community: importance of herbivore food preference and algal competitive abilities. *American Naturalist* 112: 23–39.

Magnuson, J. J., W. M. Tonn, A. Banerjee, J. Toivonen, O. Sanchez, and M. Rask. 1998. Isolation vs. extinction in the assembly of fishes in small northern lakes. *Ecology* 79: 2941–2956.

Mancinelli, G., M. L. Costantini, and L. Rossi. 2002. Cascading effects of predatory fish exclusion on the detritus-based food web of a lake littoral zone (Lake Vico, central Italy). *Oecologia* 133: 402–411.

Martinez, N. D. 1991. Artifacts or attributes? Effects of resolution on the little rock lake food web. *Ecological Monographs* 61: 367–392.

Martinez, N. D. 1993. Effect of scale on food web structure. *Science* 260: 242–243.

McCann, K. S. 2000. The diversity–stability debate. *Nature* 405: 228–233.

McQueen, D. J., M. R. S. Johannes, J. R. Post, T. J. Stewart, and D. R. Lean. 1989. Bottom-up and top-down impacts on freshwater pelagic community structure. *Ecological Monographs* 59: 289–309.

McPeek, M. A. 1998. The consequences of changing the top predator in a food web: a comparative experimental approach. *Ecological Monographs* 68: 1–23.

Menge, B. A., and J. P. Sutherland. 1976. Species diversity gradients: synthesis of the roles of predation, competition, and temporal heterogeneity. *American Naturalist* 110: 351–369.

Michiels, I. C., S. Matzak, and W. Traunspurger. 2003. Maintenance of biodiversity through predation in freshwater nematodes? *Nematology Monographs and Perspectives* 2: 1–15.

Mittelbach, G. G., C. F. Steiner, S. M. Scheiner, K. L. Gross, H. L. Reynolds, R. B. Waide, M. R. Willig et al. 2001. What is the observed relationship between species richness and productivity? *Ecology* 82: 2381–2396.

Montoya, J. M., M. A. Rodriguez, and B. A. Hawkins. 2003. Food web complexity and higher-level ecosystem services. *Ecology Letters* 6: 587–593.

Moorthi, S. 2000. Beziehung zwischen Diversität und Stabilität am Beispiel benthischer Protistengemeinschaften, Masters Thesis, Christian-Albrechts-Universität zu Kiel.

Myers, R. A., and B. Worm. 2003. Rapid worldwide depletion of predatory fish communities. *Nature* **423**: 280–283.

Naeem, S., D. R. Hahn, and G. Schuurman. 2000. Producer-decomposer co-dependency influences biodiversity effects. *Nature* **403**: 762–764.

Naeem, S., and S. Li. 1998. Consumer species richness and autotrophic biomass. *Ecology* **79**: 2603–2615.

Norberg, J. 2000. Resource-niche complementarity and autotrophic compensation determines ecosystem-level responses to increased cladoceran species richness. *Oecologia* **122**: 264–272.

Oksanen, L., S. D. Fretwell, J. Arruda, and P. Niemelä. 1981. Exploitation ecosystems in gradients of primary productivity. *American Naturalist* **118**: 240–261.

Pacala, S. W., and M. J. Crawley. 1992. Herbivores and plant diversity. *American Naturalist* **140**: 243–260.

Paine, R. T. 1966. Food web complexity and species diversity. *American Naturalist* **100**: 65–75.

Persson, L. 1999. Trophic cascades: abiding heterogeneity and the trophic level concept at the end of the road. *Oikos* **85**: 385–397.

Persson, L., A. M. De Roos, D. Claessen, P. Bystrom, J. Lovgren, S. Sjogren, R. Svanback et al. 2003. Gigantic cannibals driving a whole-lake trophic cascade. *Proceedings of the National Academy of Sciences, USA* **100**: 4035–4039.

Petchey, O. L. 2000. Prey diversity, prey composition, and predator population dynamics in experimental microcosms. *Journal of Animal Ecology* **69**: 874–882.

Petchey, O. L., P. T. McPhearson, T. M. Casey, and P. J. Morin. 1999. Environmental warming alters food-web structure and ecosystem function. *Nature* **402**: 69–72.

Polis, G. A., and D. R. Strong. 1996. Food web complexity and community dynamics. *American Naturalist* **147**: 813–846.

Post, D. M., M. L. Pace, and N. G. Hairston, Jr. 2000. Ecosystem size determines food-chain length in lakes. *Nature* **405**: 1047–1049.

Proulx, M., and A. Mazumder. 1998. Reversal of grazing impact on plant species richness in nutrient-poor vs. nutrient-rich ecosystems. *Ecology* **79**: 2581–2592.

Proulx, M., F. R. Pick, A. Mazumder, P. B. Hamilton, and D. R. S. Lean. 1996. Experimental evidence for interactive impacts of human activities on lake algal species richness. *Oikos* **76**: 191–195.

Raffaelli, D., W. H. Van der Putten, L. Persson, D. A. Wardle, O. L. Petchey, J. Koricheva, M. van der Heijden et al. 2002. Multi-trophic dynamics and ecosystem processes. In: M. Loreau, S. Naeem, and P. Inchausti (eds), *Biodiversity and Ecosystem Functioning*. Oxford University Press Oxford, pp. 147–154.

Rosenzweig, M. L., and Z. Abramsky. 1993. How are diversity and productivity related? In: R. E. Ricklefs and D. Schluter (eds), *Species Diversity in Ecological Communities*. University of Chicago Press, Chicago, USA, pp. 52–65.

Rundle, H. D., S. M. Vamosi, and D. Schluter 2003. Experimental test of predation's effect on divergent selection during character displacement in sticklebacks. *Proceedings of the National Academy of Sciences, USA* **100**: 14943–14948.

Schmid-Araya, J. M., A. G. Hildrew, A. Robertson, P. E. Schmid, and J. Winterbottom. 2002a. The importance of meiofauna in food webs: evidence from an acid stream. *Ecology* **83**: 1271–1285.

Schmid-Araya, J. M., P. E. Schmid, A. Robertson, J. Winterbottom, C. Gjerlov, and A. G. Hildrew. 2002b. Connectance in stream food webs. *Journal of Animal Ecology* **71**: 1056–1062.

Shurin, J. B. 2001. Interactive effects of predation and dispersal on zooplankton communities. *Ecology* **82**: 3404–3416.

Shurin, J. B., and E. G. Allen. 2001. Effects of competition, predation, and dispersal on species richness at local and regional scales. *American Naturalist* **158**: 624–637.

Shurin, J. B., and D. S. Srivastava. New perspectives on local and regional diversity: beyond saturation. In: M. Holyoak, R. D. Holt, and M. A. Leibold (eds), *Metacommunities* (in press).

Siemann, E., D. Tilman, J. Haarstad, and M. Ritchie. 1998. Experimental test of the dependence of arthropod diversity on plant diversity. *American Naturalist* **152**: 738–750.

Soluk, D. A., and J. S. Richardson. 1997. The role of stoneflies in enhancing growth of trout: a test of the importance of predator–predator facilitation within a stream community. *Oikos* **80**: 214–219.

Sommer, U. 1985. Comparison between steady state and non-steady state competition: experiments with natural phytoplankton. *Limnology and Oceanography* **30**: 335–346.

Sommer, U. 1999. The impact of herbivore type and grazing pressure on benthic microalgal diversity. *Ecology Letters* **2**: 65–69.

Sommer, U., F. Sommer, B. Santer, E. Zöllner, K. Jürgens, C. Jamieson, M. Boersma et al. 2003. Daphnia versus copepod impact on summer phytoplankton: functional compensation at both trophic levels. *Oecologia* **135**: 639–647.

Steiner, C. F. 2001. The effects of prey heterogeneity and consumer identity on the limitation of trophic-level biomass. *Ecology* **82**: 2495–2506.

Steiner, C. F. 2003. Keystone predator effects and grazer control of planktonic primary production. *Oikos* **101**: 569–577.

Steinman, A. D. 1996. Effects of grazers on benthic freshwater algae. In: R. J. Stevenson, M. L. Bothwell, and R. L. Lowe (eds), *Algal Ecology—Freshwater Benthic Ecosystems*, Academic Press, pp. 341–373.

Sterner, R. W., and J. J. Elser. 2002. *Ecological Stoichiometry*. Princeton University Press, Princeton, NJ.

Thebault, E., and M. Loreau 2003. Food-web constraints on biodiversity–ecosystem functioning relationships. *Proceedings of National Academy of Sciences, USA* **100**: 14949–14954.

Tilman, D. 1982. *Resource Competition and Community Structure*, Princeton University Press, Priceton, NJ.

Tilman, D., J. Knops, D. Wedin, and P. Reich. 2001. Experimental and observational studies of diversity, productivity and stability. In: A. P. Kinzig, S. W. Pacala, and D. Tilman (eds), *The Functional Consequences of Biodiversity*. Princeton University Press, Princeton, NJ, pp. 42–70.

Vadeboncoeur, Y., E. Jeppesen, M. J. Vander Zanden, H. H. Schierup, K. Christoffersen, and D. M. Lodge. 2003. From Greenland to green lakes: cultural eutrophication and the loss of benthic pathways in lakes. *Limnology and Oceanography* **48**: 1408–1418.

Vander Zanden, M. J., B. J. Shuter, N. Lester, and J. B. Rasmussen. 1999. Patterns of food chain length in lakes: a stable isotope study. *American Naturalist* **154**: 406–416.

Vinebrooke, R. D., D. W. Schindler, D. L. Findlay, M. A. Turner, M. Paterson, and K. H. Milis. 2003. Trophic dependence of ecosystem resistance and species compensation in experimentally acidified lake 302S (Canada). *Ecosystems* **6**: 101–113.

Wahl, M., M. E. Hay, and P. Enderlein. 1997. Effects of epibiosis on consumer-prey interactions. *Hydrobiologia* **355**: 49–59.

Waide, R. B., M. R. Willig, C. F. Steiner, G. Mittelbach, L. Gough, S. I. Dodson, J. P. Juday et al. 1999. The relationship between productivity and species richness. *Annual Review of Ecology and Systematics.* **30**: 257–300.

Weithoff, G., A. Lorke, and N. Walz. 2000. Effects of water-column mixing on bacteria, phytoplankton, and rotifers under different levels of herbivory in a shallow eutrophic lake. *Oecologia* **125**: 91–100.

Werner, E. E., and S. D. Peacor. 2003. A review of trait-mediated indirect interactions in ecological communities. *Ecology* **84**: 1083–1100.

Wickham, S. A. 1995. Trophic Relations between cyclopoid copepods and ciliated protists—complex interactions link the microbial and classic food Webs. *Limnology and Oceanography* **40**: 1173–1181.

Wickham, S., S. Nagel, and H. Hillebrand. 2004. Control of epibenthic ciliate communities by grazers and nutrients. *Aquatic Microbial Ecology* **35**: 153–162.

Williams, R. J., and N. D. Martinez. 2000. Simple rules yield complex food webs. *Nature* **404**: 180–183.

Williams, R. J., E. L. Berlow, J. A. Dunne, A. L. Barabasi, and N. D. Martinez. 2002. Two degrees of separation in complex food webs. *Proceedings of the National Academy of Sciences, USA* **99**: 12913–12916.

Wilson, W. G., P. Lundberg, D. P. Vazquez, J. B. Shurin M. D. Smith, W. Langford, K. L. Gross, and G. G. Mittelbach. 2003. Biodiversity and species interactions: extending Lotka–Volterra community theory. *Ecology Letters* **6**: 944–952.

Woodward, G., and A. G. Hildrew. 2002. Body-size determinants of niche overlap and intraguild predation within a complex food web. *Journal of Animal Ecology* **71**: 1063–1074.

Worm, B., and J. E. Duffy. 2003. Biodiversity, productivity and stability in real food webs. *Trends in Ecology and Evolution* **18**: 628–632

Worm, B., H., K. Lotze, C. Boström, R. Engkvist, V. Labanauskas, and U. Sommer. 1999. Marine diversity shift linked to interactions among grazers, nutrients and propagule banks. *Marine Ecology Progress Series* **185**: 309–314.

Worm, B., H. K. Lotze, and U. Sommer. 2001. Algal propagule banks modify competition, consumer and resource control on Baltic rocky shores. *Oecologia* **128**: 281–293.

Worm, B., H. K. Lotze, H. Hillebrand, and U. Sommer. 2002. Consumer versus resource control of species diversity and ecosystem functioning. *Nature* **417**: 848–851.

Yoshida, T., L. E. Jones, S. P. Ellner, G. F. Fussmann, and N. G. Hairston. 2003. Rapid evolution drives ecological dynamics in a predator-prey system. *Nature* **424**: 303–306.

Chapter 15

Almaas, E., and A.-L. Barabási, and A.-L. 2004. Power laws in biological networks. Available at: *ArXiv* <q-bio.MN/0401010>.

Bak, P. 1996. *How Nature Works: The Science of Self-Organized Criticality*. Copernicus, New York, 212 p.

Barabási, A.-L. 2002. *Linked: The New Science of Networks*. Perseus Publications, Cambridge, MA, 280 p.

Barabási, A.-L. and R. Albert. 1999. Emergence of scaling in random networks. *Science* **286**: 509–512.

Casti, J. L. 2004. Why the future happens. The Second International Biennial Seminar on the Philosophical, Methodological and Epistemological Implications of Complexity Theory. International Convention Center, La Habana, Cuba, January 7–11, 2004.

Depew, D. J., and B. H. Weber. 1994. *Darwinism Evolving: Systems Dynamics and the Geneology of Natural Selection.* MIT Press, Cambridge, MA, 588 p.

Elsasser, W. M. 1969. Acausal phenomena in physics and biology: a case for reconstruction. *American Scientist* 57(4): 502–516.

Higashi, M., and T. P. Burns. 1991. *Theoretical Studies of Ecosystems: The Network Perspective.* Cambridge University Press, Cambridge, 364 p.

Hirata, H., and R. E. Ulanowicz. 1984. Information theoretical analysis of ecological networks. *International Journal of Systems Science* 15: 261–270.

Holling, C. S. 1986. The resilience of terrestrial ecosystems: local surprise and global change. In: W. C. Clark and R. E. Munn (eds), *Sustainable Develoment of the Biosphere.* Cambridge University Press, Cambridge, pp. 292–317.

Jeong, H., B. Tombor, R. Albert, Z. N. Oltvai, and A.-L. Barabási. 2000. The large-scale organization of metabolic networks. *Nature* 407, 651–654.

Kuhn, T. S. 1962. *The Structure of Scientific Revolutions.* University of Chicago Press, Chicago, IL, 172 p.

May, R. M. 1972. Will a large complex system be stable. *Nature* 238: 413–414.

McLuhan, H. M. 1964. *Understanding Media: The Extension of Man.* McGraw-Hill, New York.

Minarik, E. M. 1957. *Little Bear.* Harper and Row, New York, 63 p.

Montoya, J. M., and R. V. Sole. 2002. Small-world patterns in food webs. *Journal of Theoretical Biology,* 214: 405–412.

Odum, H. T. 1971. *Environment, Power and Society.* Wiley, New York, p. 331.

Pimm, S. L. 1982. *Food Webs.* Chapman and Hall, London, 219 p.

Pimm, S. L., and J. H. Lawton. 1977. Number of trophic levels in ecological communities. *Nature* 268: 329–331.

Popper, K. R. 1977. The bucket and the searchlight: two theories of knowledge. In: M. Lippman (ed.), *Discovering Philosophy.* Prentice Hall, Englewood Cliffs, NJ, pp. 328–334

Popper, K. R. 1990. *A World of Propensities.* Thoemmes, Bristol 51 p.

Prigogine, I., and I. Stengers. 1984. *Order out of Chaos: Man's New Dialogue with Nature.* Bantam, New York, 349 p.

Rutledge, R. W., B. L. Basorre, and R. J. Mulholland. 1976. Ecological stability: an information theory viewpoint. *Journal of Theoretical Biology* 57: 355–371.

Ulanowicz, R. E. 1980. An hypothesis on the development of natural communities. *Journal of Theorectical Biology* 85: 223–245.

Ulanowicz, R. E. 1986. *Growth and Development: Ecosystems Phenomenology.* Springer-Verlag, New York, 203 p.

Ulanowicz, R. E. 1995. *Utricularia's* secret: the advantages of positive feedback in oligotrophic environments. *Ecological Modelling* 79: 49–57.

Ulanowicz, R. E. 1997a. *Ecology, the Ascendent Perspective.* Columbia University Press, New York, 201p.

Ulanowicz, R. E. 1997b. Limitations on the connectivity of ecosystem flow networks. In: A. Rinaldo and A. Marani (eds), *Biological Models.* Istituto Veneto de Scienze, Lettere ed Arti. Venice, pp. 125–143.

Ulanowicz, R. E. 1999. Life after Newton: an ecological metaphysic. *BioSystems* 50: 127–142.

Ulanowicz, R. E. 2001. The organic in ecology. *Ludus Vitalis* 9(15): 183–204.

Ulanowicz, R. E. 2002. The balance between adaptability and adaptation. *BioSystems* 64: 13–22.

Ulanowicz, R. E. 2004. Order and fluctuations in ecosystem dynamics. *Emergence.*

Ulanowicz, R. E. 2004. Ecosystem dynamics: a natural middle 2(2): 231–253.

Ulanowicz, R. E., and J. Norden. 1990. Symmetrical overhead in flow networks. *International Journal of Systems Science* 21(2): 429–437.

Ulanowicz, R. E., and W. F. Wolff. 1991. Ecosystem flow networks: loaded dice? *Mathematical Biosciences* 103: 45–68.

Wagensberg, J., A. Garcia, and R. V. Sole. 1990. Connectivity and information transfer in flow networks: two magic numbers in ecology? *Bulletin of Mathematical Biology* 52: 733–740.

Watts, D. J. 1999. *Small Worlds: The Dynamics of Networks Between Order and Randomness.* Princeton University Press, Princeton, NJ, 262 p.

Wulff, F., J. G. Field, and K. H. Mann (eds). 1989. Network Analysis in Marine Ecology. Springer-Verlag, Berlin, Heidleberg, New York.

Zorach, A. C., and R. E. Ulanowicz. 2003. Quantifying the complexity of flow networks: how many roles are there? *Complexity* 8(3): 68–76.

Index

Page numbers in *italics* refer to Figures.